McGraw-Hill Illustrated Telecom Dictionary

Other McGraw-Hill Telecommunications Books of Interest

ALI • *Digital Switching Systems*
ASH • *Dynamic Routing in Telecommunications Networks*
BENNER • *Fibre Channel*
BEST • *Phase-Locked Loops, Third Edition*
FEIT • *TCP/IP, Second Edition*
GALLAGHER • *Mobile Telecommunications Networking with IS-41*
GOLDBERG • *Digital Techniques in Frequency Synthesis*
GORALSKI • *Introduction to ATM Networking*
HARTE • *Cellular and PCS: The Big Picture*
HELDMAN • *Information Telecommunications*
HELDMAN • *Competitive Telecommunications*
KESSLER • *ISDN, Third Edition*
KURUPILLA • *Wireless PCS*
LACHS • *Fiber Optic Communications*
LEE • *Mobile Cellular Telecommunications, Second Edition*
LEE • *Mobile Communications Engineering, Second Edition*
LINDBERG • *Digital Broadband Networks and Services*
LOGSDON • *Mobile Communication Satellites*
MACARIO • *Cellular Radio, Second Edition*
PATTAN • *Satellite-Based Global Cellular Telecommunications*
RODDY • *Satellite Communications, Second Edition*
ROHDE ET AL. • *Communications Receivers, Second Edition*
SIMON ET AL. • *Spread Spectrum Communications Handbook*
SMITH • *Cellular Design and Optimization*
TSAKALAKIS • *PCS Network Deployment*
WINCH • *Telecommunication Transmission Systems, Second Edition*

McGraw-Hill Illustrated Telecom Dictionary

Jade Clayton

McGraw-Hill
New York San Francisco Washington, D.C. Auckland Bogotá
Caracas Lisbon London Madrid Mexico City Milan
Montreal New Delhi San Juan Singapore
Sydney Tokyo Toronto

Library of Congress Cataloging-in-Publication Data

Clayton, Jade.
McGraw-Hill illustrated telecom dictionary / Jade Clayton.
p. cm.
Includes index.
ISBN 0-07-012063-3 (paperback)
1. Telecommunication—Dictionaries. I. Title.
TK5102.C57 1998
384'.03—dc21 98-7686
CIP

1 2 3 4 5 6 7 8 9 0 DOC/DOC 9 0 3 2 1 0 9 8

ISBN 0-07-012063-3

The sponsoring editor for this book was Steve Chapman, the editing supervisor was Andrew Yoder, and the production supervisor was Tina Cameron. It was set in ITC Century Light by Lisa M. Mellott through the services of Barry E. Brown (Broker—Editing, Design and Production).

Printed and bound by R. R. Donnelley & Sons Company.

This book is printed on recycled, acid-free paper containing a minimum of 50% recycled de-inked fiber.

Dedication

This work is dedicated to the memory of Brian Blackburn and all others that have lost their life or lost someone to cancer.

Acknowledgments

The author would like to express his sincere appreciation and gratitude to the following individuals for their support during the creation of this work:

Dawn Pensiero, Kathy Thomson (my mom), Don Thomson (my dad), Mary Kirkendall (my grandmother), Johnny Clayton, Zachary Thomson, Robbie Thomson, John and Gayle Kirkendall, Drew Kirkendall, Jim and Monica Kirkendall, Velma Clayton, Paul and Tracy Anderson, David Anderson, Margurite Anderson, Todd and Holly Fraker, Brian, Claudia, Brianna and Cassidy Blackburn, Brett, Vicki and Tristin Cooper, Chad and Nicole Player, Mike and Rose Palmer, Brian and Becky Henrie, Corie Wright, Rydell Mitchell, Carrie Hughes, Tamie Jensen, Ty Mutchler, Chuck Geltz, Steve McLiesh, Scott Varley, Tricia Drake, Scott Sargent, Bryan Barr, Dave Hepworth and Chaz Benson.

A big thank-you is also given to the following companies and individuals for their material assistance, experience, training, employment, friendship, and help in gathering information for this work:

Cary Wood, Gary Callister, Allan Vollmer, Allan Wood, Brice Cox, Carol Burbidge, Dan Dick, David Hull, Davin Stillson, Dell Stillson, Dennis Moore, Don Fell, Doug Klentzman, Evan Jones, Gene Farrell, Janet Roothoff, Jeff Hatch, Merritt May, Norman Sarabia, Peggy Young, Rob Ford, Robert Palfreyman, Rod Kearl, Seth Miner, Shawn Llewelyn, Stan Hong, Steve Trauba, Tim Cassady, Troy Ostmark, Wayne Leota, and other co-workers of the American Express Technologies organization.

Jim Facer and Mario Heib of Acme Broadcasting.

Kelly Keisel, Mike Nelson and other co-workers of Microwave Tower Services.

Paul Stroh and Gerry Capps of Brooks/Worldcom Communications.

Frank Tatulli and Bob Verda of Periphonics Interactive Voice Response Systems.

Lonnie Mair of Lucent Technologies.

Phil Tonick, Frank Croan, Dave Young, Fred Kroll, Robert Kellett, Paul Backman, Bill Stieneger, Laurie Guluarte, Dick Forfar, and other co-workers of AT&T–TCG.

Lee and Kathy Guthrie, Reed Madsen, Gary Chavez, Daryle Starr, Ewe, Ken Olson, Curt Breen, John Dutt, Roland (row your Bayliner) Eysser, Lynn Montgomery, Marty Moss, Dave Smart, Randy Ipsen, Scott Everts, Sam Alvarado, Ken Morley, Kent Forbush, Kris Peterson, Anita Shoblume, Maureen Slusher, Randy Smith, Kevin Smith, Ann and Bob Smith, Jan Hembury, Scott Ross, Mike Nielsen, Steve Pryor, Kent Brown, Teresa Price, Linda Lujan, Scott Lutz, Ray (come to Jesus) Spencer, Lonnie Kresser, and other co-workers of USWest Communications.

Judith Johnson and Peggy Egbert of the Utah Public Service Commission.

Roger Harry, Tony Daniels, Dave Kahn, Larry Wang, Kevin McGarrell, Cheryl Heinz, and John Bracken of Nortel.

Mike Leyva of Blue Cross IT.

Nolan Bitters of First Security Bank IT.

Clint Smith of CCS Engineering.

Jim Holloway, Chuck Griebe, and Lynn Bayless for their Data Networking, PC and Internet Expertise.

Dan and Nyla Dick for their digital photo processing assistance.

Dawn Pensiero of Children's Medical Center of Dallas IT.

Steve Chapman and others of McGraw-Hill.

Introduction

The *McGraw-Hill Illustrated Telecom Dictionary* was written to provide convenient and easy-to-understand definitions of commonly used terminology and technology in the telecommunications industry. It is meant to be used as a reference for public telephone company professionals, telephone equipment/services vendors, telephone equipment manufacturers, telephone equipment distributors, as well as instructors and students of all levels that have subject matter relating to computer science, information systems, telecommunications, and electronics.

Because the telecommunications industry will constantly change due to improvements in technology and name changes for services and products, new terms, acronyms, and modified acronyms will constantly be created. As these new terms evolve, they will be defined in future editions of this book.

This Dictionary has more than 2000 definitions. Of these definitions, more than 250 have a picture, diagram, or chart to assist in the definition. Many of the remaining 1750 definitions refer to a diagram or picture located somewhere within the book.

0–99

0 The "in fact" standard number to dial for reaching a local phone company operator or answering service.

1-Pair Gas Lightning Protector Used in Siecor telephone network interfaces.

1PR Gas Lightning Protector

1FB A service code that defines a flat-rate business telephone line. A line where a subscriber can make unlimited local calls and not be billed extra, regardless of the number of calls or their duration.

1FR A service code that defines a flat-rate residential telephone line. A line where a subscriber can make unlimited local calls and not be billed extra, regardless of the number of calls or their duration.

1MB A service code that defines a measured-rate business telephone line. A line where the subscriber is billed either for the number of calls made or by the minute.

1MR A service code that defines a measured-rate business telephone line. A line where the subscriber is billed either for the number of calls made or by the minute.

2-Line Network Interface Old style with interchangeable lightning protectors. The white paint on the tops of the protectors indicates "gas type," rather than the carbon type.

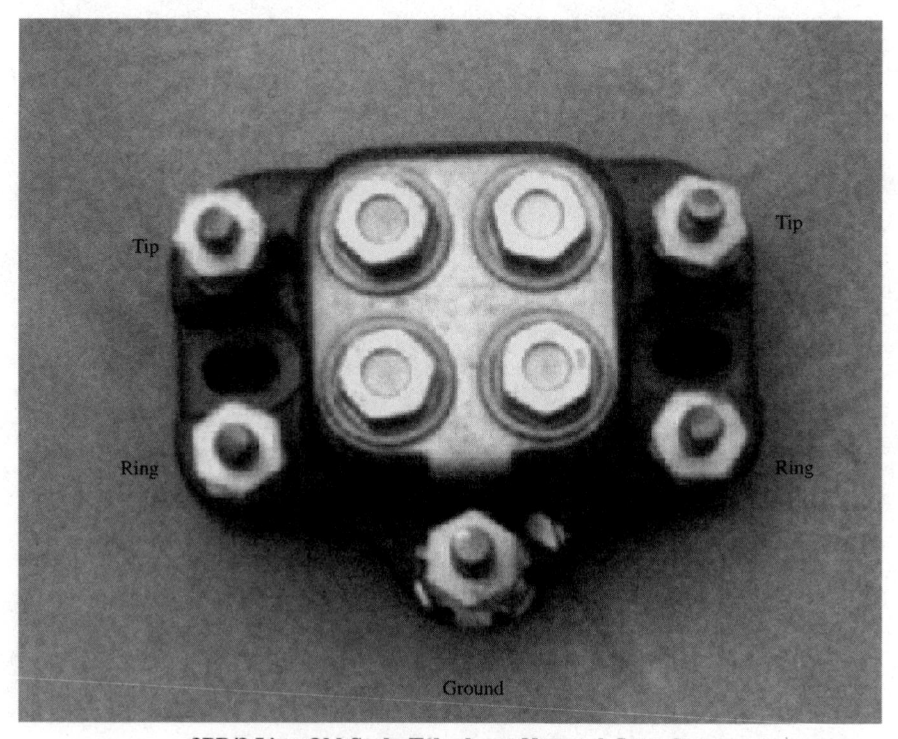

2PR/2 Line Old Style Telephone Network Interface

2FR A service code for a flat-rate party line with two subscribers. For more info, see *Selective Ringing Module* and *Party Line*.

3FR A service code for a flat-rate party line with three subscribers. For more information, see *Selective Ringing Module* and *Party Line*.

4FR A service code for a flat-rate party line with four subscribers. For more information, see *Selective Ringing Module* and *Party Line*.

4Pair Shown is 4-pair, PVC (Polyvinyl Chloride Jacketed) UTP (Unshielded Twisted Pair).

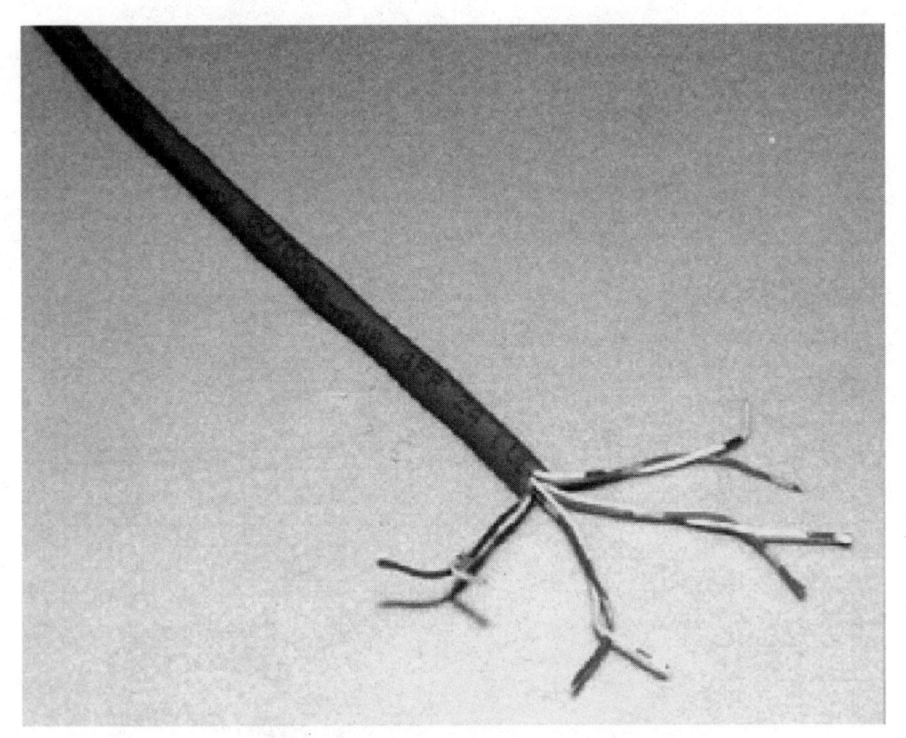

4PR PVC UTP (Unshielded Twisted Pair)

6-Pair Can A termination or splicing enclosure designed especially for 6-pair aerial or buried service wire. 6-pair cans are available with lightning protectors (protected 6-pair can).

8FR A service code for a flat-rate party line with eight subscribers. For more information, see *Selective Ringing Module* and *Party Line*.

6-Pair Can

10 Base 2 Also called *cheapernet*. 10Mb/s CSMA/CD LAN standard that is implemented over RG-58A (50 ohm) coax.

10 Base T 802.3 ethernet 10Mb/s LAN standard. See *Ethernet*.

12-Pack Coax Cable A bundle of twelve 50-ohm coaxial cables used to transport *STS-1 (Synchronous Transport Signal 1)* signals throughout a central office or node. Commonly, the cables run from a SONET carrier unit to a *DCS (Digital Cross-Connect System)*.

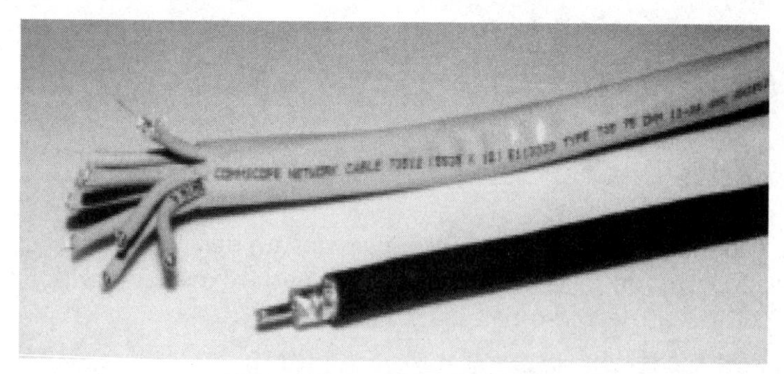

12 Pack Coax Beside RG8 Coax

25-Pair Modular Splice Used in a modular splice tool to splice PIC telephone cable.

25PR Modular Splice Unit

25PR Connector Also called an *Amphenol*, *Amp connector*, *P connector (male)*, or *C connector (female)*.

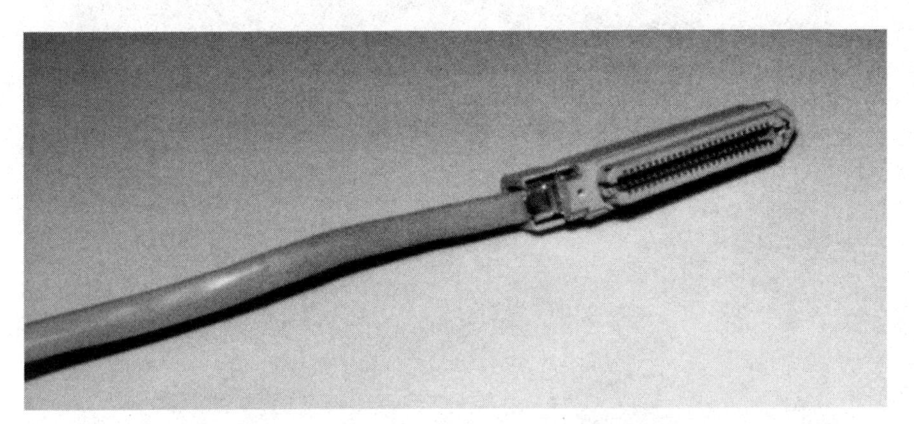

25PR Female Amp Connector (50 pin)

25PR PVC Common telephone cabling used for horizontal and vertical wiring in buildings.

25 PR PVC UTP Unshielded Twisted Pair

66 Block 66M150 termination block. It is used to terminate twisted pair wire on distribution frames and any other solid 22 to 24 wiring application.

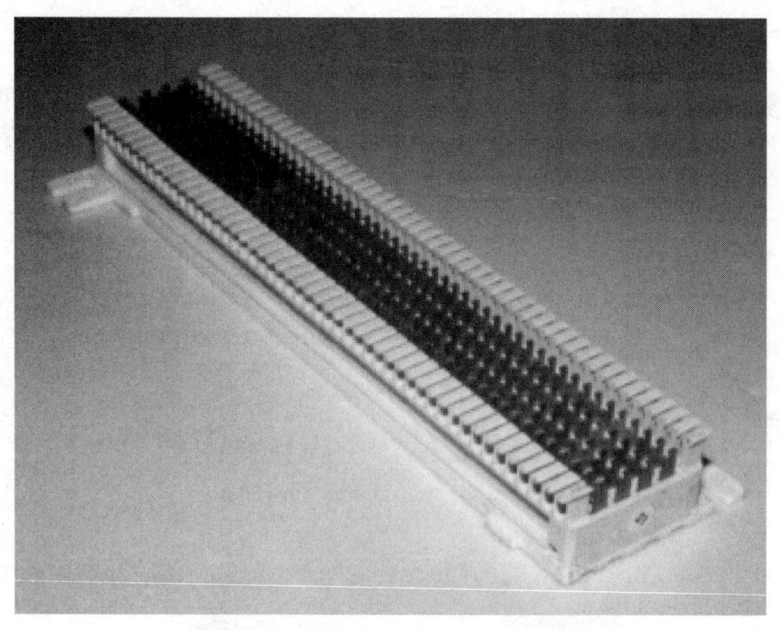

66M150 Termination Block "66 Block"

89B Bracket The bracket that is used to attach 66M150 blocks to back boards in telephone closets or distribution frames.

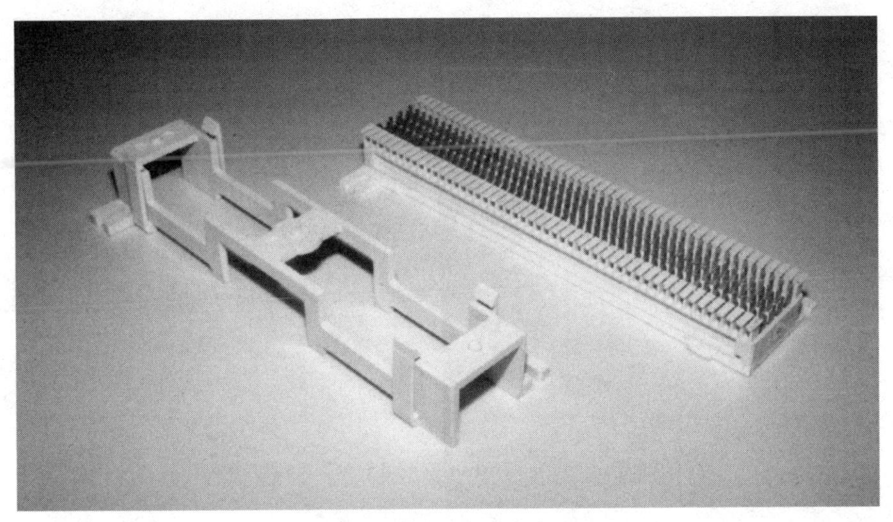

89B Bracket Beside 66M150 Block

100 Base T 802.3 ethernet 100Mb/s LAN standard. See *Ethernet*.

101B Closure A closure/housing used to protect service wire splices and inside wiring splices.

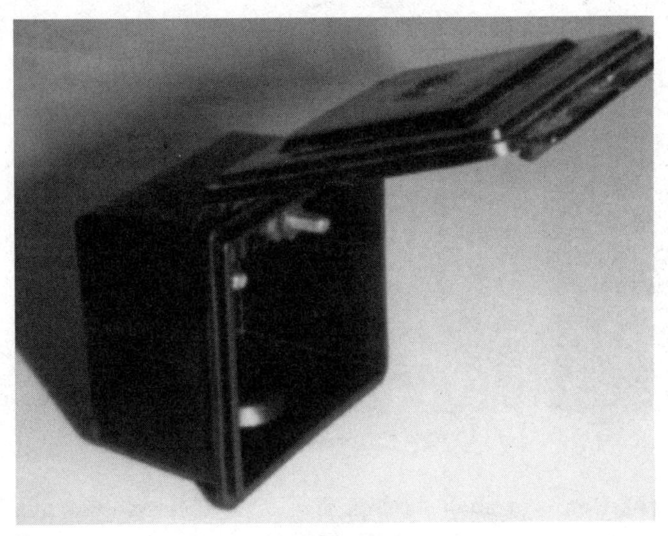

101B Closure

110 Punch Tool A tool used to terminate solid twisted-pair copper wire on AT&T 100 termination blocks.

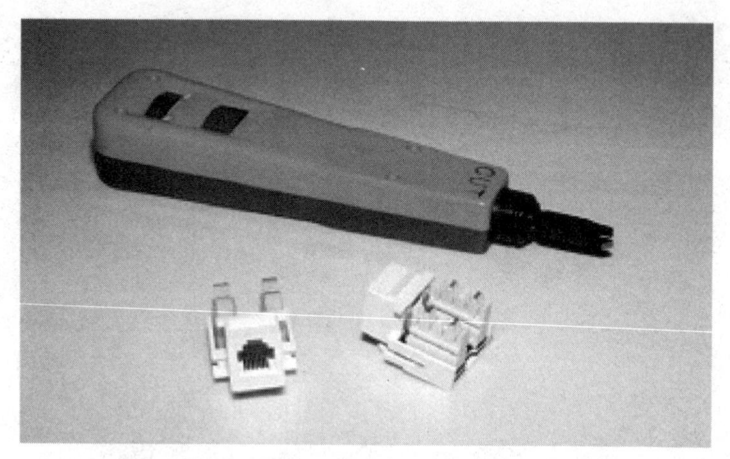

Punch Tool with 110 Blade and CAT5 RJ45 Jacks

110 Termination Block Also called *AT&T 110 blocks*. Devices used to mount twisted-pair wire so that different devices in a network can be cross connected easily.

AT&T 110 Termination Blocks

145A Test Set An analog telephone cable test set that measures the length of twisted pairs, and tests for grounds and shorts. This test set can also send a tone.

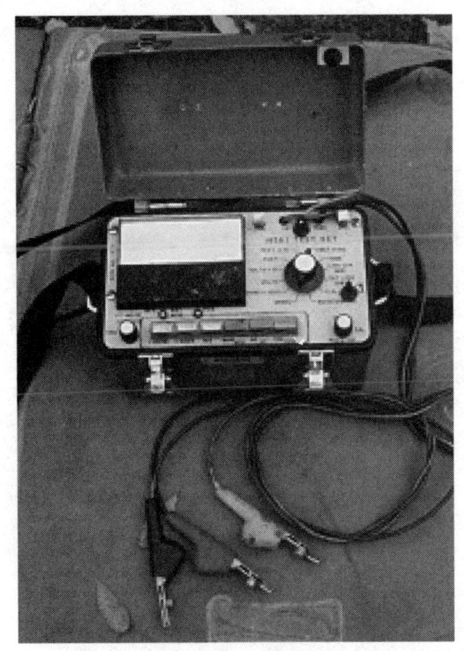

145A Test Set

258A Adapter Adapter used to connect 25-pair Amphenol cables to RJ45 patch cords.

258A Adapter—Harmonica Adapter (50 pin amp to RJ45)

267A/267C Adapter 267A adapter is a basic Y-type adapter that splits one RJ14 into two. The 267C adapter splits two lines from an RJ14 to two RJ11 jacks.

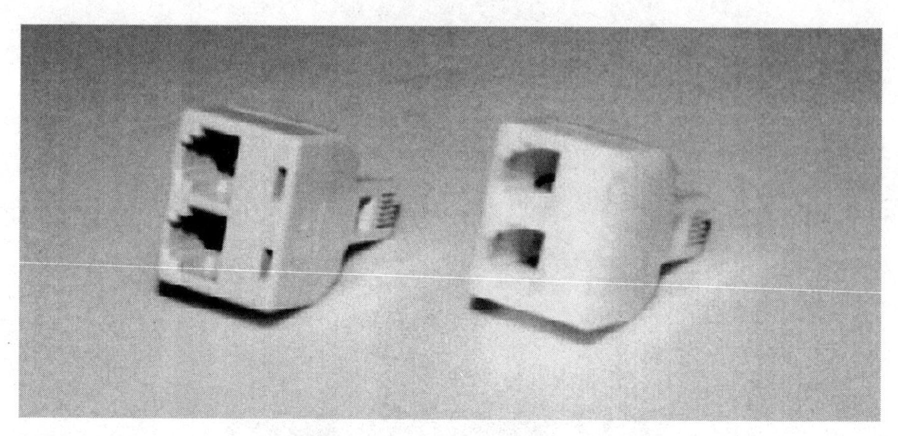

267A Adapter (left) and 267C Adapter (right)

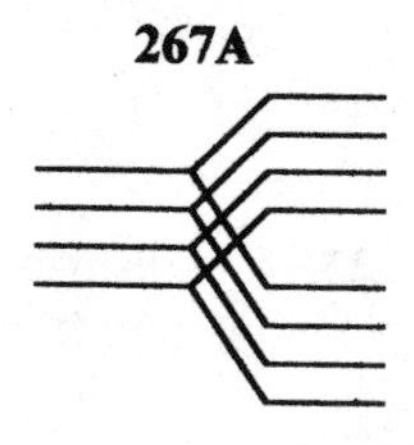

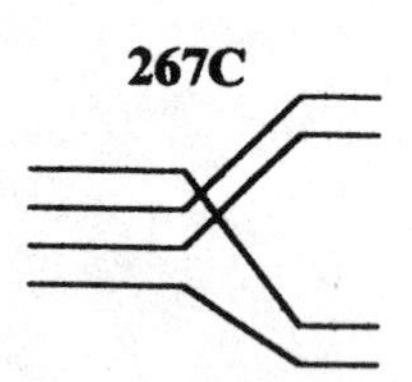

900 The "in fact" standard for services billed through telephone companies. Some 900 services include weather information, stock exchange information, and "erotic" information. Phone companies have their own individual criteria for providing 900 services to companies that wish to sell services over the telephone network. Individuals that call 900 numbers are billed a pre-determined amount for the call on a per-minute basis. Most telephone companies require that anyone selling information services on their network to inform callers of the charges in advance and to allow them a certain amount of time to hang-up before any billing begins.

911 The standard emergency service telephone number. 911 calls are not answered by the telephone company; they are answered by an emergency dispatch service. This is why 911 service has a separate charge on telephone bills.

965TD 3M Dynatel 965TD loop analyzer. Used for testing twisted-pair telephone cable. The 965TD is also a data terminal capable of accessing a database via an internal modem. The TDR (*Time Domain Reflectometer*) is another great feature of the 965TD.

3M 965T Loop Analyzer

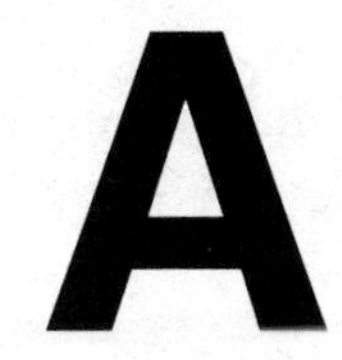

A (Amp, Ampere) A unit of electrical current flow that is equal to one volt applied to one ohm of resistance. The Ohm's Law definition of amperage is:

$$I = \frac{E}{R} \text{ or } Amps = \frac{Voltage}{Resistance}$$

The ampere can also be defined as one coulomb of charge flowing past a point in one second. One coulomb of charge is equal to 6,300,000,000,000,000,000 electrons.

AA (Automated Attendant) Most voice-mail systems come with an automated attendant built in. An automated attendant is an answering machine that asks the caller to push 1 for sales, 2 for service, etc. They are also capable of routing callers to a dial by name directory. See also *Directory Tree*.

AAL (ATM Adaptation Layer) This layer converts your data into a language or protocol that ATM can transport. Imagine that an ATM signal is a very long fast-moving train on a railroad. The train never stops, and the customers bring many different shapes and sizes of data to be transported. The AAL is the part of the train station that takes the data to be transported, packages it to fit the train cars, and loads it. Then, it gives the train car an address. It does all this without stopping the train. Each train car holds 48 bytes of data, or payload. The train car and its address are what make up the 5 bytes of overhead. This is a total of 53 bytes, which make up an ATM data frame.

AAR (Automatic Alternate Routing) A feature of some networks and protocols to reroute traffic on the fly without interrupting or corrupting traffic.

AB Switch A mechanical/manual switch used to switch a signal between two source or destination devices. For example, if you have two computers and one monitor, you could implement an AB switch to control which computer the monitor is connected to. The monitor would connect to the "C" port on the switch, and the two computers would connect to the "A" and the "B" ports. AB switches suit many applications of connectivity from computer to audio/video.

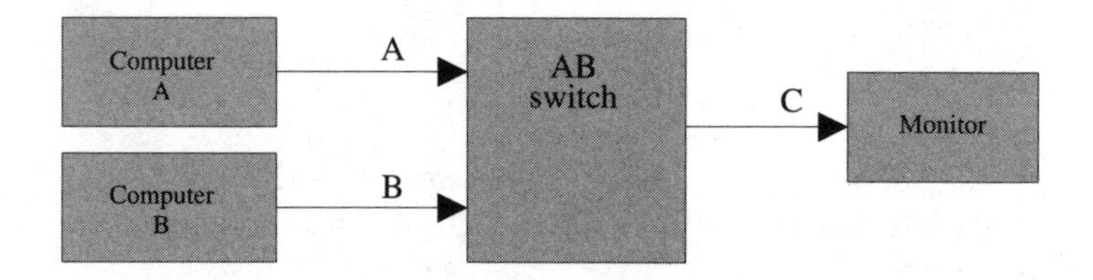

Common AB Switch

Abandoned Call When you make a call, are put on hold, then hang up before someone answers, you have abandoned the call. Customer-service call centers like to know the number of abandoned calls they have so that they know how many people to employ answering calls, etc. Believe it or not, if you hang up when calling a call center, someone that cares eventually finds out!

Ablation To burn holes into metal film with a laser. The holes represent ones and zeros for optical storage on disks.

Absorption Loss The weakening of light intensity as it travels a length of optical fiber. The unit for absorption loss is dB/Km (decibels per kilometer).

AC (Alternating Current) Alternating current is electricity that changes/alternates it's direction of flow in a steady cycle or period. The line voltage in most American homes is somewhere between 110-V and 120V AC RMS, which makes the actual peak to peak voltage about 325 V.

AC-to-DC Converter This is an electronic device that defines itself. Large-scale AC/DC converters are mostly referred to as *rectifiers*. They convert alternating current to direct current (or voltage) by incorporating a large capacitor and two or four diode rectifiers for a half wave or full wave, respectively. Almost all rectifiers have regulated output, meaning the output DC voltage is kept at a steady level, regardless of the electronic device it is providing power to. Rectifiers are also available with battery backup and redundant circuits so if a component fails the output voltage won't be disturbed. AC to DC converters are rated by input voltage requirement and output voltage/current ability.

Acceptable Angle The maximum angle that a fiber optic accepts light and doesn't reflect it away.

Access Charge (Carrier Common-Line Charge) What local phone companies charge long-distance companies to connect the far-end local portion of a call. A fee that everyone pays for every phone line to make up for subsidies that long-distance services paid to help the less-profitable local services before the divestiture of AT&T and the RBOCs (Regional Bell Operating Companies).

Access Line The physical wire between the telephone network interface (NI) and the telephone company's central office main distribution frame. An access line is usually made up of five components. An F1 (or underground) pair, a cross connect, an F2 (or aerial) pair, a service wire, and a lightning protector.

Access Link The local phone line that connects you to a central office switch and gives you access to a long-distance carrier. It's an access line with all the electronics that give you dial tone or private line communications capability.

Access Point (AP) Another name for a cross-box where telephone cables are cross connected. See also *Aerial Cross Box*.

Access Point (AP) Cross Box

Access Service Request When a special-service provider (frame relay or long distance private line) needs wire facilities from their point of presence in the city to your location, they call the local telephone company and make an access service request to provide a line that runs from your network interface to them. Many special service providers have their equipment located in the local phone company's central office as a part of a co-location agreement. When a CLEC (competitive local exchange carrier) needs to provide service where they don't have facilities, this is how they do it by using the RBOC's (Regional Bell Operating Company) wire facilities.

Access Switch This is also called a *front-end processor (FEP)*. It is where multiple services/protocols are differentiated and routed for enterprise networking services.

Access Tandem A telephone company central office or node that contains a switch in which all inter and outer area code traffic is handled. The main *LEC (Local Exchange Carrier)* central office in an area code where the hand-off for long-distance service happens.

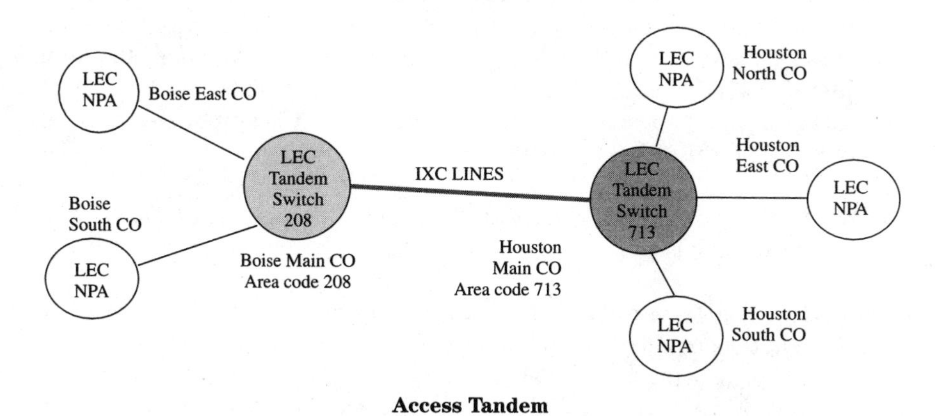

Access Tandem

Account Code In accounting for communications costs, an account code is used. If you have multiple employees in your office using multiple phone lines to make long-distance calls, figuring out who made those calls can be impossible. Some long-distance companies offer a service where employees enter an account code before their call is completed. When the phone bill comes, it is easy to see who is making long-distance calls, to where, and for how long. Most importantly, it is easy to distribute the costs among a group. Most *PBX (Private Branch Exchange)* switches have call accounting systems available that allow a telephone extension (or group of extensions) to be attached to an account code that shows how much each phone is utilized, and what calls are made and received.

ACD (Automatic Call Distributor/Distribution System) A separate footprint or built-in feature of a *PBX (Private Branch Exchange)* that equally distributes incoming calls to agents. As calls come in, they are placed into a queue (or a waiting line) for the next available agent. ACD systems are very versatile and relatively easy to program as some incorporate their own script programming feature. For incoming calls, the waiting times, pre-recorded announcements and other call treatments can be set up by the users/companies to their discretion. Some well-known ACD systems are made by ACCENT, Lucent, and Northern Telecom. See also *ACS*.

ACIS (Automatic Customer/Caller Identification) This feature comes with many *ACD (Automatic Call Distribution)* systems and enables them to make useful the DNIS signal attached to an inbound call. The *DNIS (Dialed Number Identification Service)* can be used to identify the caller or call type. If you have a call center where more than one business is served, then the Automatic Customer/Caller Identification feature can forward the call to a certain group of agents that are associated with the dialed number. It can even tell them what kind of call they are answering on the display of their phone. If a customer calls from Mexico, then the ACD ACIS feature will relate the dialed number with an ACD group that speaks Spanish. Of course, the ACD system needs to be set up or programmed by an administrator to do this.

ACK ASCII control code abbreviation for acknowledgement. Binary code is 0110000 Hex is 60.

Acoustic This term refers to the natural sound vibrations of an object or space. In telecommunications, acoustics are a concern when using hands-free or speaker-phone devices. If the acoustics of the electronic hands-free device are poor, it will vibrate or resonate when the volume level is increased. If the acoustics of the room that the device is in are bad, the device will cut in and out as it "hears" its own echo. Cloth cubicles have a good acoustic vibration-dampening effect. Wide open rooms with no ceiling tile and sheet rock walls have a poor acoustic-dampening effect.

Acquisition The process of a terrestrial-based device locking on a satellite's *GPS (Global Positioning System)* signal. Included in the process of acquiring the signal is *AGC (Automatic Gain Control)* for optimum signal level, synchronization, and processing of the data signal.

ACS (Automatic Call Sequencer) If you can't afford an ACD system, this could be the answer you have been looking for. An ACS answers the call, plays a recorded announcement, and puts the caller on hold. The calls coming in (all the calls) appear as lights on a telephone (or multiple telephones). The calls that have been on hold the longest blink the fastest (or have other signaling methods). ACS systems are designed primarily for or as a part of key systems.

Activated Return Capacity The ability of your cable TV box to send information back to the cable-TV office head end and the ability of the head end to receive the data. This information can include the ID number of the cable TV box and what station you are watching.

Active Device An active device is an electronic component that requires external power to manipulate or react to an electronic input for a desired output. Examples of active devices are: transistors, op amps, diodes, cathode ray tubes, and ICs. If it's not active, it is a passive device. Included in the passive-device category are capacitors (condensers, if you want to use a really old term) resistors, and inductors (or coils), which include transformers.

Active Vocabulary A list of words that a voice-recognition system has been programmed to recognize. Each voice-recognition system has it's own set of words that are selected to fit its application. This is done so that when a voice says "pair," the voice recognizes a word that means two, not a fruit ("pear").

ACU (Automatic Calling Unit) A device that IBM computers use to access outside dial tone for communications. It does the job of a modem, but uses its own protocols to communicate with the computer.

AD (Analog to Digital Converter, ADC) A part of a channel bank that encodes analog voice signals into a stream of binary digits. The digital to analog converter or analog to digital converter samples a caller's voice at a rate of 8000 times per second. (The sample rate for a T1 channel is 8000 times per second.) Each sample's voltage level is measured and converted to one of 256 possible sample levels. These levels are from the lowest, 0000000, to the highest, 11111111. The reason for 256 levels is because if you count in binary from 00000000 to 11111111, you end up with 256, the highest number possible with 8 bits. The bits are then transmitted one after another at a high rate of speed to their destination, where the same process happens in reverse. For a diagram, see *Analog to Digital Conversion*.

Adaptable Digital Filtering A method of conditioning twisted-pair telephone lines to carry data more efficiently up to 12,000 feet before regeneration. The adaptive digital filter can be customized to the characteristics of any given pair (that is in good condition).

Add On A PBX, Centron, or Central Office feature (also known as *three-way calling*). Some telephone stations have a button that is designated "add on." To add a third caller, you push the add-on key and dial the number of the third party, then push add-on again to bridge the calls together.

Address Signals The digits you dial on your phone pad, the phone number is actually an address signal to the local central office that you are connected to.

Addressable Programming For Pay Per View, cable TV companies use an addressable programming system. When you call the phone number to activate the Pay-Per-View movie or event, an IVR system receives your ANI signal or asks for your phone number. Then, it uses your phone number as your customer ID code for billing; in some cases, it identifies which cable-TV converter box to enable. Your cable TV box has an ID code or address code in its memory. When it receives its own address signal from the cable-TV office head end, it enables the horizontal sync for the Pay-Per-View channel for the specified time. It's called *addressable programming* because the converter box is programmed after it receives it's address, which acts like a password.

ADPCM Adaptive Differential Pulse Code Modulation

ADSL (Asymmetrical Digital Subscriber Line) This service will allow telephone companies to provide video to the home over twisted-pair telephone lines. Its current line format is T1 *AMI (Alternate Mark Inversion)*, 16Kb/s to the CO, (for control to change the channel) and 1.528Mb/s to your TV. The distance that the signal can be transmitted over twisted pairs is extended by adaptive digital filtering, which helps correct attenuation and noise.

Advance Replacement The process of getting a replacement component (card, phone, power supply, software, etc.) by calling the distributor or manufacturer and obtaining an advance-replacement reference number. When you receive your advance replacement item, you replace it in the box with the bad item, mark the box with the advance-replacement reference number and send it back. Hopefully, the replacement item doesn't go bad so that you don't have to go through all that again.

Aerial Cable Cable that is attached to power or telephone poles strung through the air. Electrical (power), telephone (fiber optic and twisted pair), and cable TV (coax) are frequently aerial. Aerial cable is attached to a steel strand with lashing wire in most cases. It is sometimes attached during manufacturing as a part of the jacket or sheath (this kind is called *figure-8 cable*). The steel strand is attached to pole with strand clamps and other pole attachments. This is all done with pole-attachment agreements with the owner of the pole, which is the power company, in most cases.

Aerial Cross Box A cross box that is mounted on a pole away from the ground. Aerial cross boxes (also called *tree stands*) are installed in areas where easement rights are narrow or in areas where vandalism is a high risk.

Aerial Cross Box (AP) "Tree Stand"

Aerial Service Wire Splice A common device used to splice aerial service wire (also called a *football* or *potato*).

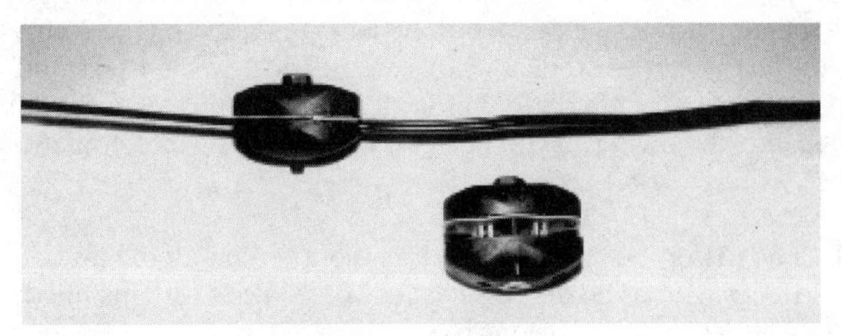

Aerial Service Wire Splice

AGC (Automatic Gain Control) To radio, this is the best thing since sliced bread. Built-in to every radio is an AGC circuit that compensates for the strength of the signal you are receiving. If your radio had no AGC and was tuned to a distant station, tuning to a local station would cause the volume to blare. The way it works is as follows: After the tuner has selected the frequency to be processed, the signal goes to an AGC circuit, which is very similar to a regular intermediate-frequency amplifier (single transistor), except that the gain (amplification) of the circuit is controlled by a level detector. The level detector samples the output voltage of the first preamp and converts it to a DC voltage that is applied to the base configuration of the AGC transistor. The DC voltage controls the bias (amplification configuration) of the AGC circuit, which directly controls how large of a signal is output to the first preamp. The entire system is designed for an optimum signal into the second preamp. When the signal is optimal, then the level of the AGC control signal is zero in most AGC circuits.

AGC is an important part of digital microwave. Some microwave links are often miles apart. When the path of the two dish antennas are aligned, the technician connects a volt meter to the AGC control signal. As the dish is rotated on its axis (azimuth), the technician watches as the AGC control signal changes. When the signal peaks, it is pointed directly at the other antenna. Even though terrestrial microwave links do not

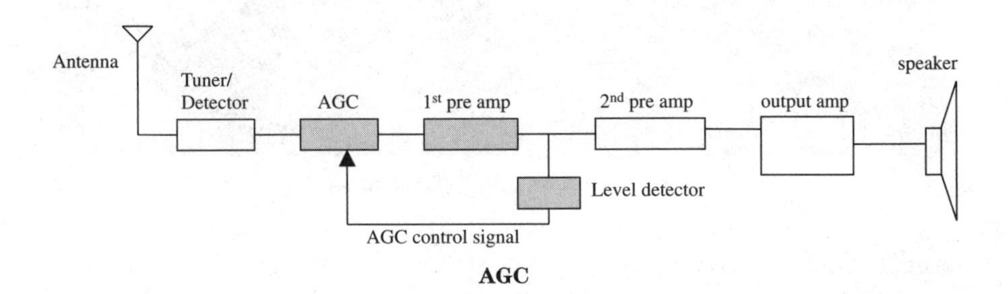

AGC

move or switch stations, they still need AGC to compensate for weather changes.

Aggregate Bandwidth The total bandwidth of a broadband circuit and all of its tributaries, including the payload and overhead. A T1 has an aggregate bandwidth of 1.544Mb/s.

Aggregation Device An ISDN adapter that can combine two B channels (64Kb/s each) together for a single channel that has twice the bandwidth (128Kb/s). These adapters can switch back and forth from aggregated to non-aggregated while the circuit is in use.

Aggregator A long-distance reseller. They sign up with a long-distance company as a reseller and all their customers are "aggregated" together for a bulk discount. The long-distance company provides the service and does the billing. The advantage to the long-distance company is that they have more people selling their long distance. The advantage to the customer is the value-added service (consultation/expertise) that the aggregator offers.

AH (Amp Hour) A battery rating for UPS system and other batteries. The amp-hour rating is derived by multiplying the amount of current that a battery can supply by the time it can supply it. It works out to be a ratio so that you can calculate how long your battery back-up system will last if the power goes out.

For example: If a battery has an amp-hour rating of 100, then it can supply 100 amps for 1 hour. Or it can supply 50 amps for 2 hours, 25 amps for 4 hours, 1 amp for 100 hours, etc.

AIN (Advanced Intelligent Network) The ability of a communications network to determine the routing or handling of a call based on the way the caller desires. AIN is used by local and long-distance companies to give customers a choice as to how they would like their calls routed. A particular trunk can be specifically programmed to route a specific path through switching centers across a geographical area. AIN is ultimately an upgrade to SS7. Some AIN trunks can be made to route to an IVR (Interactive Voice Response) system that gives the customer options for their call handling.

AIOD (Automatic Identification of Outward Dialing) This is a call-accounting system feature of PBX and some key systems that captures every number dialed by a specific telephone extension and prints it out on a report for accounting and cost-tracking purposes.

Airline Mileage The mileage between two cities that long-distance private-line pricing is based on. AT&T developed a grid coordinate system (coordinates shown in V&H table) that gives every telephone central office in the United States a vertical and horizontal grid number. To calculate the mileage between two cities, the Pythagorean theorem is used.

To calculate mileage between two cities, follow these steps:

1. Take the difference of the V coordinates and square it.
2. Take the difference of the H coordinates and square it.
3. Add the two squared numbers together.
4. Divide by 10
5. Take the square root of that number. This is the mileage.

Example: What is the airline mileage from Los Angeles, CA to New York, NY?

- The V coordinate of Los Angeles is 9213. The V coordinate of New York is 4977. The difference is 4236
- H coordinate of Los Angeles is 7878. The H coordinate of New York is 1406. The difference is 6472
- Next, square both numbers: $4236^2 = 17{,}940{,}000$. $6472^2 = 41{,}890{,}000$
- Now, add these numbers: 17,940,000 + 41,890,000 = 59,830,000
- Now, divide these numbers: 59,830,000 ÷ 10 = 5,983,000.
- Take the square root: $\sqrt{5{,}983{,}000} = 2446$.
- 2446 miles is the airline mileage between Los Angeles and New York.

Air-Pressure Cable Telephone cable that is equipped with air-pressure equipment. In many cables nitrogen is used instead of air because it is noncorrosive (air contains humidity and oxygen that corrodes copper pairs). Nitrogen is pumped into the cable and the pressure is monitored. If the cable is cut, the pressure drop notifies the telephone company of a cable problem and the nitrogen rushing out of the cable helps prevent any water from entering the cable. For a photo of an air pressure splice closure, see *Waffle Splice Closure*.

Alligator Clips Most analog test equipment comes equipped with alligator clips. For a photo, see *Bed of Nails Clips*.

All Trunks Busy You might try to make a call and get a fast busy signal or an intercept message that says "I'm sorry, all circuits are busy now. Please try your call again later." This situation can happen for a number of reasons. If you are dialing long distance, you get this message because all of the trunks that your long-distance company has between their inter-lata central office *POPs (Points of Presence)* are busy. If you are making a local call that terminates to a different local CO and you get this message, it is because all the inter-office trunks are busy. If you are calling your neighbor and you get this message, then the inter-grouping trunks within the local CO switch are all busy. Inter-grouping trunks are used in large switches to interconnect "smaller CO switch groups" within the CO switch.

ALPETH (Aluminum/Polyethylene) The sheath or jacket of an outside plant telephone cable that is used mostly for aerial applications. It is basically an aluminum wrap around the conductors, which resembles a serrated tin can, coated with ⅛" of black plastic.

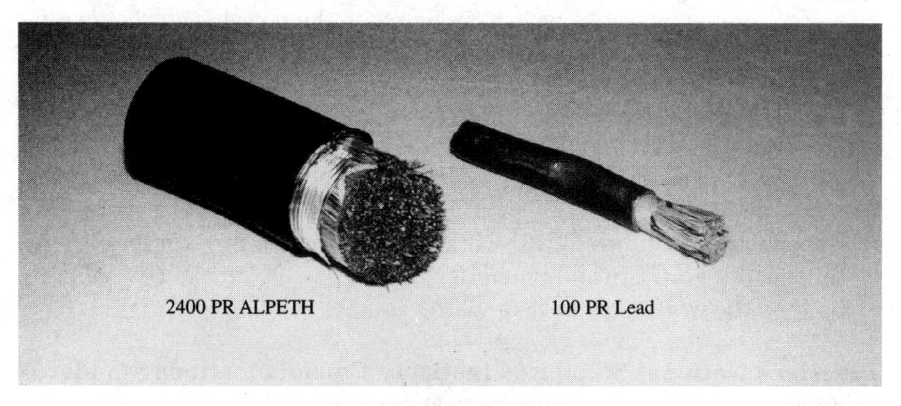

ALPETH (left) Lead (right)

Alternate Answering Position A second attendant console where the first console can forward calls if the first console attendant is absent. An alternate answering position can also be used for overflow of calls that the first attendant can't keep up with.

Alternate Routing A switch feature that enables all trunks to have alternate outgoing assignments. If the primary routing or least-cost routing is all busy or out of service, then the switch will route the call to an alternate trunk to connect the call. It is a good idea to use multiple long-distance and local services in conjunction with alternate routing in case of a service outage.

AM (Amplitude Modulation) AM is a technique of making a voice or other signal ride on (or modulate with) another frequency (the carrier frequency). For a diagram, see *Amplitude Modulation.*

AMA (Automatic Message Accounting) What RBOCS call their call tracking system for billing.

Ambient Current The result of the voltages created by random movement of electrons in a circuit when the power is off. There is always ambient voltage, which is why oscillator circuits start oscillating when the power is turned on. The natural oscillations of the electrons become filtered and amplified when the power is applied to the circuit.

Ambient Noise Noise caused by the random movement of electrons in an electronic circuit when the power is off or by the random movement of air.

Ambient Voltage Electromotive force created by the random movement/vibration of electrons in a circuit when the power is off. There is always ambient voltage, which is why oscillator circuits start oscillating when the power is turned on. The natural oscillations of the electrons become filtered and amplified when the power is applied to the circuit.

American National Standards Institute (ANSI) A nongovernmental nonprofit standards setting institute that publishes standards that industries voluntarily follow. ANSI works very hard to bring together the interests of the private and public sector. ANSI is the official U.S. member body to the world's leading standards bodies.

American National Standards Institute Communications Standards Some examples of ANSI communications standards are:

ANSI T1.110-1987 SS7	General information
ANSI T1.111-1988 SS7	Message Transfer Part (MTP)
ANSI T1.112-1988 SS7	Signaling Connection Control Part (SCCP)
ANSI T1.113-1988 SS7	ISDN user part
ANSI T1.114-1988 SS7	Transaction Capability Application Part (TCAP)
ANSI T1.206	Digital Exchanges and PBX loop-back test lines
ANSI T1.301	ANSI ADPCM standard
ANSI T1.401-1988	Interface between carriers and customer installations for voice-grade switched analog lines Loop Start and Ground Start.
ANSI T1.501-1988	Network performance/network encoding limits for 32Kb/s ADPCM
ANSI T1.601-1988	Basic access interface/electrical loops for the network side

ANSI T1.T1.Q1	Network performance standards for switched exchange and IXC
ANSI TIX9.4	SONET
ANSI X3T9.5 TPDDI	FDDI on UTP/STP
ANSI character set	ANSI 256 character set code

AMI (Alternate Mark Inversion) AMI is a line format that serves two advantages to sending a digital signal directly over a twisted pair. The feature of AMI that makes it unique is that each bit is inverted. This makes the first bit +5 V, the second -5 V, the third +5 V, etc. Alternating the bit polarity also makes the signal look like it is half the frequency to the twisted pair.

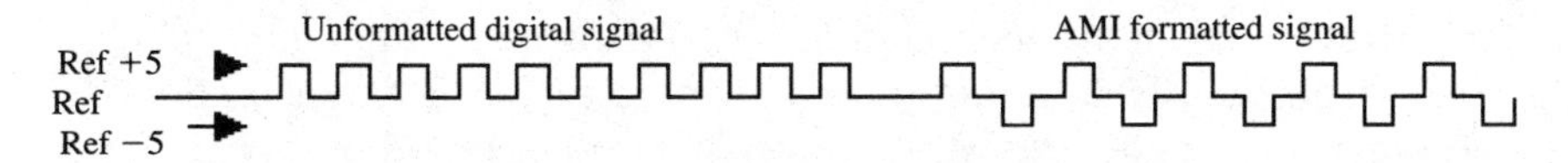

Reference voltage for T1 carrier is typically +135 Volts over the public network, so AMI actually switches from +140V to +130V

AMI (Alternate Mark Inversion)

Amp Abbreviation for Ampere, see *Ampere*.

Ampere 6,300,000,000,000,000,000 electrons moving past a point in one second (a coulomb) is equal to one ampere (also known as an *amp*) of electrical current. The shortcut/alternative to counting all the electrons as they run by is to use the Ohm's law formula and calculate the amperage instead. If you know two of the following about your circuit, voltage, resistance, or watts you can perform the calculation. The formulas are:

$$\textit{Current} \text{ (in amps)} = \frac{\textit{Voltage} \text{ (in volts)}}{\textit{Resistance} \text{ (in ohms)}}$$

or

$$\textit{Current} \text{ (in amps)} = \frac{\textit{Power} \text{ (in watts)}}{\textit{Voltage} \text{ (in volts)}}$$

or

$$\textit{Current} \text{ (in amps)} = \sqrt{\frac{\textit{Power} \text{ (in watts)}}{\textit{Resistance} \text{ (in ohms)}}}$$

Amp Hour (AH) A battery rating for UPS system and other batteries, the Amp Hour rating is derived by multiplying the amount of current that a battery can supply by the time it can supply it. It works out to be a ratio so that you can calculate how long your battery back-up system will last if the power goes out.

For example: If a battery has an amp hour rating of 100, then it can supply 100 amps for 1 hour, 50 Amps for 2 hours, 25 Amps for 4 hours, 1 amp for 100 hours, etc.

Amplified Handset A handset with a built-in amplifier for the hearing impaired. Amplified handsets can be purchased for virtually every kind of PBX telephone. Walker Electronics is a well-known manufacturer of these devices.

Amplifier An electronic circuit designed to increase an input level characteristic to a desired output level characteristic. Some amplifiers are designed to amplify the voltage level of a signal and others are designed to amplify the current of a signal flowing through a load. A typical stereo system has both of these types of amplifiers. If you are listening to a CD player, the signal (after digital to analog processing) is fed to a voltage amplifier to increase its ability to drive a current amplifier, which amplifies the current that is driven through the loudspeaker. Both of these amplifiers (voltage and current) combined make a power amplifier, hence power (in watts) is a function of voltage and current. Amplifiers are usually rated by the amount of power that they are capable of producing in a loudspeaker. Peak power is calculated by using the peak value of a sinusoidal waveform. RMS (root mean square) power is calculated by using the RMS value of the sinusoid waveform, which mathematically works out to be 70.7% of the peak value. Many amplifier manufacturers use the peak-power rating because it looks better. If you compare a JVC amplifier that is rated at 71 watts RMS output, it is the same output as the 100-watt peak-power "other brand" amplifier. High-quality audio amplifiers are usually rated in RMS power.

Amplitude The peak or peak-to-peak amplitude of a signal measured in volts. The AC signal below has a peak amplitude of 10 volts or a peak-to-peak amplitude of 20 volts.

Amplitude Modulation AM is a technique of making a voice or other signal ride on (or modulate with) another frequency (the carrier frequency).

AMPS (Advanced Mobile Phone System) This is the cellular/PCS network as we know it today. The first mobile phone system was called *MTS*

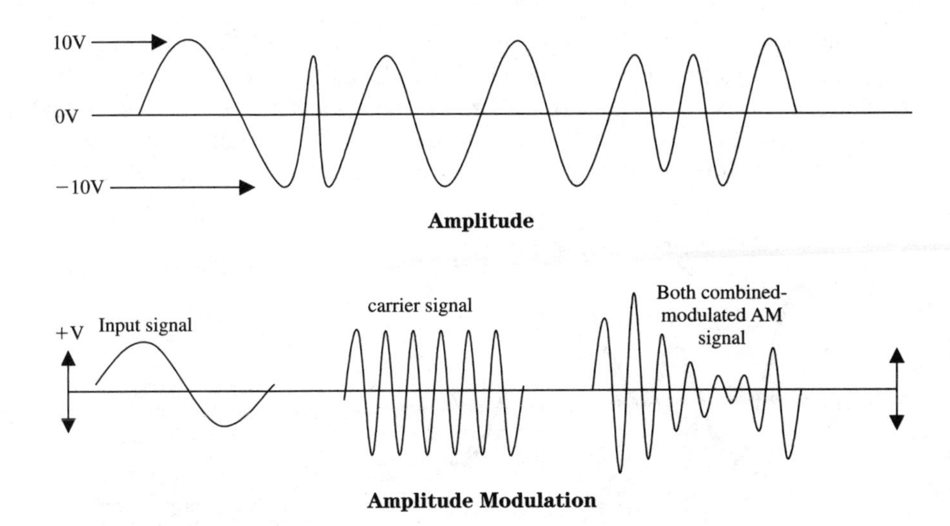

Amplitude

Amplitude Modulation

(Mobile Telephone System) and was developed during World War II. To make a call on an MTS system, a human operator was needed to hand-off/connect the call. In the early 60's *IMTS (Improved Mobile Telephone System)* evolved. IMTS did not require a human operator to connect a call from one mobile phone to another, but calls could only be made within one cell. In 1983, the implementation of the *AMPS (Advanced Mobile Phone System)* began. AMPS allows callers to call from one mobile phone to another, from one cell to another, and connect calls between the land-based network and the mobile network without the need for an operator.

Analog A signal having an infinite number of levels per cycle, in contrast to digital, which has only two possible levels per cycle (i.e., on or off).

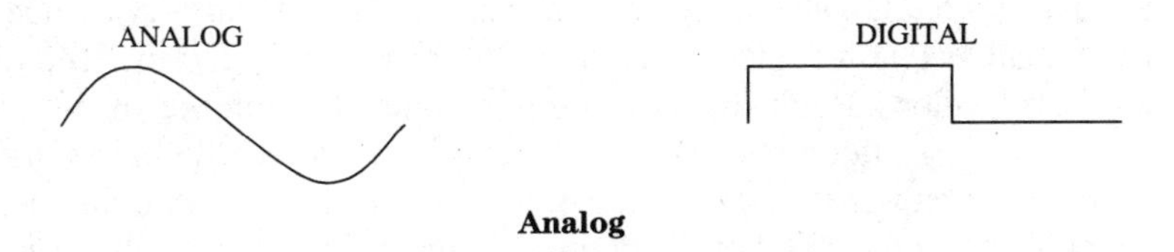

Analog

Analog to Digital Conversion (Digital to Analog Converter/Analog to Digital Converter) A part of a channel bank that performs the function of encoding analog voice signals into a stream of binary digits. The analog-to-digital converter samples a caller's voice at a rate of 8000 times per second. (The sample rate for a T1 channel is 8000 times per second) Each sample's voltage level is measured and converted to one of 256 possible sample levels. These levels are from the lowest, 0000000, to the highest 11111111. It has 256 levels because if you count in binary from 00000000 to 11111111, you end up with 256, that is the highest number possible with 8 bits. The bits are then transmitted one after another at a high rate of speed to their destination, where the same process happens in reverse.

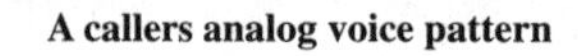

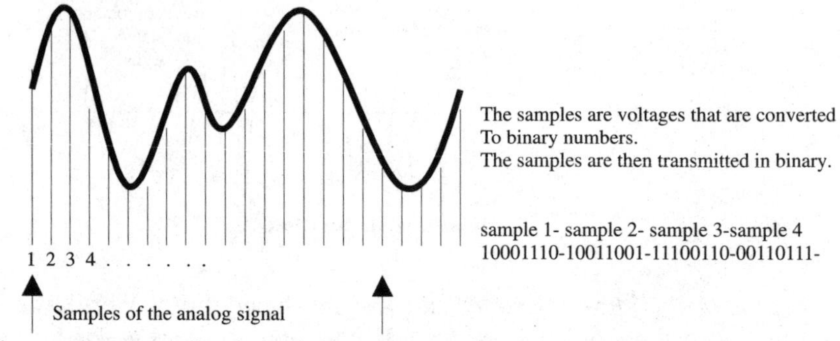

The samples are voltages that are converted
To binary numbers.
The samples are then transmitted in binary.
sample 1- sample 2- sample 3-sample 4
10001110-10011001-11100110-00110111-
1 2 3 4 . . . . . .
Samples of the analog signal

The binary numbers are actually voltages, the binary bit stream of four samples is shown as voltages below.

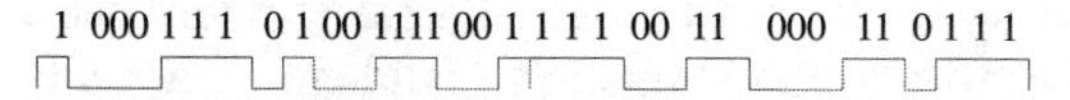

At the other end of the transmission line, the binary code is converted back into its original analog form using the same process in reverse.

Analog to Digital Conversion

ANI (Automatic Number Identification) ANI is also called Caller Identification or Caller ID. This feature, offered by local phone companies, sends the phone number (and often the name of the caller) down the phone line in a digital data packet between the first and second ring. To receive the data, a subscriber that has signed up for the service needs to have a caller ID unit (also called a *caller-ID box*) plugged into the phone line. The caller-ID unit displays the name and the number of the calling party for each incoming call. Caller ID only works if the caller and the called party's phone service is fed out of a central office that has caller-ID capability. If the central office does not have caller-ID capability, then the display will read "out of area" to the called party. If the called party does not have caller service they will get a display that says "no data sent."

Annex A A frame relay standard extension that outlines the provisioning of a *Local Management Interface (LMI)* that goes between the customer's equipment and the frame relay network. The LMI will provide network monitoring and status through the transmission overhead.

Annex D The second frame relay standard extension that outlines the provisioning of a Local Management Interface (LMI) that goes between the customer's equipment and the frame relay network. The LMI will provide network monitoring and status through the transmission overhead.

Annular Ring A marking around a cable to show length—some are in meters and some are in feet.

Anonymous Call Rejection A feature that can be provided by the local phone company that will not complete anonymous calls to your line. You can also reject anonymous calls by installing a caller-ID unit that has anonymous call rejection built-in to it.

ANSI (American National Standards Institute) A nongovernmental nonprofit standards-setting institute that publishes standards that industries voluntarily follow. ANSI works very hard to bring together the interests of the private and public sector. ANSI is the official U.S. member body to the world's leading standards bodies.

ANSI Standards See *American National Standards Institute Communications Standards* for some examples of their communications standards.

Answer Back A command that a local data terminal sends to a computer or device across a network asking it to send its address so that the local terminal can verify that it has connected to the correct computer.

Answer Supervision The ability of a central office to know when a long-distance call has been answered so that the call can be billed. This feature is a combination of the SS7 network and CO switch software that integrates with the telephone company's call accounting and billing system. Anyone who bills people for phone calls should have this system, but some do not. Hotels are famous for not having answer supervision. If you make a call from your hotel room and the end you are calling rings for more than 30 seconds (eight rings) you will probably be billed for the call. The hotel's PBX has no way to know if anyone picked up the phone on the other end, so it starts billing for the call after a certain time. The hotel's PBX is supervising and billing according to you being off-hook and the digits you dialed, not according to when the "answerer" picked up the line. To this day, some long-distance companies don't have answer-supervision capability.

Antenna A device for receiving and transmitting electromagnetic signals. The optimal antenna for a given transmission or reception of a frequency has a length equal to the wavelength (or a usable fraction) of that frequency. CB radio antennas are very long in comparison to the antenna on your PCS cellular phone. CB radio is transmitted at frequencies that are low, which have a long wavelength and PCS cellular is transmitted at high frequencies, which have shorter wavelengths. See also *Yagi Antenna*, *Parabolic Dish Antenna*, and *Monopole Antenna*.

Antennas are available in many varieties that are designed to manipulate the incoming or outgoing signal.

- *Single pole, full wavelength* For picking up weak signals or transmitting over long distances.
- *Single-pole half or quarter wavelength* Smaller size and for receiving nearby stations.
- *Dipole* For directional applications. Rabbit ears, for example, are directional.
- *Phased array* For picking very distant signals.
- *Parabolic dish* For focusing a signal from or to one direction.
- *Vertical Loop* For low-noise, directional reception. Common for UHF TV.
- *Horizontal Loop* For low-noise, directional reception.
- *Multielement* For low-noise, directional reception or transmission.
- *Monopole* Used in cellular and PCS applications.

Antenna Gain The two types of antenna gain are *field gain*, which is expressed in volts, and *power gain*, which can be expressed in decibels. The important thing to remember about antennas and their characteristics is that they are built to be an integral part of the transmitter. An antenna by itself is nothing and a transmitter by itself is nothing. An antenna's ratings determine what type of transmitter it should be matched with and vice-versa.

Anti-Static Materials coated or manufactured with semiconductive materials makes them anti-static, which is good for *CMOS (Complementary Metal-Oxide Semiconductor)* components. CMOS components are highly sensitive to static discharges and static fields (ESD). Your body can easily hold a static charge of 40,000 V on a dry day. About 25,000 V is required to get a static shock from a door knob. Exposure to static electricity can ruin a CMOS component instantly or just weaken it, which would cause it to fail unpredictably. CMOS components include microprocessor chips, transistors, RAM and ROM chips, and many others. If a CMOS device must be used in an ESD hazard area, TTL components are used as an alternative. TTL components are not as static sensitive, but, they are not as fast, not as small, use more electricity, and produce more heat.

AOSP (Alternate Operator Service Provider) A long-distance company that works like the old days, when a live operator would assist you with your call. Some calling-card companies incorporate this in their service. AOSP service is great, and it's a good thing because you pay for it.

AP (Access Point) Another name for a cross-box where telephone cables are cross connected. For pictures see *Access Point* and *Aerial Cross Box.*

APD (Avalanche Photo Diode) A device used as a light-to-electricity converter and signal amplifier at the same time. They are incorporated in optoelectronic circuits used in fiber-optic terminating applications.

Application Layer The seventh and highest layer of the OSI communications model. The applications layer is the function of connecting an application file or program to a communications protocol. The latest model or guideline for communications protocols is the OSI (open systems interconnect). It is the best model so far because all of the layers or functions work independently of each other. Older proprietary communications models are shown below along side the OSI. For a diagram of the OSI, SNA, and DNA layers, see *Open Systems Interconnection.*

Applications Processor An add-on to a PBX (Private Branch Exchange) system or CO (Central Office) switch that expands its ability to provide extended services or process additional protocols. An example of an applications processor is a voice-mail system, ACD, frame relay or ISDN interface. Physically, the applications processor is often an additional shelf, module, or card that interfaces into the PBX system's bus architecture.

Area Code An area code is a three-digit code that designates a toll center in the North American Numbering Plan. To call outside of your toll center, you first dial 1, then the area code for the toll center or "area" you wish to call. See Appendix C for a listing of area codes by area. See Appendix D for a listing of area codes by number.

ASCII (American Standard Code for Information Interchange) ASCII is a code developed by ANSI. It is a seven-bit character and command code. More than one variation of ASCII exists, such as extended ASCII, which is eight bit and has many more characters. ASCII is the standard code that PCs use and is the code that is transmitted into your computer every time you push a key on your keyboard.

Least significant bits (hexadecimal)	Most significant bits (hexadecimal) 000 (0)	001 (1)	010 (2)	011 (3)	100 (4)	101 (5)	110 (6)	111 (7)
0000 (0)	NUL	DLE	SP	0	@	P	`	p
0001 (1)	SOH	DC1	!	1	A	Q	a	q
0010 (2)	STX	DC2	"	2	B	R	b	r
0011 (3)	ETX	DC3	#	3	C	S	c	s
0100 (4)	EOT	DC4	$	4	D	T	d	t
0101 (5)	ENQ	NAK	%	5	E	U	e	u
0110 (6)	ACK	SYN	&	6	F	V	f	v
0111 (7)	BEL	ETB	'	7	G	W	g	w
1000 (8)	BS	CAN	(	8	H	X	h	x
1001 (9)	HT	EM	)	9	I	Y	I	y
1010 (A)	LF	SUB	*	:	J	Z	j	z
1011 (B)	VT	ESC	+	;	K	[	k	{
1100 (C)	FF	FS	,	<	L	\	l	
1101 (D)	CR	GS	-	=	M	]	m	}
1110 (E)	SOH	RS	.	>	N		n	~
1111 (F)	SI	US	/	?	O		o	DEL

DEFINITIONS OF ASCII CONTROL CODE ABBREVIATIONS

ACK - ACKNOWLEDGE
BEL - BELL
BS - BACKSPACE
CAN - CANCEL
CR - CARRIAGE RETURN
DC - DIRECT CONTROL
DEL - DELETE IDLE
DLE - DATA LINK ESCAPE
EM - END OF MEDIUM
ENQ - ENQUIRY
EOT - END OF TRANSMISSION
ESC - ESCAPE
ETB - END OF TRANSMISSION BLOCK
ETX - END OF TEXT
FF - FORM FEED
FS - FORM SEPARATOR
GS - GROUP SEPARATOR
HT - HORIZONTAL TAB
LF - LINE FEED
NAK - NEGATIVE ACKNOWLEDGE
NUL - NULL
RS - RECORD SEPARATOR
SI - SHIFT IN
SO - SHIFT OUT
SOH - START OF HEADING
STX - START OF TEXT
SUB - SUBSTITUTE
SYN - SYNCHRONOUS IDLE
US - UNIT SEPARATOR
VT - VERTICAL TAB

ASCII

ASIC (Application Specific Integrated Circuit) A proprietary integrated circuit created to do a specific job. An example is Motorola manufacturing a chip to do multiplexing specifically for a MUX that is made by Nortel. Motorola manufactures the chips for Nortel and stamps Nortel's name on them. Then, only Nortel uses the technology and no one else.

ASR (Access Service Request) If a special service provider (frame relay or long-distance private line) needs wire facilities from their point of presence in the city to your location, they call the local telephone company and make an access service request to provide a line that runs from your network interface to them. Many special service providers have their equipment located in the local phone company's central office as a part of a co-location agreement. When a CLEC needs to provide service where they don't have facilities, they do it by using the RBOCS wire facilities.

Asymmetric Communications transmission that is full or half duplex, where one direction is very fast, compared to the other. Cable TV is an example of asymmetrical communication. The cable TV head end sends massive amounts of video and audio information down a coax one way, and the cable TV set-top decoder boxes send small amounts of ID and status information the other way back to the head end over the same coaxial connection. Sometimes asymmetrical channels are referred to as "upstream" for slow and "downstream" for fast.

Asymmetrical Digital Subscriber Line (ADSL) This "service in the making" will allow telephone companies to provide video to the home over twisted-pair telephone lines. Its current line format is T1 *AMI (Alternate Mark Inversion)*, 16Kb/s to the CO, (for control to change the channel) and 1.528Mb/s to your TV. The distance that the signal can be transmitted over twisted pairs is extended by adaptive digital filtering, which helps correct attenuation and noise.

Asynchronous To communicate without external timing and to have each communicating device work at its own speed. People talk asynchronously. Even though one person talks very fast and another very slowly, their brains still receive the conveyed messages and respond. Modems and FAX machines are asynchronous.

Asynchronous Transfer Mode (ATM) An ANSI and CCITT standard communications protocol. ATM is a frame-format communications protocol whereby data is transmitted and received 53 bytes or octets at a time. There are 48 customer bytes (Payload) and 5 bytes for control and addressing. Four things make ATM special.

1. It is capable of carrying delay-sensitive transmissions without delay (such as speech, music, or video).
2. Many ATM channels can be concatenated to provide more bandwidth (carrying capacity).
3. It is future-compatible to *BISDN (Broadband Integrated Services Digital Network)* over SONET.
4. It has the ability to carry many different types of data at the same time: LAN, video, voice, and anything else that is capable of being digitized.

ATM FRAME FORMAT

48 byte payload	5 byte overhead

Asynchronous Transfer Mode (ATM)

Another good thing about ATM is that the overhead is a bit more than 10%, which is an improvement over other transport methods. To put ATM into a simple picture, imagine that you have a computer network LAN signal and a video signal. You want to send the signals that your computer and video are generating to other computers and TVs.

Now, take the scenario a step further. Imagine that your computer LAN signals are motorcycles and the video signals are cars. ATM would then be large trucks that carry the motorcycles and cars to their destination and back, linking your communications gap with two great things in mind. Your cost is lower in contrast to many small circuits and you only buy one piece of gear to connect everything (in contrast to many different terminal adapters, CSU/DSU, etc.). ATM service is not yet available everywhere.

AT (Access Tandem) A telephone company central office or node that contains a switch in which all inter and outer area-code traffic is handled. The main LEC central office in an area code, where the hand-off for long-distance service happens. For a diagram, see *Access Tandem.*

ATM (Asynchronous Transfer Mode) An ANSI and CCITT standard communications protocol. ATM is a frame-format communications protocol whereby data is transmitted and received 53 bytes or octets at a time. There are 48 customer bytes (Payload) and 5 bytes for control and addressing. For more information, see *Asynchronous Transfer Mode.*

Attendant Another name for a PBX operator. The person who connects outgoing calls and/or answers, screens and directs incoming calls in a polite mannerly way. If they don't, they would soon be replaced by an auto-attendant.

Attenuation Reduction of a signal's voltage level as it travels down a line, measured in decibels. Attenuation is also called *loss*, because some signal is always lost through resistance and reactance. Optical light-wave signals are also attenuated when they traverse through a fiber-optic because of impurities in the fiber optic, and the fact that light intensity decreases with distance.

Attenuator An attenuator is also called a *pad*, *T pad*, or *H pad*. It is a device that reduces the voltage level of a signal without changing it's impedance. Attenuators are frequently used on telephone lines that terminate to customers close to the central office so that the volume in the handset does not hurt their ears. Attenuators are also made for fiber-optic applications. A fiber-optic attenuator works like your sunglasses, it reduces the level of light entering your eyes so that you can see more effectively.

Audio Sound. Signal frequencies that if amplified and applied to a loudspeaker can be heard. These frequencies range from 20 Hz to 17 kHz.

Audio Frequency Signal frequencies that, if amplified and applied to a loudspeaker, can be heard. These frequencies range from 20 Hz to 17 kHz.

Auger A device that looks like a giant drill bit, which is used for boring holes into the ground for telephone or power poles. Some utility construction vehicles are equipped with augers.

AUI (Autonomous Unit Interface) A 15-pin connector used in *CTI (Computer Telephony Integration)* applications. For a picture, see *DB15*.

Auto Baud A term that refers to a modem or other communicating device's ability to match or adapt to the bit transmission rate of the device at the far end.

Auto Dialer A device that automatically dials preprogrammed digits when the line it is attached to breaks a dial tone. Auto dialers are used to program long-distance carrier access codes so that people don't have to deal with the confusion of which long-distance company to access and when. The person making a call pushes a key on their phone that accesses a long-distance trunk equipped with an autodialer that dials the access code, then the person making the call dials in the number they want to call. Auto dialers are also used to make a phone dedicated to one phone number or directory extension. In the city where I live there is a restaurant with no waiters or waitresses. When the patrons to the restaurant are seated, they find a phone at their table. The phone has no dial pad. When the phone is picked up, an autodialer instantly dials the order-taker's extension and the patrons place their order.

Automated Attendant The machine that answers the line and plays a message that says: "Thank you for calling company X. To speak to a person in sales press one, to speak to a repair person press two." Advantages of auto attendants are that you can have one number advertised for multiple departments and not have to have a full-time person directing calls.

Automated Voice Response Not to be confused with *integrated* voice response, an *automated* voice-response system or network is a way of guiding callers to a department, agent or pre-recorded information. Automated-attendant/voice-mail systems that have directory trees programmed into them are becoming the most popular ways of accomplishing this. Directory trees are set-up by the voice-mail administrator. Directory trees are capable of being as long and complicated as the caller can tolerate. For example, the caller hears a pre-recorded message that prompts them to dial "1" for information about skin ailments, dial "2" for flu-like symptoms or dial "3" for head pain. Options 1, 2, and 3 then branch out into a tree:

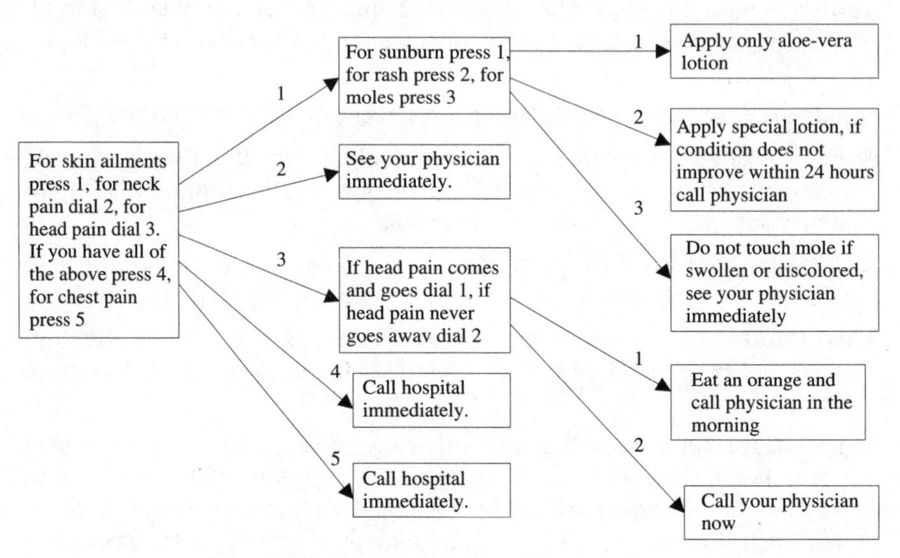

Sample Information Mailbox/Directory Tree

Automatic Call Distributor (ACD system) A separate system or built-in feature of a PBX that equally distributes incoming calls to agents. As calls come in, they are placed into a queue (or a waiting line) for the next available agent. ACD systems are very versatile and relatively easy to program because some incorporate their own script programming language. For incoming calls, the waiting times, pre-recorded announcements and other call treatments can be set up by the users/companies to their discretion. Some well-known ACD systems are made by ACCENT and Northern Telecom.

Automatic Gain Control see *AGC*.

Automatic Number Identification (ANI) A service provided by local and long-distance telephone companies that sends the name and number of the calling telephone to a display attached to the called phone line. In an in-band signaled phone line or residential telephone line, the ANI signal packet comes as a miniature data burst between the first and the second ring.

Avalanche Photodiode A device used as a light-to-electricity converter and signal amplifier at the same time. They are incorporated in optoelectronic circuits used in fiber-optic terminating applications.

Avalanching Avalanching occurs when a PN diode or transistor junction is reverse biased (reverse positive and negative) with enough voltage to force it to conduct in the wrong direction. Diodes and transistors have an avalanche voltage rating. When this voltage is exceeded, the device avalanches. The avalanche is actually a sudden steady rush of current that causes lots of heat. This usually damages the device (severely). Some devices (avalanche photodiode, SCR, etc.) are designed to use the avalanche effect in a useful way. They switch "on" or conduct when the reverse or gate voltage applied to them reaches a certain level.

AWG (American Wire Gauge) A measurement standard for copper wire. The gauge rating is the thickness of a solid copper wire. The larger the gauge, the smaller the wire. Most telephone wire is 19 AWG at the largest and 26 AWG at the smallest. Cat 3 is commonly 24 AWG. The electrical wire in your home is probably 12 AWG.

Azimuth In directional radio transmission, the azimuth is the direction in degrees (bearing) that an antenna is transmitting to its far-end counterpart. Azimuth is also the manual tracking adjustment on a magnetic tape head, such as in a cassette player. When your cassette player sounds muffled, the heads could be dirty or the azimuth could need adjustment. If a directional radio signal seems weak after a large bird crashed into the dish, the azimuth might need to be realigned.

B3ZS (Bipolar 3-Zero Substitution) A line-coding/data-transmission format. Transmission formats are used to prevent too many consecutive zeros from being transmitted. If too many zeros go down the line in a row, the transmission line effectively becomes a flat line, with no timing.

B8ZS (Binary 8-zero substitution) A line-coding/data-transmission format. Transmission formats are used to prevent too many consecutive zeros from being transmitted. If too many zeros go down the line in a row, the transmission line effectively becomes a flat line, with no timing. If a sequence of eight bits are detected prior to being transmitted, they are replaced with a different pre-determined byte that is not all zeros.

B 911 This is also known as *Basic 911 emergency service*. It is the older version of the 911 system and it is what public service commissions want phone companies to phase out. B 911 does not provide automatic location information. In many cases (depending on the CO switch/911 software serving the customer that makes the 911 call), it does not provide automatic number identification.

B Channel The "bearer" channel of an ISDN circuit. It carries 64Kbp/s of end-user data. The other ISDN channel is referred to as the *D* or *data* channel, which is 16Kbp/s and carries phone company signaling along with the other stuff that makes the ISDN circuit work. The two categories of ISDN are the *BRI* (*Basic Rate Interface* 2B channels and 1D channel) and the *PRI* (*Primary Rate Interface* 23B channels and 1D channel).

For a relational diagram of the two types of ISDN lines, see *Integrated Services Digital Network*.

B Connector A wire-splicing connector for splicing twisted pair wire, also called *beans*. Beans are shaped like a plastic tube that is about as big around as a drinking straw, but only an inch long. They have metal teeth inside them so that when two wires are crimped inside, they make a good connection and don't slide out. Beans can also have a water-retardant jelly inside them as well.

B Washer, Curved The washer used between a telephone pole and a strand clamp when installing pole attachments.

Curved B Washer

Back Bone The part of a communications network that connects main nodes, central offices, or LANs. The backbone usually has its own high-speed protocol, such as switched token ring or FDDI for LAN interconnections, and SONET for central-office and main-node interconnections.

Back-Feed Pull When it is difficult to get large cable pulling equipment into end locations of a cable installation, outside plant construction personnel use a technique called a *back-feed pull*. If there is a vault or hand hole that the cable passes through between the end points, the cable will be fed in two parts. The first section of cable will be fed one direction from the mid-point. After the first half is fed, the remaining cable is unreeled and fed through the opposite direction to other end point.

Back Haul A long-distance service term. Sometimes you can save money by creatively routing long-distance calls. *Back haul* is a routing term that means routing a call past its destination and then back. Many new

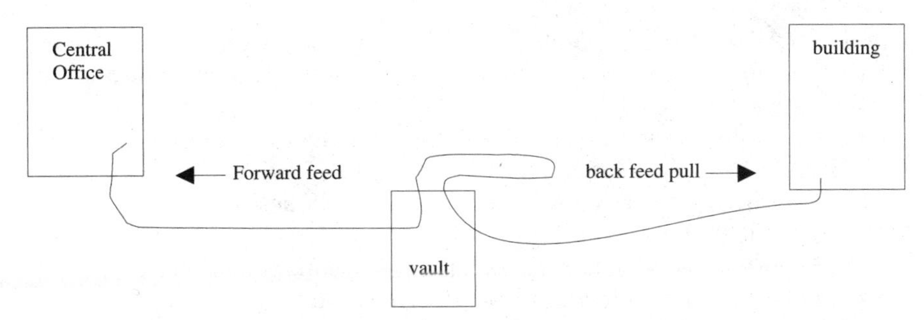

New cable installation using a back feed pull

long-distance companies are providing less-expensive long-distance services to specific cities and conglomerate long-distance companies take advantage of this by back hauling phone calls. Airline passengers are often back hauled because it is less expensive to have a lay-over in another nearby larger city—even though it is beyond your destination.

Back Hoe Fade The cutting of a buried fiber optic that connects two communications nodes. It is called a *fade* because in some equipment architectures, not all communications are cut off—they are either divided or rerouted. To some, this is a "fade" in communications service, not a complete outage.

Back Plane The main electronic PC board in a communications equipment cabinet that has slots or connectors for circuit cards to be plugged into. The back plane in almost all key service units and PBX card cages is where the power bus, CPU bus, and the control bus are located.

Northern Telecom Meridian PBX backplane

Backward Channel The channel that flows upstream in an asymmetrical transmission. An asymmetrical communications transmission that is characterized by one direction being very fast compared to the other. Cable TV is an example of asymmetrical communication. The cable TV head end sends massive amounts of video and audio information down a coax one way and the cable TV set-top decoder boxes send small amounts of ID and status information the other way back to the head end over the same coaxial connection. Sometimes asymmetrical channels are referred to as "upstream" for the slow channel and "downstream" for the fast channel, or "forward" for the fast channel and "backward" for the slow channel.

Bad Line Button A button on an attendant console that enables an attendant to busy out a trunk that is in trouble (such as noise or static). Some bad line buttons don't busy out the trunk; they just mark it so that it can be identified from the rest when the local phone company is called to fix the problem. Sometimes the trouble is not in the trunk, but at least when you call the phone company, you can tell them which line you are having trouble with.

Baffle An enclosure for the back of a loudspeaker that improves its sound quality by improving its acoustic profile. Baffles are made of thin, flexible rubber foam and are frequently used on ceiling intercom speakers and automobile speakers. They usually cost about $6.00 to $10.00 each, depending on the size of the speaker.

Balance, Circuit An electronic circuit that can be active (using external power) or passive (using only capacitors, resistors, and inductors) that is attached to a twisted copper pair to even out the electronic characteristics of both wires. Balance is very important in a transmission line. If the two wires or pair is not balanced, noise is created on the line, which interferes with the transmission signal.

Balancing Network See *Balance, Circuit.*

Band, Citizens A low-power two-way transmission radio band in the United States. There are actually two types or bands of CB radio. They are 26.965 MHz to 27.225 MHz and 462.55 MHz and 469.95 MHz

Band, Frequency The following are the defined boundaries of radio frequency bands:

ELF	Extremely Low Frequency	below 300 Hz
ILF	Infra Low Frequency	300 to 3000 Hz

VLF	Very Low frequency	3 kHz to 30 kHz
LF	Low Frequency	30 kHz to 300 kHz
MF	Medium Frequency	300 kHz to 3000 kHz
HF	High Frequency	3 MHz to 30 MHz
VHF	Very High Frequency	30 MHz to 300 MHz
UHF	Ultra High Frequency	300 MHz to 3000 MHz
SHF	Super High Frequency	3 GHz to 30 GHz
EHF	Extremely High Frequency	30 GHz to 300 GHz
THF	Tremendously High Frequency	300 GHz to 3000 GHz

Band, Marking A label placed around an insulated wire or fiber optic for identification. Some are printed on during manufacture and some are attached during installation.

Band-Elimination Filter A band-elimination filter is a circuit used to pass a certain range of frequencies away from specific equipment or devices to ground, or somewhere they are wanted. Radio-frequency interference is easily eliminated with the correct band-elimination filter. Band-elimination filters are also called *RFI (Radio Frequency Interference)* filters, RFI suppressors and *EMI (Electromagnetic Interference)* suppressors. They work by providing an easier path for noise to go through, rather than your electronic device (such as your telephone or modem). Many RFI filters are on the market, and some are adjustable to different frequencies. Some are modular instead of hard wired into the NI or jack so that you can plug them right into your phone line.

Band-Pass Filter A band-pass filter is used in frequency-division multiplexing, as well as the equalizer in your stereo. It is usually a capacitor/resistor/inductor network that has a resonant frequency and a rating of how well it passes a band of frequencies and blocks out others, called the *Q* (quality) of the circuit. The resonant frequency of the circuit is the frequency that the circuit will pass.

Band-Stop Filter See *band-elimination filter.*

Bandwidth The difference in frequency between the top end of a channel and the bottom end. A good example of a bandwidth is sound. If you are listening to a sound, such as music, you notice the different pitches. All of these pitches or tones of sound are actually audio information that your ear can process. The sounds are actually vibrations. Bass tones vibrate at a slow rate, about 20 to 700 vibrations per second. Treble tones vibrate faster, from 3000 vibrations per second to 17,000 vibrations per second. The total bandwidth (vibration range) that you are listening to is about 17,000 – 20 = 16,980 vibrations per second. This is the range or bandwidth of human hearing.

Banjo This is also called a *beaver tail.* It is used by technicians to connect other devices to modular jack wiring for testing purposes. *Banjo* is a trademark of Harris Dracon Division.

Harris Dracon "Banjo" RJ14 6 conductor (left)
RJ45 8 conductor (right)

Bantam Connector/Plug A standard plug and jack that is used to interface test equipment with digital circuits (DS1, DS3, STS-1) that are wired to DSX patch panels.

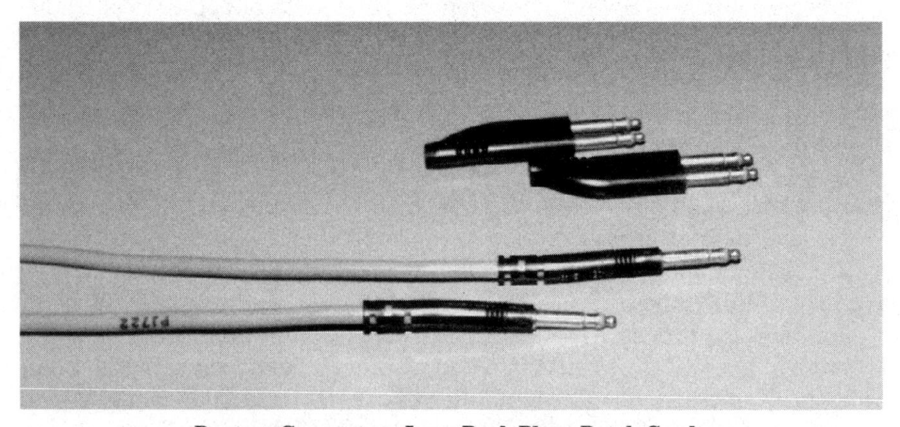

Bantam Connectors Loop Back Plugs Patch Cords

Barge In When an attendant or operator adds themselves on to a line that is already in use. See also *Busy Override*.

Barrel Connector A connector that connects two male coaxial F connectors together.

Base Station A device that connects wireless communications to the land-line phone network.

Baseband The opposite of broadband. Baseband is the transmission of one signal over a media or carrier. Telephone conversations in themselves are baseband. Applying 24 telephone conversations to a T1 carrier is a broadband application. Because T1 has more than one channel within one transmission, it is broadband.

Baseband Modem A modem that modulates and demodulates a single digital data transmission to and from another modem. Baseband modems can be used over plain twisted-pair wire with no telephone service. In this application, they are better known as *short-haul modems*, which extend peripheral devices (such as printers) 50 to 1000 feet from their host.

BASIC (Beginners All-Purpose Symbolic Instruction Code) A type of programming language that has many variations, depending on the developer. Microsoft Visual Basic is a newer version of this style of programming.

Basic Rate Interface The small size ISDN line (the other size is a primary rate interface). It consists of two bearer or "B" channels and one data or "D" channel. The B channels are 64Kbp/s each. With the appropriate service package from the phone company and correct terminal adapter, you can talk on one B channel while using your computer modem on the other B channel. When your phone conversation ends and you hang up, the terminal adapter will send a message back to the phone company through the D channel that connects both B channels together for a total transmission bandwidth of 128Kbp/s for your computer automatically.

Battery A device that converts chemical energy to electrical energy. Batteries of over 1.5 V (nominally) are composed of cells, each cell being a smaller battery that is equal to 1.5 V in electrical potential. A 12-V lead-acid automotive-type battery is comprised of eight 1.5-V cells in series that add up to 12 V. Some batteries are re-chargeable, depending on the two chemically interacting materials that the battery is made of.

Battery Back Up There are two different types of battery back-up systems. There are rectified power sources, which continuously charge batteries that power a system. This system is used by telephone companies for their central offices. When the power goes out, the charging on the batteries stops, but the system still runs because it's running on the batteries. The other type of battery back up is a UPS (uninterruptable power supply), which is always on standby. When the power goes out, the UPS converts the DC battery power to AC power to run the system. Which system you use depends on the type of power that your phone system requires. If you have the option of running on DC, the rectified battery back-up is far superior in reliability, and they are available in many different sizes. Reltec is a well-known manufacturer of these systems.

Baud Rate The actual bit rate on a communications line. Not to be confused with **bit rate**, which includes data compression and other transmission enhancements.

Baum An impedance-matching transformer used in RF applications. Fifty to 75 ohms is typical.

BCD (Binary Coded Decimal) A four-bit code that represents the numbers zero through nine in binary. It is basically implemented as a short cut for entering many binary numbers into a machine-language program. Logic circuitry decodes the BCD to binary for the microprocessor. The code is

Decimal	BCD
0	0000
1	0001
2	0010
3	0011
4	0100
5	0101
6	0110
7	0111
8	1000
9	1001

BCM (Bit-Compression Multiplexer) A multiplexer that increases bandwidth by encoding data bits into a special format.

Beacon The name of a special frame sent by token-ring equipment that notifies other nodes in the ring that there has been a major failure. Other nodes receiving the beacon work in conjunction with each other to reroute transmission around the failure.

Bean A wire-splicing connector for splicing twisted-pair wire. Beans look like a plastic tube that is about as big around as a drinking straw, but only an inch long. They have metal teeth inside them so that when two wires are crimped inside, they make a good connection and don't slide out. Beans can also have a water-retardant jelly inside them as well. For a photo, see *Plain B Wire Connectors*.

Beaver Tail Another name for a Harris Dracon Division Banjo or a similar break-out device. The Banjo is a device for connecting test equipment to the wiring in modular jacks. For a photo, see *Banjo*.

Bed-Of-Nails Clip A test clip similar to an alligator clip, except that it has a section of very sharp needle-like objects bunched together that poke through a wire's insulation when the clip is applied. These clips achieve a good connection for testing without stripping the insulation off the wire.

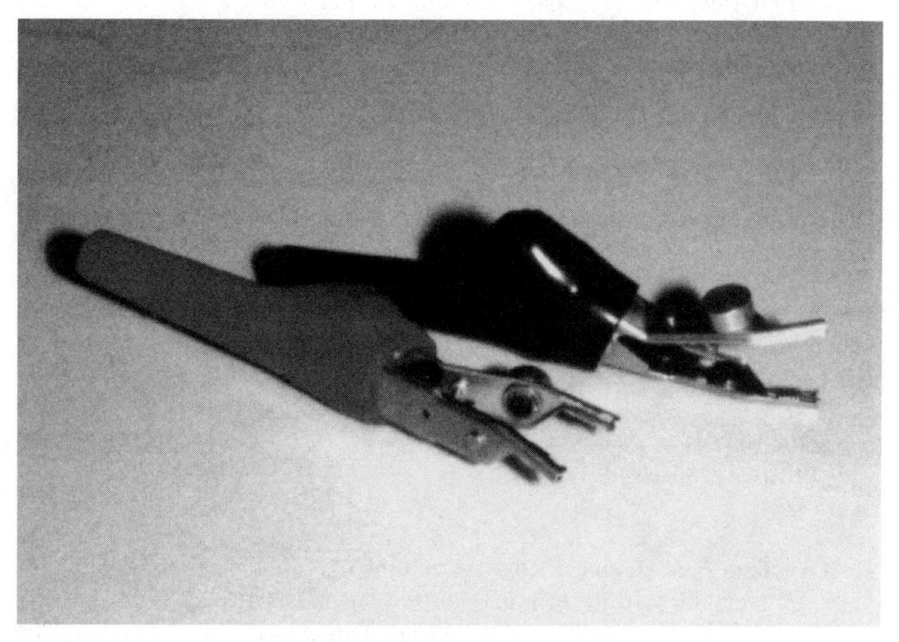

Alligator/Bed of Nails Clips

BEL This is an ASCII control-code abbreviation for bell. The Binary code is 0111000 Hex is 70.

Bell Operating Company (BOC) or Regional Holding Company At the time of divestiture, there were 22 BOCs, grouped into seven Regional Bell Operating Companies (RBOCs).

BOCs:

Bell Telephone Company of Nevada
Illinois Bell Telephone Company
Indiana Bell Telephone Company
Michigan Bell Telephone Company
New England Telephone and Telegraph Company
US West Communications Company
South Central Bell Telephone Company
Southern Bell Telephone and Telegraph Company
Cincinnati Bell Company
Mountain Bell Telephone Company
Mountain States Telephone and Telegraph Company
Southwestern Bell Telephone Company
The Chesapeake and Potomac Telephone Company of Maryland
The Bell Telephone Company of Pennsylvania
The Chesapeake and Potomac Telephone Company of Virginia
The Chesapeake and Potomac Telephone Company of West Virginia
The Diamond State Telephone Company
The Ohio Bell Telephone Company
The Pacific Telephone and Telegraph Company
New Jersey Bell Telephone Company
Wisconsin Telephone Company

RBOCs:

Ameritech
Bell Atlantic
Bell South
NYNEX
Pacific Telesis
Southwestern Bell
US West

Bell System Practices (BSPs) A volume of standards that explain how to do everything from terminate an RJ11 jack to install a central office. They even had a standard on how to collect a past-due phone bill. The BSPs were a pre-1984 (divestiture) tool for operating phone companies. They are no longer widely embraced by the RBOCS or AT&T. New equipment manufacturers have their own instructions for operating and installing their products, and each RBOC has its own way of operating a communications company.

Bend Loss The loss of transmission in a fiber-optic or twisted-pair cable because of a bend. Bending fiber-optic cable causes the light traversing through it to reflect outward, instead of down the core of the fiber. A bend in a twisted-pair wire causes the dielectric or insulation to change its electrical properties, which results in a loss of signal.

Bending Radius The smallest or tightest bend that a Cat5, Cat7 or fiber can withstand under a tensile-pulling force without damaging its transmission characteristics.

BER (Bit Error Rate) A way to measure data-transmission integrity. The bit error rate is a ratio of bad bits to good bits.

BERT (Bit Error Rate Test) A way to measure data transmission integrity. The test gives a result as a ratio of bad bits to good bits.

Beta A way of referring to a test site or test product. If a new revision is released by a manufacturer, the first sites that it is installed at is referred to a *beta site.*

Bidirectional Bus A bus that connects devices that clock bits in as well as out of their shift registers. The devices that a bus is connected to make it unidirectional or bidirectional, not the bus itself. All buses are merely a group of parallel conductors that connect the shift registers of components, as well as the power for the devices.

Billed Telephone Number The number that is the regarded as the billing account number on a phone bill. Sometimes when a customer calls a phone company for service, the customer-service representative will ask the customer for the billed telephone number because that is the number that all the other customer's phone numbers and charges are referenced to. This method is used so that a customer doesn't get a phone bill for every individual phone line they have.

Binary A number system that counts with only two digits, 0 and 1. We are all more familiar with the arabic base ten, which counts with ten digits: 0, 1, 2, 3, 4, 5, 6, 7, 8, and 9. The following is a list of numbers and their binary equivalent. For a larger table of binary numbers, see Appendix E.

0	0000
1	0001
2	0010
3	0011
4	0100
5	0101
6	0110
7	0111
8	1000
9	1001

10	1010
11	1011
12	1100
13	1101
14	1110
15	1111

Binary Coded Decimal (BCD) A four-bit code for representing the numbers zero through nine in binary. It is basically implemented as a short cut for entering many binary numbers into a machine-language program. Logic circuitry decodes the BCD to binary for the microprocessor. The code is:

decimal	BCD
0	0000
1	0001
2	0010
3	0011
4	0100
5	0101
6	0110
7	0111
8	1000
9	1001

Bindary Coded Decimal (BCD)

Binary-to-Decimal Conversion For a conversion table of Binary to Decimal and Hexadecimal, see *Appendix E*.

Binary-to-Hexadecimal Conversion For a conversion table for binary to decimal and hexadecimal, see *Appendix E*.

Binder A method of separating groups of 25 pairs in a twisted-pair cable with counts of more than 25. Colored plastic ribbon binds, designates and separates each group of 25 pairs. The first binder group is white/blue pairs 1 to 25, the second is white/orange pairs 26 to 50, the third is white/green pairs 51 to 75, the fourth is white/brown pairs 76 to 100, etc. If you would like to see the entire list of binder groups and their associated pairs, see *Color Code, Twisted Pair*.

Binder Group A method of separating groups of 25 pairs in a twisted-pair cable with counts of more than 25. See *Binder*.

Binding Post A reference used to identify where twisted copper pairs are terminated in access points, cross boxes, and terminals. Physically a binding post is a pair of teeth on a 66M150 block or a pair of 9/16" lugs. Each binding post has a number. When a technician looks for a specific pair in a cable (called a *cable pair*), they refer to documents that list the pairs and which binding posts they are spliced to.

BIOS Basic Input Output System residing in a PC. It contains the shift registers (dynamic RAM) used as buffers for sending bits to the specific hardware that they are intended for.

Bipolar 1. A copper twisted-pair transmission method (or line format), where bits that are transmitted are alternated positive and negative. This transmission technique increases the distance a transmission can travel on a twisted pair. 2. A transistor.

BISDN (Broadband Integrated Services Digital Network) The future standard of communications service. BISDN is working its way up to be a platform (or higher-level service), supported by and a combination of ATM, SONET, and ISDN. The idea is that with these three technologies together, a customer could have bandwidth on demand. The good thing about bandwidth on demand is that you only pay for what you use. Currently, you can purchase a BRI or PRI from a local phone company and they charge you a price for that service. That price is based on the fact that the service is dedicated to you 24 hours a day, 365 days a year. BISDN services will be shared in the same way that long-distance dial tone is today (time-share). You will be able to dial up a PRI or two, watch some HDTV programs, or download some data, then hang-up. You will then be billed only for the time you used the service (or bandwidth).

Bit (Binary Digit) A unit of data that is represented as a one or a zero. Inside most data devices, a bit is physically a positive 5 volts or 0 volts.

Bit Error Rate (BER) A way to measure data transmission integrity. The bit error rate is a ratio of bad bits to good bits.

Bit Interleaving A simple way to time-division multiplex by interleaving individual bits, instead of bytes or packets. Timing of the two ends is not as complex with this method. Used in X.25 and HDLC, not in T1 or T3.

Bit Oriented Communications protocols that use bits to represent control information in contrast to bytes. A byte can mean different things in the different varieties of character sets that are transmitted. Bit-oriented protocols are not character code sensitive.

Bit Parity A way to check that transmitted data is not corrupted or distorted during the transmission. The way parity works is as follows. Take a bit stream that will be transmitted, add all the bits as binary numbers mathematically, and the resulting number is odd or even. Add a 1 at the end of the stream if the number is even and a 0 if the number is odd. When the bits are received at the other end, they are added up and compared to the last bit. If they add up to be an even number, then the last bit should be a 1. If they add up to an odd number, then the last bit should be a 0. If the case for either does not hold true, then the receiving end sends a request to retransmit the stream of bits. They are retransmitted, with the parity bit attached all over again.

For example, a computer sends a bit stream of 10101011. Simply adding the bits gives a sum of 1 + 0 + 1 + 0 + 1 + 0 + 1 + 1 = 5. This is an odd number, so add a 0 to the end of the stream to make it 101010110. The bits are received at the other end, added together, and compared to the parity bit the same way. There are new and more sophisticated ways of checking for errors in data transmission, such as cyclic redundancy checking.

Bit Rate The average net number of bits being transmitted over a communications line in a second, including compression and encoding techniques, as well as retransmission of corrupted data.

Bit Robbing Bit robbing is usually known as *in-band signaling*. The practice of taking a bit here and there in the beginning and end of a digital transmission for use in the overhead of the transmission equipment. Bit robbing is bad when the signals being multiplexed into the transmission are data. Robbing a bit from a data stream severely corrupts data. Bit robbing is a technique reserved for multiplexing multiple voice circuits onto a T1. Circuits intended to transmit data use out of band signaling or clear-channel signaling.

Bit Stream A series of voltage pulses that represent a binary code. A serial data transport. A bit stream can exist on a transmission line or within the electronics of a data device.

Bits Clock A device that provides a timing pulse in the form of a 1-0-1-0-1-0-1-0 bit stream. Bits clocks are used extensively in SONET networks. The bits clock provides the timing pulse that everything in the network synchronizes itself to.

Bits Per Second The average net number of bits being transmitted over a communications line in a second, including compression and encoding techniques, as well as retransmission of corrupted data.

Black Box Usually a device that converts or routes one type of data or signal applied to the input to a desired useful output for a specific application. One company called "Black Box Corporation" specializes in the manufacture of these specialized devices.

Blended Agent An agent in a call center that receives calls from outside customers. When the incoming call rate slows down, it makes outgoing calls.

Blended Call Center A call center that receives calls from customers and also calls customers. Sometimes agents are dedicated to either inbound or outbound calls.

BLF (Busy Lamp Field) A part of or an add-on module to a phone or console that allows the user to see multiple extensions and if that extension is in use (or busy).

Blocked Call A call that cannot be completed because the Central Office or PBX switching capacity is full at the time the call was attempted. Blocking can occur at any point in a network where a call is switched (from CO to CO or from local to long distance). The caller with a blocked call either hears a fast busy or an intercept message that says "I'm sorry, all circuits are busy now. Please try your call again later."

Blocking When a central office or PBX has fully utilized its capacity to connect calls, it blocks them. Callers trying to call in or out of a switch that is blocking calls will get a fast busy signal.

BNC A type of connector used on all different types of coax. It is keyed so that it locks into place and it has better transmission characteristics than an "F" connector.

BOC (Bell Operating Company or Regional Holding Company) At the time of divestiture, there were 22 BOCs, grouped into seven *Regional Bell Operating Companies (RBOCs)*. For a listing of the BOCs and RBOCs, see *Bell Operating Company*.

Body Belt Used by communications/power/construction personnel to harness themselves to telephone/power poles or tower structures. This is also called a *safety belt* or a *climbing belt*. For a photo, see *safety belt*.

Bond 1. What telephone company construction personnel call the connection between the sheath of a telephone cable and an electrical ground. 2. An electrical connection.

Bonding In ISDN the joining of two 64Kbp/s B channels together for one 128Kbp/s channel.

Boot To restart a computer or CPU-based system by physically turning it off, then back on, which resets the CPU. This is also called *bootstrap*.

Bounce To re-start a T1 or other digital service circuit.

BPAD (Bisynchronous Packet Assembler Dissasembler) A hardware-based device that inserts bytes into packet frames and vise-versa in packet multiplexing/transmission equipment.

BPS (Bits Per Second) The average net number of bits being transmitted over a communications line in a second including compression and encoding techniques, as well as retransmission of corrupted data.

BPSK (Binary Phase-Shift Keying) A method of transmitting binary bits in a form of frequency shift, or FM that is the same concept as frequency-shift keying. The difference is that with BPSK, you change the phase of the frequency, instead of the frequency itself. The two are shown in the diagram. If you look closely, you can see the changes in the waveforms that represent the switch from a one to a zero value.

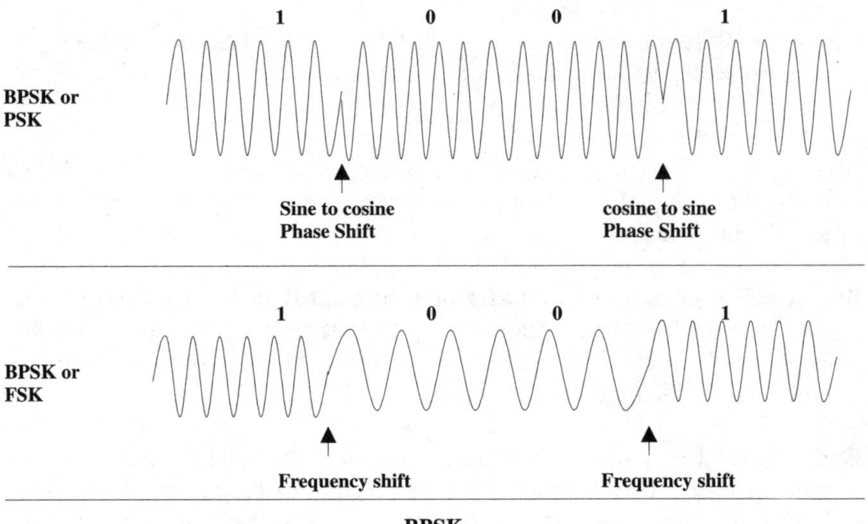

BPSK

Breakdown Voltage The voltage at which insulation in a cable or an electronic device fails.

Break-Out Box A test device that plugs into a data cable (i.e., RS232) and provides easy test access for each wire in the cable.

BRI (Basic Rate Interface) The small-size *ISDN* (*Integrated Services Digital Network*) line (the other size is a Primary Rate Interface). It is made up of two bearer or "B" channels and one data or "D" channel. The B channels are 64Kbp/s each. With the appropriate service package from the phone company and correct terminal adapter, you can talk on one B channel while using your computer modem on the other B channel. When your phone conversation ends and you hang up, the terminal adapter will send a message back to the phone company through the D channel that connects both B channels together for a total transmission bandwidth of 128Kbp/s for your computer automatically. For a diagram, see *ISDN*.

Bridge A bridge is a device that converts a data transmission from one media or protocol to another. If you want to extend an Ethernet connection beyond the 300ft CAT 5 UTP limit, you would have to "bridge" the transmission to a different media that CAT 5 that has less loss over long distances, like fiber-optic cable. The bridge would convert the electric data signal into an optical signal and vice versa.

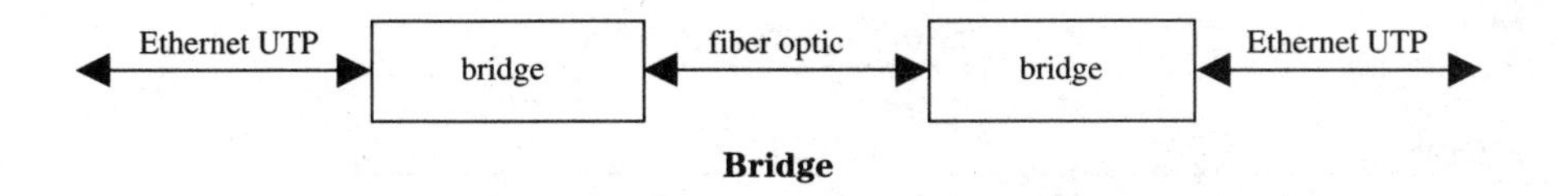

Bridge

Bridge Clip A metal clip (sometimes plastic insulated) used to electronically connect or bridge across the left side of a 66M150 block with the right side.

Bridge Tap A Y splice in a copper twisted-pair communications cable. It gets its name because the first splice is a straight through splice, and the second splice connects the wires by cutting or "tapping" them into the first splice. A bridge tap adds flexibility to telephone plant. A telephone company never knows which customer is going to use lots of pairs for their service and who will use only a few. To remedy the situation without installing 100 pairs of copper to each individual building, the telephone company installs the same 100 pairs into three or four buildings. This is done by bridge tapping the cable splices. The bad thing about bridge tapping is that it lengthens the pair. Length is bad for digital services (additional loss, and reactance), such as T1 and ISDN, so the bridge taps must be disconnected before these services are installed. On a cable drawing, bridge taps are shown as arrows and telephone cable is shown as a line.

Bridge Clip

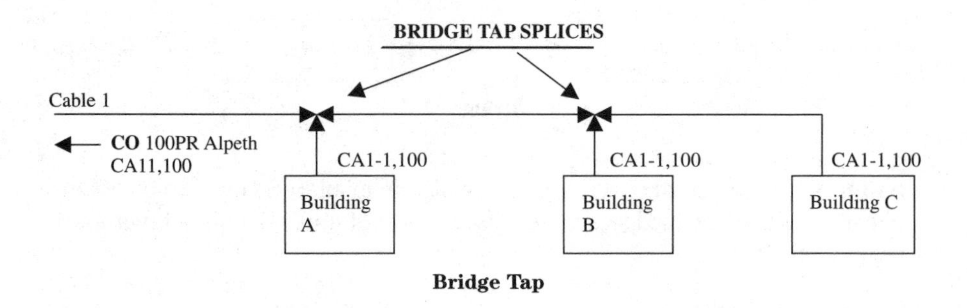

Bridge Tap

Broadband Incorporating more than one channel into a communications transmission. T1 is a broadband communications protocol because it carries 24 conversations over four wires. Cable TV is also broadband because it carries many TV channels over one coax.

Broadband Integrated Services Digital Network (BISDN) The future standard of communications service. BISDN is still evolving as a future platform (or higher-level service) supported by and a combination of ATM, SONET, and ISDN. The idea is that with these three technologies together a customer could have bandwidth on demand, and the good thing about bandwidth on demand is that you only pay for what you use. Currently, you can purchase a BRI or PRI from a local phone company and they charge you a price for that service. That price is based on the fact that the service and the money it cost to provide it is dedicated to you 24 hours a day, 365 days a year. With BISDN services will be shared the same way long-distance dial tone is today (time-share). You will be able to dial up a PRI or two, watch some HDTV programs or download some massive data, then hang-up. You will then be billed only for the time you used the service (or bandwidth).

Broadcast To send information in any form to more than one place.

Brouter A device in a LAN, MAN, or WAN that performs the functions of a bridge and a router at the same time. A bridge converts a signal from one transmission media to another or one protocol to another. An example would be a device that converts an electrical signal to an optical signal so that a data transmission can be taken from an Ethernet LAN to a distant PC. A router is a device that routes data intended to be sent to certain devices to those devices. A common application for a router is to send Internet traffic only to the Internet and keep local traffic local. This increases the efficiency of the Internet connection because it is only receiving and transmitting traffic intended for it instead of all the traffic on the network. The following drawing depicts where routers, brouters, bridges, and a hub would be implemented in a small Ethernet LAN to MAN network.

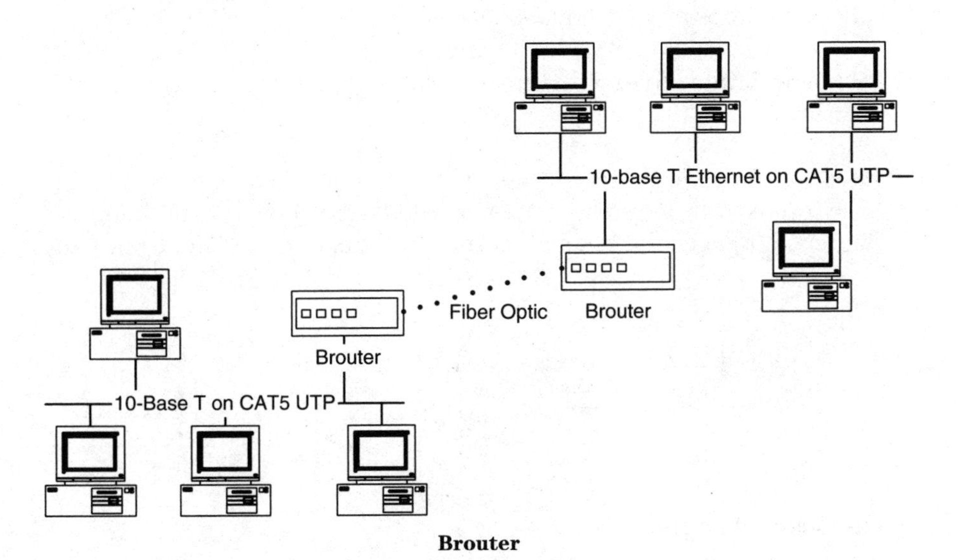

Brouter

BS An ASCII control code abbreviation for backspace. The binary code is 1000000 and the hex is 80.

BSP (Bell System Practice) A volume of standards that explain how to do everything from terminate an RJ11 jack to install a central office. They even had a standard on how to collect a past-due phone bill. The BSPs were a pre-1984 (divestiture) tool for operating phone companies. They are no longer widely embraced by the RBOCS or AT&T. New equipment manufacturers have their own instructions for operating and installing their products, and each RBOC has its own way of operating a communications company.

BSS (Base Station System) A wireless communications device that manages radio traffic and bandwidth between a group of base transceiver stations.

BTA (Basic Trading Area) Geographical boundaries defined within a cellular radio license.

BTN (Billed Telephone Number) The number that is regarded as the billing account number on a phone bill. Sometimes when a customer calls a phone company for service, the customer-service representative will ask the customer for the billed telephone number because that is the number that all the other customer's phone numbers and charges are referenced to. This method is used so that a customer doesn't get a phone bill for every individual phone line they have.

BTS (Base Transceiver Station) A station that transmits mobile radio signals.

Buffer A temporary storage (memory) device for data. A buffer is basically a box with a bunch of RAM inside it. A common application for buffers is to collect a stream of data and temporarily store it until an-

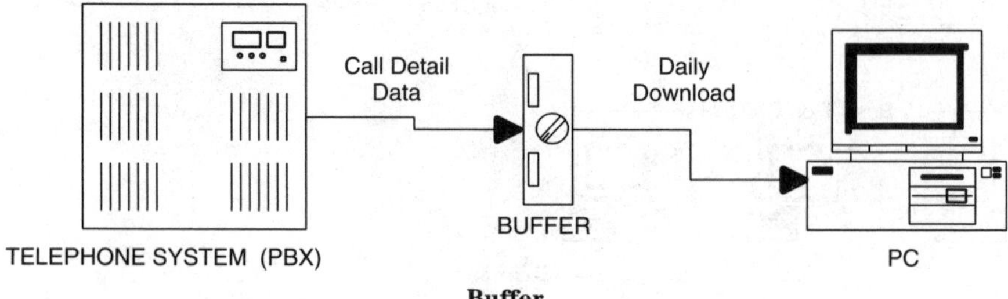

Buffer

other device, such as a PC or server asks the buffer to download it. This is useful when the PC, server or LAN could be out of service for a period of time. When the server or PC is returned to service it just asks for the data from the buffer and it is downloaded. The buffer is then empty and ready to receive more data.

Building Entrance Agreement The piece of paper that all building management companies are hounded for since the eruption of the CLECS (Competitive Local Exchange Carriers). A building entrance agreement gives a telephone company or other utility the privilege to construct communications facilities into their building and to occupy their own space for equipment, power for the equipment, and access to said equipment. Even though having a CLEC present in a building is a great advantage for tenants, smart building management companies use their position to their own advantage. They only allow the CLEC to construct the facilities into the building the way they want, when they want. The facilities (fiber, cable, conduit, not electronics) when completed belong to the building. The building management then charges rent back to the CLEC for the use of the facilities that they paid to have designed and constructed. The CLECs must agree to all of the building management's terms because without building entrance agreements the CLECs can only exist by using another phone companies facilities (usually an RBOCs) to provide service.

Buried Cable Terminal Where buried service wires are fed from between the feeder cable and the standard network interface. For a photo of a *Buried Terminal, see Pedestal.*

Buried Service Wire Splice A special watertight splice that is filled with an encapsulant. Common types of these splices are made by Keptel and Communications Technology Corporation.

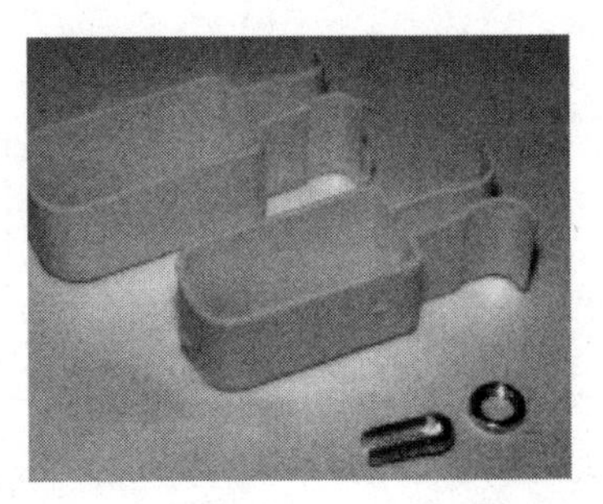

Buried Service Wire Splice

Bus Topology A LAN physical topology, which means the way that the individual devices within the LAN are physically connected. The method at which the devices connected to the LAN access the media (coax is most common for this topology) is called the *logical topology*. The bus topology is a wire (UTP) that behaves as a street that connects a number of PCs. When a PC wants to access the network or wire to a server or another PC, it looks at the street to see if there is no traffic. If there is no traffic, then the PC (or server) sends data down the wire (or street). The data has an address attached to it and all the other devices connected to the network see the address. If the address belongs to a certain device, that device reads the data attached to the address. This scheme of sending data is called *Ethernet*. One of the inefficiencies of Ethernet is something called a *collision*. A collision happens when two or more devices look at the network (the wire) and see that it is clear at the same time, then attempt to send data at the same time. The data that is transmitted by the multiple devices collides and becomes corrupted. The devices on the network sense the collision and try to send their data again when the network is clear. This method of control is classified as a contention-based protocol because all the devices on the network contend for its use. All of the Ethernet protocols are contention based. See also *CSMA/CD*, *CSMA/CA*, and *Ring Topology*.

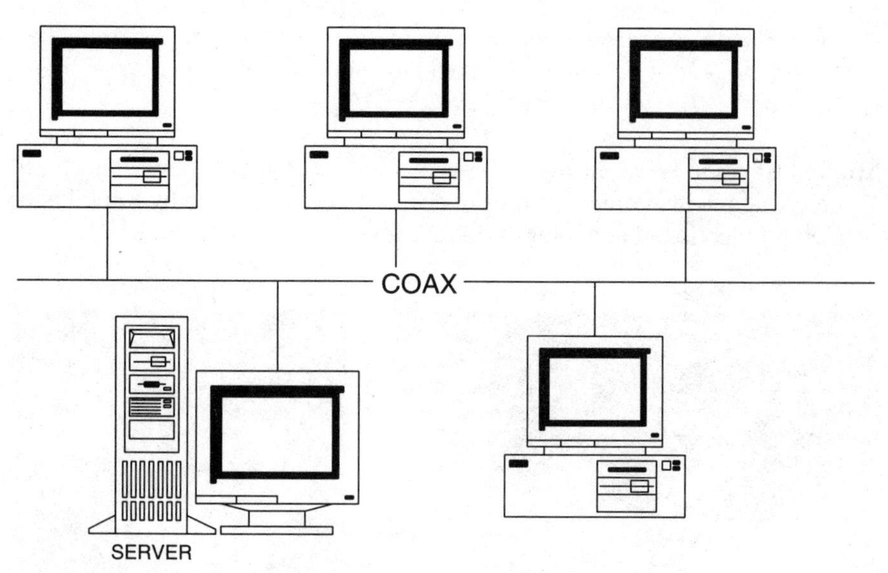

Bus Ethernet Physical Topology

Busy Hour The hour in the day or month when a central office or PBX connects the most calls. The busy hour is an important factor in designing a switch for blocking.

Busy Lamp Field The lights on an attendant console that indicate who is on their line (what lines are busy) and who is not. Some busy lamp fields are an add-on module that can be attached to a phone.

Busy Out A temporary fix or condition of a phone service. To "busy a line out of a hunt sequence." If a business phone line becomes defective and it is in a hunt or roll over sequence, calls will not hunt or roll past this line. Say that you have four lines coming into your business. The first line is the main number and if that first line is busy, then calls come in on the second line, etc. If line one goes bad then it can't be called, so it can't be busy. Because it is not busy, then calls will not hunt or rotate to the next three lines. When you call the phone company repair service they busy out the bad line, which makes it look busy to the network. Your calls then start coming in on the other three lines. When a repair technician finishes with repairing the problem on the bad line, he has it unbusied.

Busy Override When an attendant or operator adds themselves on to a line that is already in use. This is also called a "barge in." If a local phone company operator barges in on a phone conversation the people on the line will hear a beep tone, then subsequent short beep tones as long as the operator is connected. The operator can converse with the two parties on the line after barging in and pass on urgent information. This is a common feature of PBX systems. A PBX system can be programmed to not warn the people in the middle of their call that another person is listening. This feature is often used to monitor the quality of customer service in call centers.

Busy Signal There are two types of busy signals. The most common type is a *slow busy*, which means that the number you are trying to call is being used. The other type of busy signal is a *fast busy*, which means the phone company central office could not understand the digits that you dialed or the actual phone network is too busy to take your call. Many fast busy signals are being replaced with "intercept messages," which tell the dialer what the problem is. Examples of intercept messages are: "I'm sorry, all circuits are busy now. Please try your call again later." And "The number dialed is out of service. Please check the number and try your call again".

Busy Verification The local telephone company test to see if a particular phone number (or circuit, or loop as they call them) is busy. The test is run by a *DATU (Direct Access Test Unit)* in the central office. When the test comes back, it says the line is "in use busy speech" or "ROH *(Ringer Off Hook)*," which means that the handset has been taken off the hook and left there.

Butt Set A test telephone set used by telephone installation and repair personnel. Instead of a plug on the end of the cord, it has a pair of alligator/bed-of-nails clips. For a photo, see *Craft Test Set*.

Bypass Trunk Group A method of connecting one central office to another without going through a tandem. This method reduces tandem traffic and reduces blockage between the two offices that have bypass trunk groups installed. A bypass trunk group is shown in the following diagram.

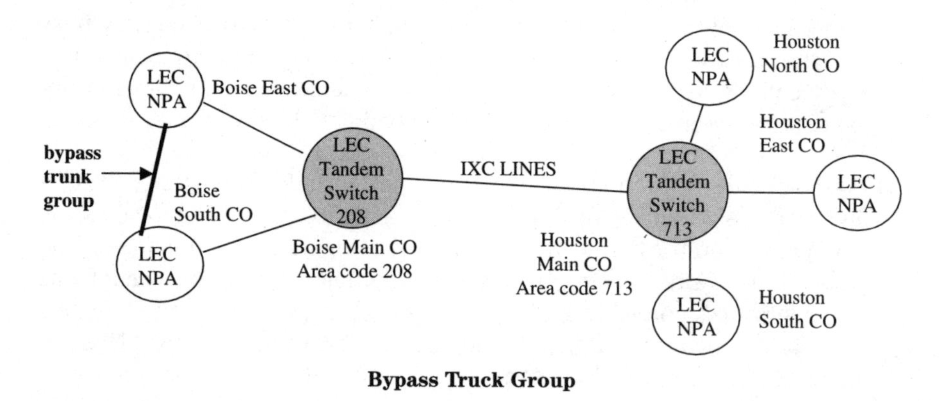

Bypass Truck Group

Byte Eight bits, also known as an octet.

10101010 – A mathematical representation of one byte.

– An electronic representation of one byte.

Byte

Byte Stuffing Byte stuffing is what some communications protocols do to make data more suitable for transmission. If a customer wants to send 30 bytes, but one transmission frame carries 48 bytes, then the transmission equipment adds 18 bytes so that the frame is full. Think of byte stuffing as the styrofoam peanuts or wadded-up paper you put in a box when you ship something through the mail.

C

C+ A high-level programming language.

C++ A high-level programming language.

C7 The European version of SS7. SS7 and C7 are not the same protocol even though they perform the same functions. Gateway switches (class 1 central offices) convert the two different international standards when calls are handed through.

C Connector Also called a *female amp connector* or *25-pair female connector*. For a photo of a C connector, see *25-Pair Connector*. The male version is called a *P connector*.

C Drop Clamp A clamp used to fasten aerial service wire to homes/buildings.

C Programming Language A high-level programming language that is somewhat similar in application to the Basic programming language.

C Wire Wire that is strengthened with steel for "long-span" aerial plant applications. Some C wire is non insulated, so it is also called *open wire*. Open wire better fits the application as it is used in wide open or very rural areas. The old telegraph system was an open-wire system.

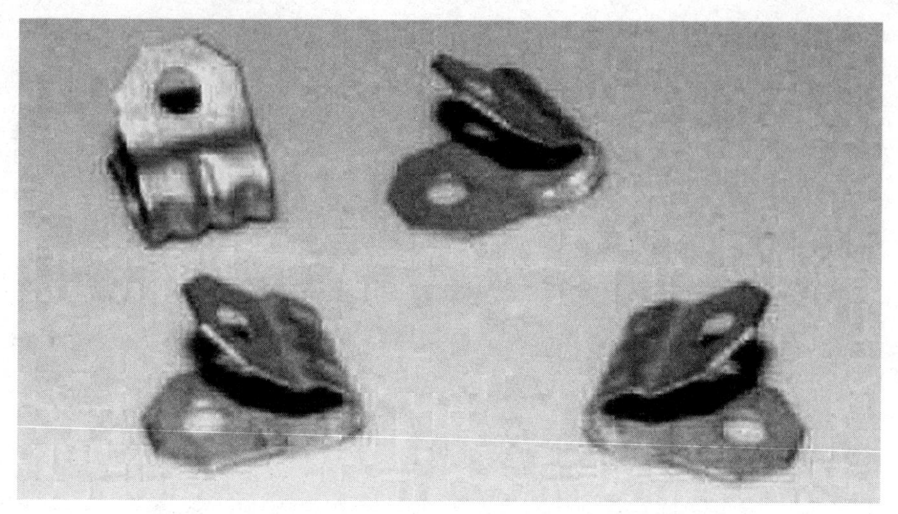

C Drop Clamps and S Clamps

Cable The general name for copper-based media to transport electrical voice, data, and video signals. It can be twisted pair or coax, and indoors or outdoors. In general, two types of cable are installed indoors, PVC and Plenum. Many types of cable are used outdoors. The basic difference in outdoor cables is how many PVC and aluminum sheaths are around the conductors and if the cable is filled with a water-proofing jelly.

Cable Act of 1984 An act passed by The United States Congress in 1984 that deregulated almost all of the Cable TV industry and was eventually superseded by another act in 1992. After the 1984 Act was passed, the FCC had control of cable TV in only the following six areas: 1. Each CATV system had to be registered with the FCC prior to any operations. 2. Enforcement of all subscribers having access to an AB switch so that they could easily switch from CATV to regular TV programming. 3. Rebroadcast of local television stations without any alteration or deletion. 4. Non-duplication of local broadcast regular TV programs. 5. Fines and/or imprisonment for broadcasting indecent material. 6. Licensing for receive-only earth stations for satellite delivered via pay cable.

Cable Knife A knife used by communications cable installers to strip ALPETH (aluminum/polyethylene) jacketed cable.

Cable Mapping The tracking of installed cable and pairs in a network. Sometimes an actual map is used and sometimes a tracking system of cable, pair, address and binding post is used. The RBOCs use both.

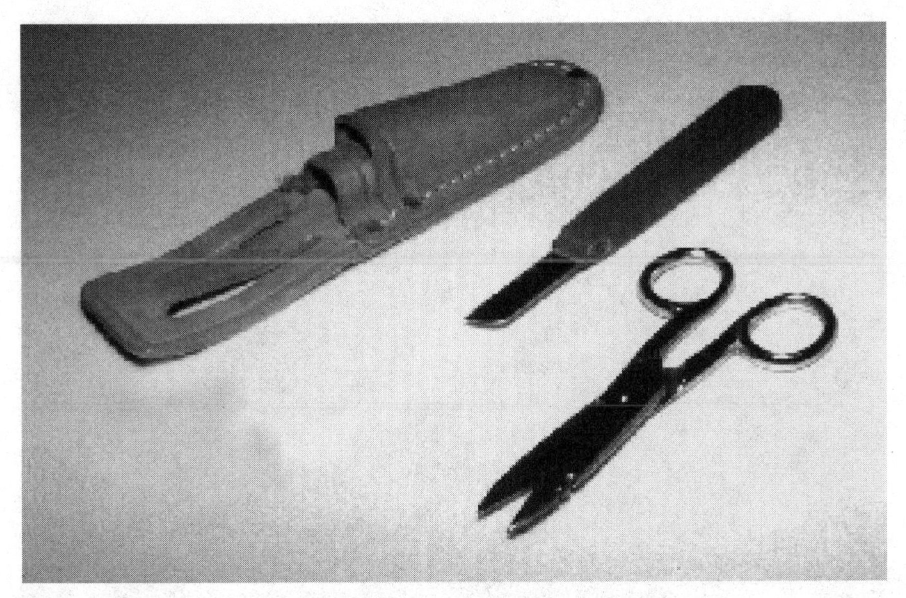

Cable Splicer's Snips and Sheath Knife

Cable Modem A modem that is used on cable TV lines so that Internet service providers can be dialed up over a TV.

Cable Plant A term that refers to a communications utility's twisted pair and/or coax network that winds through towns and neighborhoods. It includes terminals, pedestals, cross boxes, and vaults.

Cable, Riser A twisted-pair cable (usually several hundred pairs) distribution system that progresses from the telephone company Demarc or point of entrance in a building to each floor of that building.

Cable Span The cable suspended in the air between two telephone or power poles.

Cable Stripper Several tools are used to strip the jackets off of ALPETH and lead-jacketed telephone cable. The most popular is a cable knife and snips (for a photo, see *Cable Knife*). An alternative tool is shown.

Cable TV (Community Antenna Television) A cable-TV company simply has satellite dishes in a central location that pick up TV signals from around the country. They retransmit those channels down a coax that branches out through a geographical area where subscribers to the

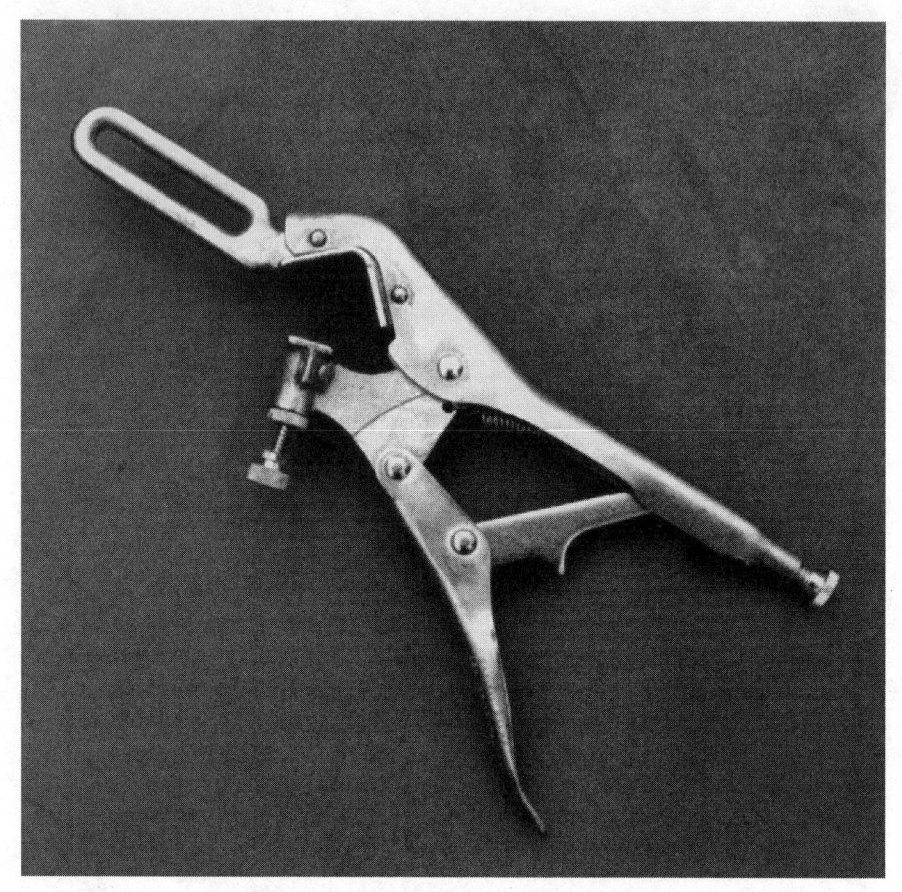

Lead/alpeth Cable Jacket Stripper Tool

antenna service can be connected. Some cable-TV companies add their own information channels to their broadcast and local advertising between programming on selected channels.

Cable Vault Formerly known as a *manhole*. A concrete enclosure that is usually under ground and varies in size from a small crate to a large room. They are specifically designed for the housing and easy access to telephone and cable-TV splices. Cable vaults are extremely dangerous places because of lack of oxygen and accumulated carbon monoxide. Natural gas, and other dangerous gasses have a tendency to build up inside them.

Cable Weight A specification of outside plant cable used by outside plant engineers when figuring how long a span can be, and how big of a messenger to use.

Call-Accounting System A computer (usually a dedicated PC) that connects to a switch via a serial data port and monitors the details of every phone call made through that switch. The call details are stored as call records. With the appropriate software, the records can be retrieved, sorted, processed, and queried to almost any specific nature that the call-accounting system administrator desires. These systems are used by hotels to track all the calls you make from your room so that they can bill you appropriately. They are also used by companies to do bill-back reports for individual departments within the company.

Call Accounting System

Call Announcement A way of transferring a call to another extension. A feature of PBX and key systems where you have a chance to talk to the person you are transferring the call to before you make the transfer. The way it works on most systems is: You are talking to a person that you want to transfer, you push the transfer button on your phone, dial the extension you would like to transfer the call to, and you hear it ring. When the person at the extension you have called answers the phone, you can notify them that you are transferring a very important client to them. You then push the transfer button again to complete the transfer. If the person you want to transfer to does not answer their phone, you can switch back to the call you wanted to transfer and speak to that person again. The other way of transferring calls is called a *blind transfer*, where you simply transfer the caller to another extension with no intervention or control of the call.

Call Attempt An uncompleted call because of blocking, where callers cannot get through because all lines are busy. This is a statistic of central office switches as well as PBX systems.

Call Blending The mix of typical incoming calls in a call center with outgoing calls made by an auto dialer with call-blending capability. Call centers blend calls to fully utilize their employees working at the time. If inbound calls are slow, the auto dialer increases the number of outgoing call attempts so it can connect people that answer with agents. Auto dialers are usually programmed with a predetermined list of numbers to dial, usually for customers that are expecting follow-up on their services from the company making the call.

Call Detail Recording (CDR) The initial function of a call-accounting system is to receive detailed information on telephone calls connected through a PBX switch and store them in memory. This function is called *call detail recording*. Call details include number dialed and duration of the call for outbound calls, and the trunk ID (or phone number) and call duration for inbound calls. Each call event (transfer, connect, disconnect, etc.) gets a time stamp.

Call Duration The length of time a phone call lasts from the time both ends are off-hook, until the time that one end hangs-up.

Call Forwarding A service offered by local phone companies to their subscribers and a feature of PBX systems that allows a user to make calls dialed to their phone ring to a different phone or phone number.

Call Hand Off When your call is transferred from one cellular site (or cellular transmitter) to another.

Call Letters The station identification letters assigned to broadcast radio and TV stations by the FCC. Some examples are: WKND, WLIS, WPPR, WKRP, KXRK, and CKLW.

Call Menu A recorded message that gives callers options to choose from by using their dial pad. See also *Auto Attendant.*

Call Mix A statistical account of calls in a system. This information is useful for determining architecture upgrades and troubleshooting. An example of a call mix would be 40% DID calls, 30% autoattendant-routed calls, and 30% zero out to human-attendant calls. From this call mix, you can see exactly how each percentage of your calls are handled.

Call Packet A packet of data that contains X.25 *Switched Virtual Circuit (SVC)* addressing information and other overhead.

Call Park A PBX feature that allows a call to be placed on hold in a way that it can be picked up from any extension in the office. The way it works is: The attendant or anyone else who wishes to park the call presses a park key on their telephone. The display on the phone then shows the extension that the call is parked in (park extensions are imaginary, just reference numbers). In this example, the display said "parked in 60." The person that parked the call then pages the person that they wish to pick up or receive the call over a loudspeaker "Johnny please pick up park 60." Johnny hears the page, goes to the nearest telephone set, and presses the park key and then 60. He is immediately connected with the calling party that was parked by the attendant that answered the call.

Call Pick Up The ability to answer a ringing phone that is not yours. You hear a phone in the next office ringing, pick up the receiver and press the pick-up key on your telephone, then enter the extension number of the phone you wish to intercept the call from. Your phone is immediately connected with the call and you say "Hello, John Doe's (or name of the person's extension) line."

Call Pick-Up Group A group of telephones that receive or "ring" when a certain number is called. An example of a pick-up group would be all of the phones on all the desks that ring simultaneously when a 1-800 hotline is called. When someone picks up the handset of any one of those ringing phones, they answer the call.

Call Queuing The function of placing incoming calls on hold and in line for the next available call-answering agent. Call queuing is a function of *Automatic Call Distribution (ACD)* systems. The queue is an extension number within the ACD system that calls are transferred to. An ACD system integrates with a PBX system.

Call Record One of the many reports generated by a call-accounting system. A call record details a telephone call made or received by a telephone or extension by the number dialed, incoming trunk ID, duration of the call, and the time of connection and disconnection.

Call Restrictor A device that can be attached to a phone line or trunk that prevents certain numbers from being dialed through or that only allows a certain group of numbers to be dialed. When installing or purchasing a call restictor, it is a good idea to be sure that 911 is dialable under any application.

Call-Request Packet A packet that is sent by the originating DTE equipment in a frame-type data transmission that requests a network terminal number, network facilities, and call user data (or X.29 control information).

Call Return Also called *last call return*. A service offered by local telephone companies that enables telephone customers to return a call they missed by dialing *NN (the two numbers after the * depend on the local company). This service is handy for those times when you are running to the phone and the caller hangs up right when you pick up the receiver. All you have to do is dial the *NN code, you hear a recording that tells you the phone number of the last caller, and gives you the option of ringing them back by pushing 1 or just hanging up. Using this service can cost 50 to 75 cents each time you use it up to a maximum amount per month, usually about $6.00.

Call Second One phone call for one second. This is the smallest unit of telephone switch traffic. One hundred call seconds is equal to a *Centum Call Second (CCS)*. A one-hour call is equal to 36 CCS, which is one Erlang. The Erlang is the standard unit of measure for telephone-switch traffic.

Call Setup Time The time from when you go off-hook, dial a number, the phone network checks to see if the number you are calling is busy, a path is established between central offices, the other end rings and is picked up. Even though call setup costs money, customers making long-distance calls don't pay for the call setup time.

Call Trace A service offered by local phone companies. If you receive a malicious or obscene telephone call, you can have the call traced by immediately dialing *57 after the call. This only works for the last call received. After the *57 is pressed, a recorded message is played that gives further instructions. The call is tagged in the local telephone company's call detail-recording log. The caller is usually warned for a first offense. The person that made the report never finds out who made the obscene call, but is not surprised to see how quickly the obscene calls cease. *57 is the North American Standard for last-call trace service.

Call Transfer A feature of PBX systems that allows users to transfer a conversation or call connection to another extension. The feature is usually executed by pressing a transfer key while on the line with the person or party they wish to transfer, dialing the digits of the extension they wish to transfer to, then pressing the transfer key again.

Call Waiting A feature offered by local phone companies that allows someone that is talking on their phone line to receive another incoming call by briefly pressing the switch hook. The person knows they are getting another call because they hear a short beep or click on the line. If a person does not have call waiting, the caller that is trying to reach them while they are on the phone will receive a busy signal, as opposed to a ring when they have this service.

Caller ID (Caller Identification) Also known as *ANI (Automatic Number Identification)*. A feature offered by local phone companies that sends the phone number (and often the name of the caller) down the phone line in a digital data packet between the first and second ring. To receive the data, a subscriber that has signed up for the service needs to have a caller-ID unit (also called a caller-ID box) plugged into the phone line. The caller-ID unit displays the name and number of the calling party for each incoming call. Caller ID only works if the caller and the called party's phone service is fed out of a central office that has caller ID capability. If the central office does not have caller-ID capability, the display will read "out of area" to the called party. If the called party does not have caller service, they will get a display that says "no data sent."

Caller Identification (Caller ID) See Caller ID.

Caller-Independent Voice Recognition A voice recognition system that recognizes a certain number of words, rather than a specific voice.

Camp (Camp On) A way of placing incoming callers on hold. A feature of PBX systems. If you are trying to call someone and they are on their phone, you or a PBX attendant can put your call on hold in a way that when the person you are calling hangs up, your call rings through to their phone instantly.

CAN The ASCII control-code abbreviation for cancel. The binary code is 100001 and the hex is 81.

CAP (Competitive-Access Provider) A company that offers private line services in competition with the *RBOC's (Regional Bell Operating Companies)* in providing private line access to long-distance carriers. The private line service offered by a CAP can carry a long-distance company's dial tone. A CAP shouldn't be confused with a *CLEC (Competitive Local Exchange Carrier)*, which not only provides private-line service, but also provides their own switched dial-tone services with their own switches. Some CAP companies are Electric Light Wave, Teleport Communications Group (the first CAP founded in NY City), Brooks Communications, and Metropolitan Fiber Systems. Some of these companies operate as CLECs in certain cities.

Capacitance A measure in farads of a capacitor. See *Capacitor*.

Capacitive Coupling In audio amplifiers, the different stages of amplification are linked by capacitors. The capacitors prevent the DC transistor bias voltage from passing onto and interfering with the next amplifier, yet it allows the AC audio signal to pass through and be further amplified. Some very high end and very expensive audio amplifiers are direct coupled, which means all the different amplification stages are connected by only a conductor and are integrated with each other. The transistors in direct-coupled amplifiers are usually biased with other active devices instead of resistors. They are very complicated and expensive, in contrast to capacitive-coupled amplifiers, but the low-end frequency-response approaches 0 Hz.

Capacitor A capacitor is an electronic device that has two special properties. It only allows alternating current to pass through it (blocks DC current) and it can store an electric charge. One of the many applications of capacitors is to filter alternating current (AC) out of DC power supplies and rectifiers. This is done by placing a capacitor from the DC output to ground. The capacitor appears as an easier path to voltage fluctuations and RFI, and an impossible path to direct current (DC). Physically, a capacitor is two plates of metal separated by an insulator (mylar is common). The physical size of a 1-F capacitor would be two sheets of tin foil the size of a football field insulated (or separated) by a thin sheet of mylar. The farad is a huge unit of capacitance. This is why most capacitors are microfarads (μF) in value.

Schematic symbols for capacitors:

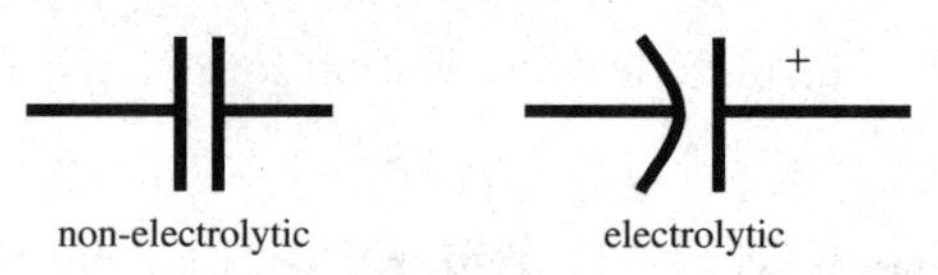

Capacitor

Capcode An ID code for a pager. The code is usually labeled on the outside of the pager and is the actual address ID code for that pager. When the ID code is broadcast, your pager receives the information that follows, then it beeps (or vibrates) and displays the transmitted information.

Carbon 1. A semiconductor that is used to make electronic components, such as resistors and microphones. 2. A name for a lightning protector. Carbon-type protectors are being replaced with gas-type protectors.

Carbon Transmitter A microphone in a handset that is made from small grains of carbon packed into the shape of a diaphragm. The way these microphones work is that as the sound waves from your mouth strike the carbon in the receiver (microphone) they vibrate the carbon. These vibrations in the carbon change its electronic resistance in conjunction with your voice, converting your voice into electronic fluctuations, which traverse down the phone line. This is an older type of microphone. The most common problem with carbon microphones is that after a period of time, the fine carbon grains stick to themselves or become settled. A few good whacks usually fixes the problem. When a telephone customer with this problem saw the telephone man pound the handset with the handle of screwdriver and say "there you go, all fixed," they probably did not know that this was a standard procedure for repairing defective carbon receivers.

Card Cage A box frame that has one open side for inserting electronic circuit cards and a back plane on the opposite side of the opening that the circuit cards plug into. For a photo of a card cage, see *Backplane*.

Carrier Band A range of frequency that is used by a specific transmission system. The carrier band for a T1 is 1.544 Mb/s. The carrier band for a DS0 is 64 kb/s.

Carrier Common Line Charge What local phone companies charge long-distance companies to connect the far-end local portion of a call. Also called an *access charge*. A fee that everyone pays for every phone line to make up for subsidies that long-distance services paid to help the less profitable local services before the divestiture of AT&T and the RBOC's (Regional Bell Operating Companies).

Carrier Detect (CD) Most modems have a little red LED with "CD" next to it. When that light is on, your modem is connected to another modem or communications device that it can communicate with.

Carrier Failure Alarm (CFA) A notification that timing has been lost in a digital transmission because of excessive zeros in the transmission.

When a carrier failure alarm occurs, all of the calls and data on that transmission are dropped until the carrier equipment regains timing.

Carrier Frequency In radio and television communications, the frequency that carries the audio or TV signal. A radio carrier frequency is specified by its location on the dial. For instance, the radio station at 1590 on your AM radio dial rides on a 1590-kHz carrier.

Carrier ID Code The code that is entered by a long-distance caller that wishes to bypass the preselected long-distance company. Each local telephone line has a preselected long-distance company that the subscriber chooses when the phone line is ordered and initially put into service. If you would like to bypass this long-distance company and use another, you dial 1, then 0, then the carrier ID code, which is three digits long. After you enter the carrier ID code, you dial the area code and the seven-digit number. Carrier-ID codes are available in almost every exchange for long-distance equal access.

Carrier, Long Distance Also known as a *long-haul company* or *long-distance company*. The four big long-distance companies in the United States are AT&T, MCI, SPRINT, and Worldcom.

Carrier Loss In T1 transmissions, a carrier loss occurs when too many consecutive zeros are transmitted or when a component of the T1 circuit fails. In other transmissions, carrier loss is simply an unintentional loss of signal, regardless of the reason.

Carrier-Provided Loop A carrier-provided loop is a local phone line that is bought by a long-distance company and re-sold as a part of a WAN service. In most WAN services, the long-distance portion of the service is billed separately from the local portion, just like your residential (phone in your home) service.

Carrier Sense Multiple Access (CSMA) An Ethernet LAN protocol. In *local-area networks (LANs)* with CSMA, PCs check the network to see if it is clear before transmitting. They do this because if more than one PC sends data at the same time, the data gets garbled and is meaningless to the other PCs. This simultaneous data transmission is called a *collision*. The two other types of LAN protocols that are advanced versions of this one are: *Carrier Sense Multiple Access/Collision Avoidance (CSMA/CA)* and *Carrier Sense Multiple Access/Collision Detection*. In Ethernet networks, PCs sense and transmit hundreds of times per second. If the network looks clear for a tiny fraction of a second, the PC will try to transmit.

Carrier Sense Multiple Access/Collision Avoidance (CSMA/CA) An Ethernet LAN protocol. In *Local-Area Networks (LANs)* with CSMA/CD, PCs check the network to see if it is clear before transmitting. If the network is clear, it sends a jam signal, then waits a specified time to allow all the other PCs to receive it. It transmits its data and sends a clear signal. They do this because if more than one PC sends data at the same time, the data gets garbled and is meaningless to the other PCs. This simultaneous data transmission is called a *collision*. In Ethernet networks, PCs sense and transmit hundreds of times per second. If the network looks clear for a tiny fraction of a second, the PC will try to transmit.

Carrier Sense Multiple Access/Collision Detection (CSMA/CD) An Ethernet LAN protocol. In *Local-Area Networks (LANs)* with CSMA/CD, PCs check the network to see if it is clear before transmitting. If it is clear, it transmits its data. They do this because if more than one PC sends data at the same time, the data gets garbled and is meaningless to the other PCs. This simultaneous data transmission is called a *collision*. CSMA/CD senses these collisions and attempts to retransmit the same data again when the network is clear again. In Ethernet networks, PCs sense and transmit hundreds of times per second. If the network looks clear for a tiny fraction of a second, the PC will try to transmit.

Carrier Shift A change in frequency. Carrier shift is also a way of transmitting binary ones and zeros over a phone line or radio carrier, which is called *frequency-shift keying*.

Carrier Signal A signal that carries another signal. In telecommunications, the word *carrier* has a broader meaning than in the broadcast radio or TV industry. In telecommunications, a carrier can simply be a digital signal connecting two modems or a T1 circuit. The data in itself is the carrier. In radio, TV, or cable TV, a carrier is a continuous unchanging waveform (ac sine wave) of a specific frequency. The sound or video sent over the carrier signal changes or alters the once unchanging carrier. In simple terms, the sound or video "rides" within or on the carrier. See *AM* for more radio carrier information.

Cascaded Amplifier An amplifier that consists of two or more amplifiers coupled together. Almost all consumer audio electronics products have more than one stage of amplification or cascaded amplifier circuits inside them. This design is very common in all kinds of amplifiers.

Cascaded Stars A LAN physical topology where star networks are connected together.

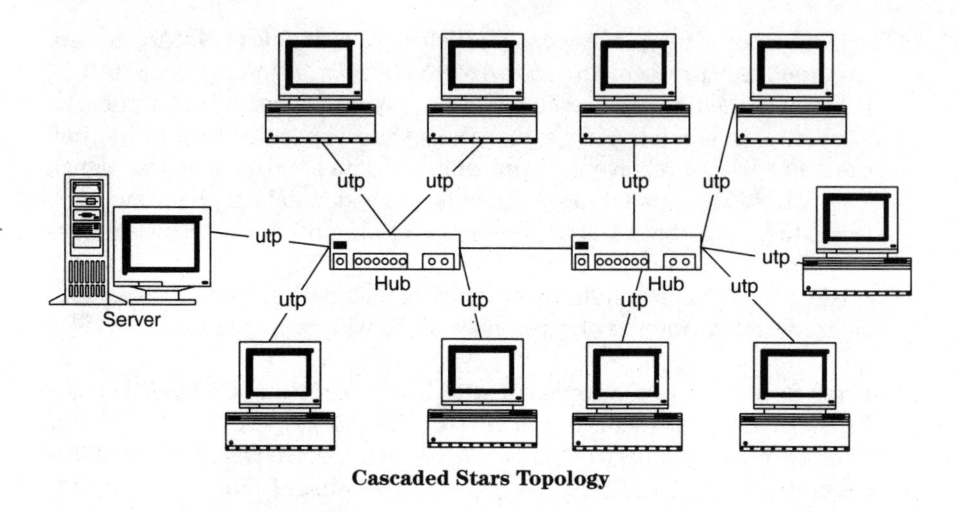

Cascaded Stars Topology

Case Sensitive When a computer's software recognizes a difference between upper and lower case alphabet symbols. Sometimes passwords are case sensitive.

CAT 1 Category 1. Any wire other than phone wire used for transmission, except coax. CAT 1 is nontwisted and can be any AWG. Applications include audio speaker wire, telephone quad conductor (four wire, red-green-black-yellow), electrical, doorbell, thermostat, and other control wire.

CAT 2 Category 2. Twisted-pair wire of 22 to 26 AWG. UTP or STP digital or data-transport media good to speeds of up to 1.5 MHz at 300 feet. The nominal impedance is 100 Ω (±10%). Typical applications include analog telephone and lesser analog transmission and control.

CAT 3 Category 3. Twisted-pair wire of 22 to 24 AWG. UTP or STP digital or data-transport media that is good to speeds of up to 16 MHz at 300 feet. Nominal impedance is 100 Ω (±10%). Typical applications include analog telephone, 10 base-T, T1 (on conditioned pairs), and 4/16 token ring.

CAT 4 Category 4. Twisted-pair wire of 22 to 24 AWG. UTP or STP digital or data-transport media that is good to speeds of up to 20 MHz at 300 feet. Nominal impedance is 100 Ω (±10%). Typical applications include analog voice, 10 base-T, token ring, and T1.

CAT 5 Category 5. Twisted-pair wire of 22 to 24 AWG UTP or STP, where each pair of wire within the sheath has a different number of twists per foot. Digital or data-transport media that is good to speeds of up to 100 MHz at 300 feet. Nominal impedance is 100 Ω (±10%). Typical applications include 10 base-T, 100 base-T, token ring, switched token ring, ATM, and T1.

CAT 5 UTP Plenum

CAT 7 Category 7. Twisted-pair wire 22 to 24 AWG. Each pair of wire placed side by side within a sheath has a different number of twists per foot. The unusual flat shape of CAT 7 makes it very distinguishable from other twisted-pair wire types. *UTP* or *STP (Unshielded Twisted Pair* or *Shielded Twisted Pair*) digital or data-transport media that are good to speeds of up to 250 MHz at 300 feet. Nominal impedance is 100 Ω (±10%). Typical applications include 10 base-T, 100 base-T, token ring, switched token ring, ATM, T1, T3, and STS-1.

CAT 7 Plenum UTP

Category 3 Twisted Pair (CAT 3) See *CAT 3*.

Category 4 Twisted Pair (CAT 4) See *CAT 4*.

Category 5 Twisted Pair (CAT 5) See *CAT 5*.

Category 7 Twisted Pair (CAT 7) See *CAT 7*.

Cathode The more negative end of a diode or other electronic device, such as a vacuum tube. The screen of your TV set and monitor are actually the cathodes of large vacuum tubes. A diagram of rectifier symbols and the cathode locations follow.

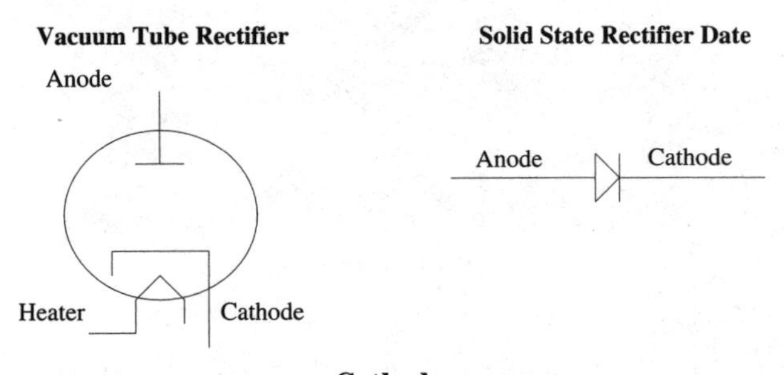

Cathode

Cathode-Ray Tube (CRT) The real name for a TV or monitor screen.

CATV (Community Antenna Television) Better known as *cable TV*. A cable-TV company receives satellite and terrestrial TV signals. They retransmit those channels down a coax that branches out through a geographical area, where subscribers to the antenna service can be connected. Some cable-TV companies add their own information channels to their broadcast and local advertising between programming on selected channels.

CB (Citizens Band) A frequency bands from 26.965 MHz to 27.225 MHz and 462.55 MHz to 469.95 MHz that are set aside for unlicensed two-way communications. CB radios are limited to a transmission power of 4 Watts.

CBR (Constant Bit Rate) A type of data transmission that is nonbursty. CBR transmissions are needed for undistorted audio and video transmission. The SLC 96-pair gain system and ATM provide constant bit rates.

CCC (Clear-Channel Coding, Clear-Channel Capability) A reference to a type of T1 service. A clear-channel T1 is formatted for out-of-band signaling, which means there is no bit robbing and the dial tone, hook flashes, and DTMF digits are sent over the 24th (the last) channel in the T1 circuit. Clear-channel signaling is usually best for data circuits and in-band signaling is best for voice circuits. In a clear-channel circuit, all 64 kb/s in each channel of the T1 instead of 56 kb/s (except for the 24th, which is 100% dedicated to signaling for the other 23 channels) are available to the end user.

CCFL (Cold-Cathode Fluorescent Lamp) A technology used to light LCD screens. The Dynatel 965T has a CCFL built in so that technicians can see the display in the dark.

CCH (Connections Per Circuit Hour) The number of connections or calls completed at a switching point per hour.

CCITT (Consultative Committee on International Telegraphy and Telephony) The CCITT is one of the four parts of the ITU (International Telecommunications Union), which is based in Switzerland. The CCITT makes recommendations for the manufacture and interoperability of telecommunications equipment. The recommendations are not enforced by anything other than the peer pressure of the industry and the fact that following standards greatly improves the chances for a product's success.

CCS (Centum Call Second) A centum call second is 100 seconds of telephone conversation. 36 centum call seconds is one Erlang, which is one call hour (one hour of phone conversation). Erlangs are measurements of telephone switch traffic.

CCS7 (Common-Channel Signaling No. 7) ISDN version of SS7. An out-of-band signaling system between central offices throughout the telephone network that carries information and signaling for each phone call (such as billing, ANI, and ringing), as well as information about each central office (such as trunks busy or blocking and routing information). CCS7 is an uncommonly used in North America, Malaysia, and Japan.

CCTV (Closed-Circuit Television) Usually CCTV is a network of security cameras that terminate into a video processor, which displays the camera images on one or more video monitors.

CD (Carrier Detect) Most modems have a little red LED. When that light is on, your modem is connected to another modem or communications device that it can understand.

CDDI (Copper Distributed-Data Interface) The twisted-pair version of *Fiber Distributed-Data Interface (FDDI)*. Pronounce them the way they look, "fiddy" and "siddy." These two token-passing systems are intended to be backbone applications for LAN environments. CDDI is capable of transmission speeds of 100 Mb/s. For more information on the way it works, see the original version, *FDDI*.

CD-I (Compact-Disc Interactive) A 5.7" compact-disc format that is used for interactive home entertainment through a TV. An example is a video game.

CDMA (Code Division Multiple Access) Also called *spread-spectrum radio*. This technique makes the PCS form of cellular phone service work. CDMA converts your voice into a digital signal, adds an address (the ID code of the destination you are conversing with) to each digital voice packet, scrambles it, and transmits it into the air. The best thing about CDMA is that 20 times the number of calls can be placed within the same amount of bandwidth. Also, it can carry digital information along with your voice. When it changes transmission from one antenna (cell site or geographical cell) to another, the old site checks that the new one has picked you up before disconnecting. The only disadvantage of PCS is a lower clarity in transmission. Voices sound digitized and distorted, in comparison to the static interference that plagues the older analog cellular technology.

CDPD (Cellular Digital Packet Data System) A data packet transfer standard for sending data over cellular that was developed by the *CTIA (Cellular Telephone Industry Association)* in 1993. It is offered by some cellular companies. With CDPD-compatible equipment, you can send data over cellular the same way you send data over a land-based data-packet service (such as frame relay). For the cellular telephone company (operator), the CDPD equipment is physically and functionally separate from the cellular switching equipment, but it shares the cell site and radio spectrum.

CDR (Call Detail Recording) See *Call Detail Recording*.

CD-R (Compact-Disc Recordable) A compact-disc format that is capable of being recorded on one time. A special drive is needed to do the recording.

CD-ROM (Compact-Disc Read-Only Memory) A read-only nonmagnetic data storage device in the form of a reflective disk that is 4.7 inches in diameter. Data stored can be audio, video, data, or a combination of these. Maximum storage is typically 650MB. The CD-ROM is read by a laser diode. When the CD-ROM is manufactured, tiny reflective spots are burned into the surface of the disk.

Here is a simple description of how a CD-ROM works: Imagine that you are driving a car through a tunnel with the headlights pointing straight up. On the ceiling of the tunnel are mirrors. Every time you pass a mirror, you see a flash of light. The flashes of light and periods of darkness would be ones and zeros. The tunnel is a track on the CD-ROM disk. Your headlights would be the laser diode and your eyes would be the optical receiver.

CD-V (Compact-Disc Video) A disk that is about 3 inches in diameter that is capable of storing about five minutes of audio and video.

CED (Called Equipment Identification Tone) A 2100-Hz tone with which a fax machine answers a call.

Ceiling Distribution System Also called a *ceiling rack*. It consists of rows of ladder-shaped iron (usually painted gray), supported above electronic equipment by more iron posts bolted to the floor. It is used as a safe, out-of-the-way place to mount the cables that connect the electronic equipment below. They are used in telecommunications central offices and large computing environments.

Cell A geographical area in cellular communications. Each cell consists of a cell site. A cell site consists of an antenna, a hut, and a doghouse (the doghouse contains the transmitting electronics and is in the hut).

Cell Site A cell site consists of an antenna, a hut, and a doghouse (the doghouse contains the transmitting electronics and is in the hut). A cell site is the transmit and receive center for a geographical area, called a *cell*.

Cell Switching The process of handing a call from one cellular broadcast site or antenna to another without interrupting the call. This process is controlled by a *Mobile Telephone Switching Office (MTSO)*, to which all the cell sites within a region are connected.

Cellular A wireless local telephone service that operates by dividing a geographical area into sections (*cells*). Each cell has its own transmitter/receiver that tracks and operates with cellular telephones within its area. The dimensions of a cell can range from several hundred feet to several miles.

Diagram of a cell layout for a geographical area

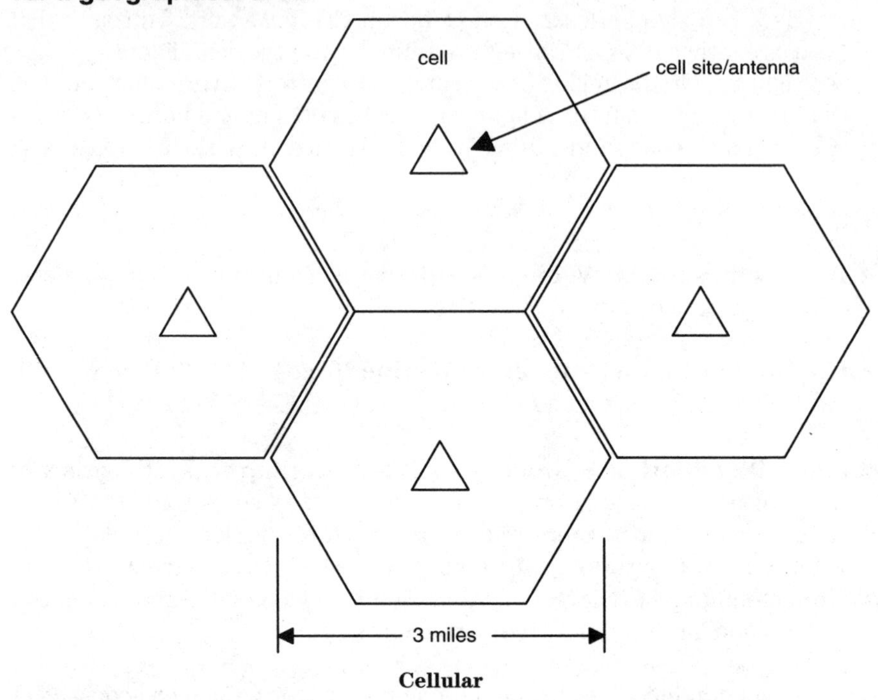

Cellular

Cellular Data-Link Control (CDLC) A protocol for sending data over cellular. The equipment is not integrated with the cellular-service provider. It works the same way a modem does for a regular phone line. It error checks and retransmits corrupted data.

Cellular Digital Packet Service (CDPD) A service offered by some cellular companies. With CDPD-compatible equipment, you can send data over cellular the same way you send data over a land-based data-packet service (such as frame relay).

CEMH (Controlled Environment Man Hole) The new nondiscriminatory name for this is *CEV (Controlled-Environment Vault)*. It contains heating and cooling equipment and communications electronics, unlike plain old vaults that only contain splices.

Centel A company that was bought by Sprint in 1992.

Central Office A building that houses a telecommunications switching or trafficking system. Typical switching systems installed in central offices in North America are Lucent Technologies' 5ESS and Northern Telecom's DMS family of switches. There are five classes of central offices and five major parts to a central office. As a whole these parts are referred to as *inside plant*.

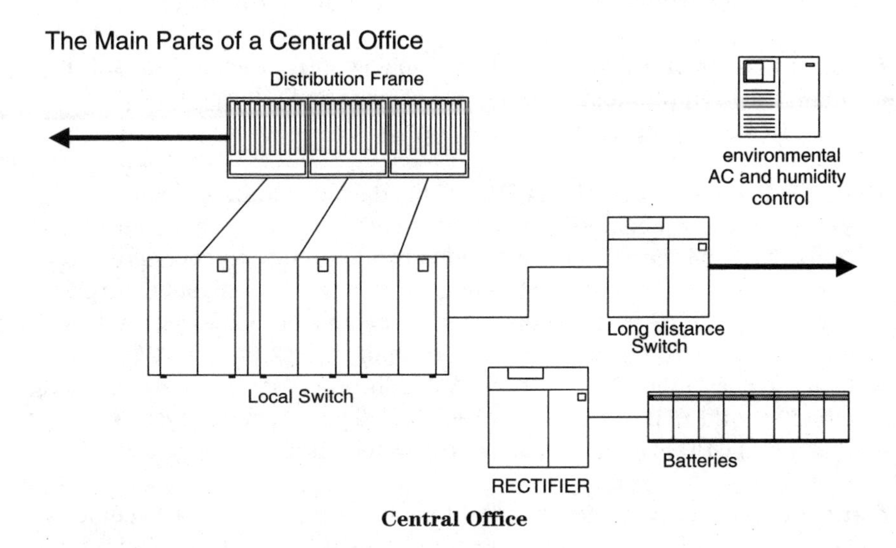

Central Office

Central Office Battery In theory, "central office battery" is –48 Volts. In reality, it is –52 Volts. The deviation is because of the difference between rectifier output voltage and the true battery voltage. The batteries that the central offices are powered from are arrays of 12-V batteries in series and parallel. These 12-V batteries actually output 12.7 V. If four 12.7-V batteries are placed in series, they add up to 50.8 V. The rectifiers in the central office that charge the batteries have an output voltage of –52 V. This is a difference of 1.2 Volts, which is the trickle charge for the batteries. Ultimately, if the power is on at the central office, you are getting –52 V, which is the output of the rectifiers that power the switching system and charge the batteries at the same time. If the street power is out and the back-up generator is not running, the real central office battery voltage is –50.8 V.

Central-Office Code The address of a central office. The second three digits of your phone number (including the area code). It is also referred

Area Code (NPA) – Central Office Code – Extension

805 – 555 –1998

Central Office Code

to as the *NXX*. It defines an exchange area, which is the boundary area of a certain central office.

Central-Office Trunk A communications path between central offices. A central office trunks are usually multiplexed into T1 formats.

Central Processing Unit (CPU) The main controller of a data, telecommunications, multiplexing or any other electronic device that performs or carries out complex instructions or tasks.

Central Processing Unit (CPU) The device within a computer (or switch or other machine that performs complex tasks) that controls the transfer of the individual instructions from one device connected to its bus (the data or I/O bus) to another, such as ROM, RAM, subcontrollers, decoders, and I/O ports. Some communications equipment manufacturers actually call a certain card or portion of the system the *CPU*. That is because they include all of the RAM, sub processors, buffers, clocking circuitry, and ROM as a part of the CPU. This is OK because we know that a real CPU is actually a small integrated circuit.

Centrex A service provide by local telephone companies that mimics an on-premises PBX. The customer purchases a block of telephone numbers (e.g., 555-1000 to 555-1999), then every telephone on the customer's premise is connected to the telephone company as an individual phone line. Each line is associated with one of the numbers in the customer's block. The telephone company then programs those specific lines to route calls as desired by the customer. Voice mail can also be incorporated into Centrex.

Centrex LAN A service that uses your modem, the phone company's wire, and the phone company's switch to connect equipment in an office or campus environment. The signal from a peripheral in your office goes all the way to the phone company's central office, then back, just to connect to a computer in the next room. Ethernet or token ring is a much less expensive and better-performing way to connect your LAN in the long run.

Centum Call Second See *CCS*.

Cesium Clock A clock that is used to synchronize communications equipment (i.e., SONET transport) by providing a perfectly steady output pulse (a very fast one), the same way that a metronome provides steady timing for a musical band. Its base is a factor of the atomic vibrations of the element Cesium.

CEV (Controlled Environment Vault) A vault that is designed to have electronics in it. The environment inside the vault must be kept at a certain temperature and humidity.

CFA (Carrier Failure Alarm) A notification that timing has been lost in a digital transmission because of excessive zeros in its transmission. When a carrier failure alarm occurs, all the calls and all the data on that transmission are dropped until the carrier equipment regains timing.

CGSA (Cellular Geographic Service Area) The geographical area that a cellular company provides service, which means their cellular radio waves can be received within this area.

Channel One segment or time-slot in a broadband communications transmission.

Channel Capacity The maximum number of bits per second that can be carried by a channel. The channel capacity of a DS0 within band signaling is 56Kb/s, the channel capacity of a DS0 with out-of-band signaling is 64Kb/s.

Channel Loop-back A method of testing a digital service line, such as a T1, where the receive channel is connected into the transmit channel (sometimes with only a pair of wires) at the far end. The signal can then be tested at the originating location and analyzed for errors. Equipment is made where the loop-back can be performed via remote control. One example of this equipment is a smart jack.

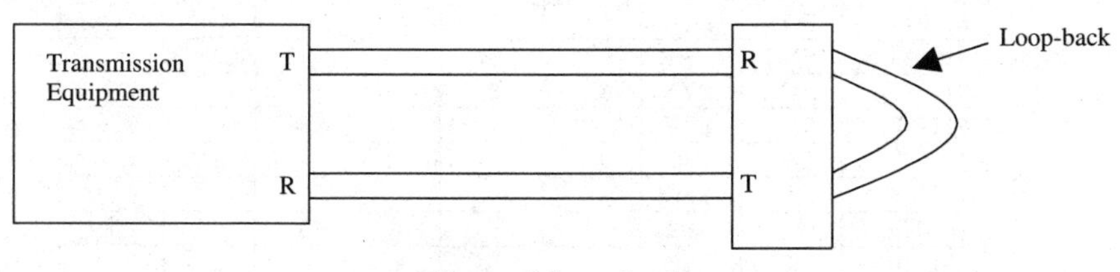

Channel Loop-back

Channel Service Unit (CSU) This is also called a *CSU/DSU (Channel Service Unit/Data Service Unit)*. A CSU is a hardware device that can come in many shapes and sizes. Rack-mount, shelf-mount, and standalone CSUs are available. A CSU/DSU has three main functions. The first function is to act as a demarcation point for a T1 (DS1) service from a local communications company. The second function is to provide line-format and line-code conversion (B8ZS to AMI, SF or D4 to ESF, 135 to 0 V) between the public network and the customer-premises equipment, if necessary. The third function is to provide maintenance or alarm services and loop-back for isolating problems with the T1 line or customer's equipment. For a photo, see *CSU/DSU*.

Channel Termination Also called a *chanterm*, is a cross connect that links the transmit and receive of two devices. Channel terminations are used to connect private-line services through a central office. Many local phone companies charge an additional amount of money for each channel termination that a private line has. If a private line goes from one building to another building across town, it probably passes through two or three central offices to get there. Each connection through a central office requires a channel termination.

Chanterm See *Channel Termination.*

Character A number, letter, or symbol that is represented by a binary code. See *Character Code* for more information.

Character Code A code in binary numbers that represents the Alphabet and other symbols. The table shows ASCII character codes.

Least significant bits (hexadecimal)	Most significant bits (hexadecimal)							
	000 (0)	001 (1)	010 (2)	011 (3)	100 (4)	101 (5)	110 (6)	111 (7)
0000 (0)	NUL	DLE	SP	0	@	P	`	p
0001 (1)	SOH	DC1	!	1	A	Q	a	q
0010 (2)	STX	DC2	"	2	B	R	b	r
0011 (3)	ETX	DC3	#	3	C	S	c	s
0100 (4)	EOT	DC4	$	4	D	T	d	t
0101 (5)	ENQ	NAK	%	5	E	U	e	u
0110 (6)	ACK	SYN	&	6	F	V	f	v
0111 (7)	BEL	ETB	'	7	G	W	g	w
1000 (8)	BS	CAN	(	8	H	X	h	x
1001 (9)	HT	EM	)	9	I	Y	I	y
1010 (A)	LF	SUB	*	:	J	Z	j	z
1011 (B)	VT	ESC	+	;	K	[	k	{
1100 (C)	FF	FS	,	<	L	\	l	
1101 (D)	CR	GS	-	=	M	]	m	}
1110 (E)	SOH	RS	.	>	N		n	~
1111 (F)	SI	US	/	?	O		o	DEL

Character Code

Characteristic Impedance The impedance or AC resistance of a transmission media such as CAT 5 twisted pair. The characteristic impedance of CAT 5 twisted pair is 100 ohms. This means that when a CAT 5 twisted pair is terminated (or connected to) a device that also has an impedance of 100 ohms, The twisted pair, regardless of its physical length, "looks" infinitely long to the circuit. The usefulness of this is that the voltage-to-current ratio is the same all the way down the line. So, if you have 2 V of signal at 20 mA at the beginning of the twisted pair, that ratio equals 2 ÷ 0.02 = 100. That's the same as the characteristic impedance of the twisted pair, imagine that. For a photo of different types of coax, see *Coax*. Here are some values of characteristic impedance for common physical media:

X-mission Media	Characteristic Impedance (Z)
RG-6 coax	75 ohms
RG-8 coax	50 ohms
RG-58 coax	50 ohms
RG-59 coax	75 ohms
RG-62 coax	93 ohms
Cat 5 UTP	100 ohms
Cat 7 UTP	100 ohms
3IBM type 1 Data	150 ohms

Character Set The letters and numbers on computer keyboards. Different standards apply to how the letters and numbers are converted to binary code. The most widely embraced standard for PCs is ASCII

Chip An integrated electronic component. *Integration* refers to many circuits integrated into one small device.

Choke Also called a *choke coil* or *RF choke*. A coil of wire manufactured with the intent of being a filter that reduces the passing of high frequencies through it. For a photo, see *RF Choke*.

Choke Coil See *Choke*.

Circuit 1. Another name for a phone line. There are many types of circuits: digital, analog, T1, and ISDN circuits. 2. An electronic device that receives a given input and converts it into a desirable output. For instance, a TV converts a transmission input into a picture and sound. A TV can be regarded as one giant circuit or many small circuits.

Circuit Board Any form of electronics parts placement that is on a flat surface.

Circuit Breaker An electronic or electrical device installed in a power-distribution system that disconnects (or turns off) the power when the specified current rating of the circuit breaker is exceeded. In most cases, when the specified current in a circuit is exceeded, there is a problem. Circuit breakers disconnect the power before the "problem" causes other problems (such as fire or power-supply overloading).

Circuit ID Code The part of a CCS or SS7 signaling message that identifies (gives a name to) the circuit that is being established between two points.

Circuit Switching Also referred to as *line switching*. A long time ago, the line switching was performed by operators sitting in front of a "cord board." When a caller wanted to make a call, they would pick up the receiver and an operator would ask "what number, please?" The caller would then tell the operator what number to connect them to, and the operator would plug a cord into the line and connect it to the line (or loop) associated with the number they wanted to call. Today, line switching is performed by "switches." *Switches* (also called *PBX* and *key systems*) are electronic machines that work similar to the operator. Instead of speaking with the operator, the switch and caller "signal" each other to accomplish the switching function. When the caller goes off hook, the central office switch sends a dial tone. The dial tone is a signal to the caller to "enter number please." The caller enters digits that signal the number to which they would like to be connected. The central office then sends a "ring" signal to the party being called and a "ring simulation" signal to the calling party, which lets them know that the central-office switch received the digits ok and is signaling the party being called. The party being called then picks up the handset, which activates the switchook and places a 1000-ohm short on the line. The 1000-ohm short signals the central office that the party is ready to receive the call. The central office switch then stops the ring signaling and "switches" the talk paths into place for the two callers.

Citizens Band (CB) See *CB*.

Cladding One of the two glass sections of a fiber optic.

CLAS (Centrex Line-Assignment Service) A feature offered with centrex service that allows customers to dial into the telephone company's line-assignment computer system and make changes to where extensions are located, ringing and hunt groups, and which extension numbers or phone numbers are in service.

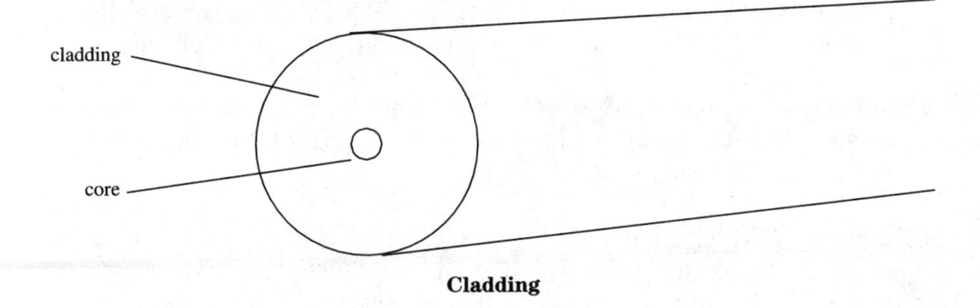

Cladding

Class 1 Central Office A centermost point in a large logical network hierarchy of long-distance central office switches.

Class 2 Central Office An older term for an network hierarchy of switch points. Central offices are now becoming so well connected in so many ways that they are rapidly becoming either a class-5,4 or a class-1 central office.

Class 3 Central Office An older term for an older network hierarchy of switch points. Central offices are now becoming so well connected in so many ways that they are rapidly becoming either a class-5,4 or a class-1 central office.

Class 4 Central Office A tandem central office or main switch center for an area code. That might also perform class-5 end-connection functions.

Class 5 Central Office A local telephone company central office that connects to end customers.

Class of Service (COS) A type of telephone service or telephone line purchased from a telephone company. Some class-of-service examples follow:

- *1FR* One flat rate, residential. What most residential customers have.
- *1MR* One measured rate, residential. Where the line has a low monthly fee, but each call beyond a certain number costs up to an additional 10 cents.
- *1FB* One flat rate, business. What most small-business customers have.
- *1F4* A four-user party line.

Clear Channel See *CCC*.

Cleaving A term used in the splicing fiber optic that means to cut the end of the fiber clean at 90 degrees, with minimal rough edges, in preparation for a fusion splice. Where a mechanical splice is involved, the end of the fiber optic is hand-smoothed with a polishing puck before splicing.

Client-Server Environment A type of network environment with requesters (clients) and providers (servers). A service requested could be for processing, a file, or an application.

Climbers What telecommunications and power company personnel wear to climb wooden telephone and power poles. The official name for these devices are *lineman's climbers.* They are also called *spurs*, *hooks*, and *gaffs*. They consist of a steel shank with straps so that it can be strapped to a person's leg. On the inside of the shank is a spike that is used to stab into the pole. The climbers illustrated were manufactured by Buckingham Mfg. Inc.

Lineman's Climbers

Climbing Belt Used by communications/power/construction personnel to harness themselves to telephone/power poles or tower structures. Also called a *safety belt* and *body belt*. For a photo, see *Safety Belt*.

Clipper A circuit that takes a sine-wave input, like that of a timing oscillator and converts it into a square-wave form that can be used as a clock signal for a digital device. A clipper circuit is simply an amplifier that is over driven into saturation, which causes a form of distortion (in this case, useful) called *clipping*.

Clipping You can experience clipping by turning up the volume on an inexpensive stereo and noticing that at a certain point, the sound reproduction is unclear. First, the bass tones are affected because they take the most energy to reproduce. At this point in increasing the volume, the sound level is not getting any louder, the signal is only becoming more corrupted. Technically, *clipping* is a form of signal distortion where the amplification of a signal in volts exceeds the saturation bias voltage of the transistor in an amplifier. In other words, the peak and negative peaks of the output waveform are not included because the amplifier is not capable of amplifying the input anymore.

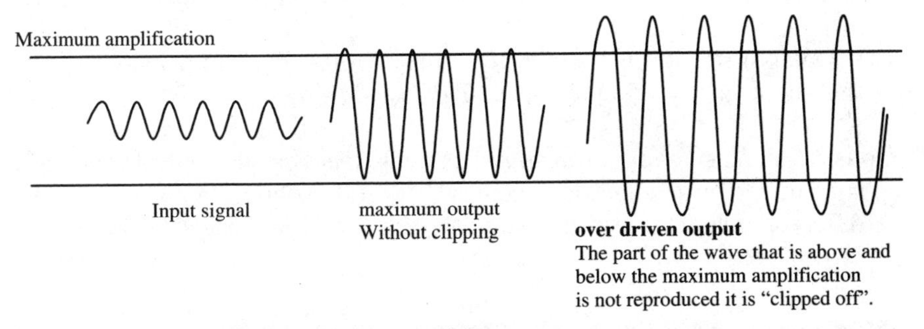
Maximum amplification
Input signal
maximum output
Without clipping
over driven output
The part of the wave that is above and
below the maximum amplification
is not reproduced it is "clipped off".

Clipping

CLLI Code An 11-digit alphanumeric code that identifies physical locations in the phone network. Almost every significant building in the United States has a CLLI code. It is pronounced "silly code."

Clock A device that provides timing pulses for communications equipment or devices within a computer the same way that a metronome provides a steady time for a musical band.

Clock Bias The difference between a clock's output and true universal time. Simply put, how far off a piece of equipment's timing is. It is usually measured in positive or negative microseconds.

Clocking See *Clock*.

Clone Fraud The crime of finding cellular ID codes by monitoring cellular transmissions (with expensive cellular equipment), then copying the code to another cellular phone and making calls with it. The airtime for the copied phone (the clone) is billed to the original phone.

Closure A casing, pedestal, or cabinet used to house open ends or splices in outside cable plant. Closures mostly refer to splices. Different names for buried splice closures or enclosures are Xaga and Cold-N-Close. A popular aerial splice closure is the TRAC closure.

CMOS (Complementary Metal-Oxide Semiconductor) The reason why many computer and other high-speed components are static sensitive. Complementary Metal-Oxide Semiconductor's largest advantage over TTL is their low power consumption (less than 1/10 of TTL), they switch on without drawing very much current, in contrast to TTL technology. Because very little current is drawn, very little power is consumed and very little heat is given off. This allows the devices to be much smaller.

CO An abbreviation for an RBOC Central Office. CLEC's usually refer to their central switching offices as *Type One Nodes*.

Coax A shielded copper transmission media that has one central conductor surrounded by a dielectric. It comes in several varieties. For a listing of the different characteristic impedances of coax types, see *Characteristic Impedance*.

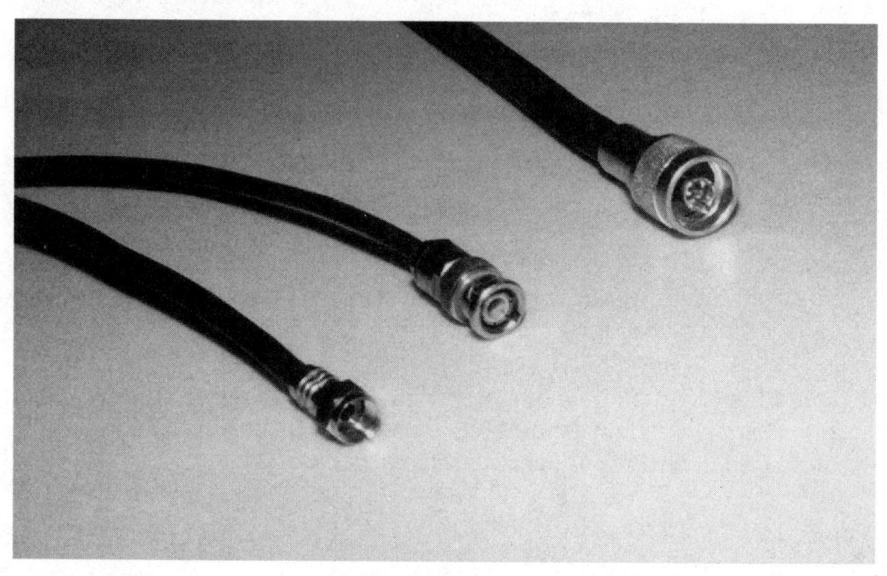

Coax/Connector Types

Coaxial Cable See *Coax.*

COBOL (Common Business Oriented Language) A computer programming language.

Code Blocking A feature of a telecommunications switch that enables it to restrict specific extensions from dialing long distance or just a specific group of area codes. Local telephone companies use this feature to restrict a customer's long-distance calling ability from their phone if they have a poor credit rating and were not able to make a deposit for their phone service. This feature is frequently used in PBX applications to prevent employees from making any long-distance calls.

CODEC (Coder Decoder) Another name for an analog-to-digital (and digital-to-analog) converter.

Coherent Light Light that consists of only one frequency or very close to one frequency. Coherent light looks to the human eye as a very pure color. Lasers and *LEDs* (*Light-Emitting Diodes*, like the one that lights when the hard drive in your computer is running) emit light that is very close to being coherent. A light bulb emits noncoherent light, which consists of many colors and wavelengths.

Coil In telecommunications, a *coil* refers to a *load coil*, which is a voice-amplifying device for twisted-pair wire. A load coil is usually placed on each twisted pair used for a voice line every 3000 feet past a central office. Coils are usually located in vaults, with twisted-pair splices. Other coils, used for other applications are usually referred to as *choke coils*.

Coin-Operated Telephone A telephone that can be installed, operated, and maintained by a local phone company or purchased from a telephone-equipment distributor, connected to a local phone line and maintained by a private individual or company other than the local phone company. Many different kinds of coin-operated phones are manufactured by different companies, just the same as the different single-line phones you can purchase for your own home.

Collision Detection A part of *CSMA/CD (Carrier Sense Multiple Access/ Collision Detection)*, An ethernet LAN protocol. In LANs with CSMA/CD, PCs check the network to see if it is clear before transmitting. If it is clear, it transmits its data. They do this because if more than one PC sends data at the same time, the data gets garbled and is meaningless to the other PCs. This simultaneous data transmission is called a *collision*. CSMA/CD senses these collisions and attempts to retransmit the same data again when the network is clear again. In ethernet networks, PCs are sensing and transmitting hundreds of times per second. If the network looks clear for a tiny fraction of a second, the PC will try to transmit.

Colocation, Physical A colocation is an interconnection agreement and a physical place where telephone companies hand-off calls and services to each other. This is usually done between a CLEC and an RBOC. The CLEC installs and maintains interconnection equipment usually consisting of optical carrier (SONET) equipment and a digital cross-connect system. There are other types of colocations. Alarm companies like to have their alarm signaling equipment located in the local central office for the security and convenience of connecting alarm circuits. Long-distance companies colocate with local telephone companies, as well.

Colocation, Virtual A co-location is an interconnection agreement and a physical place where telephone companies hand-off calls and services to each other. This is usually done between a CLEC and an RBOC. A virtual colocation is when telephone company A (the CLEC) requests that their phone company's network be connected to telephone company B's (the RBOC's) network. Telephone company B charges company A lots of money. Company B owns, installs, and maintains the equipment. To company A, the interconnection is virtual because they never physically do anything to it when and after it is installed.

Color Code The three communications color codes are the twisted-pair cable color code *(PIC)*, the fiber-optic color code, and the resistor color code. For a listing of these color codes, see Appendix F.

Combination Trunk A *DID (Direct Inward Dial)* and *DOD (Direct Outward Dial)* trunk all in one. These trunks are basically the same as the phone service in your home. They can be dialed out, and they can ring in.

Command Set Parts of a program within a switch or PBX's software. A command is a set of instructions written in a program and attached to a command. A command is a simple entry that a user makes to instruct the PBX to perform a function, such as "Enable Trunk Port 8." When this command is entered, a command set within the software is activated and trunk port 8 looks for calls to come in.

Commercial Building Telecom Wiring Standard A wiring standard for buildings. The standard states that all wire on each floor of a building is run and terminated to a single location. All wire run between floors is fed by a riser that terminates at each floor where the wire runs terminate. All the riser cables must originate at a single point where the local telephone company point of presence (NI) is located.

Committed Information Rate The rate in Kb/s or Mb/s that a communications company will guarantee over a frame relay circuit that they provide to their customer. If you purchase a frame relay circuit, there is a place on the order agreement that you state the rate of information that you want to transmit. The choices are 16Kb/s, 32Kb/s, 48Kb/s, and 64Kb/s. If you enter 32Kb/s for your committed information rate, you will pay more for your service than the 16Kb/s Committed Information Rate. You have the capability to transmit and are permitted to transmit at rates up to the full 64Kb/s, unless the frame relay network becomes congested. Then, only your committed rate of 32Kb/s will get through. This is similar to purchasing the use of lanes on the freeway. If the freeway has very few cars travelling on it, then there is no sense in buying the right for multiple lanes because no one is using them anyway. However, if the freeway is congested, then you are getting your money's worth with your lane rights. You never know how congested a frame relay network will be, or when and where bottlenecks will occur in the network.

Common Audible Ringer A loud ringer. A bell connected to a telephone line that is in a noisy or wide-open area. When the phone rings, the loud bell also rings.

Common Channel Signaling A telephone network architecture that uses SS7 for its major signaling functions.

Common, Electrical Not to be confused with ground. *Common* is a reference point, and is ungrounded. It is usually a signal return or DC reference coupling for transmission circuits.

Common, Return See *Common, Electrical.*

Communications Controller Also called a *front-end processor* or *proxy server*. It has the capability of receiving multiple data communications transmissions of different protocols and converting all of those protocols to one common protocol, then routing the data to its destination. You could think of a communications controller as a language translator.

Communications Protocol The method that a communications circuit or link exchanges information. If you are a customer, there isn't a whole lot to worry about with protocol. Just imagine that protocol is just another word for "dialect." If one piece of gear "speaks Chinese," don't try to hook it up to a piece of gear that "speaks English." Some examples of communications protocols are ISDN, frame relay, V.32, and ethernet. They are all very different.

Communications Satellite Usually a geostationary piece of electronic equipment that stays in the same location, relative to the earth's surface, only at a distance or "altitude" of 22,000 miles. They are used to overcome the curvature of the earth for radio-transmission applications.

Communications Workers of America The organized labor union of the Regional Bell Operating companies. A top craft (or top skill level) network technician earns about $20.00 per hour, according to the most recent wage scales in the CWA contract handbook.

Community-Antenna Television (CATV, Cable TV) See *Cable TV*.

Compact Disc (CD) See *CD*.

Complementary Metal-Oxide Semiconductor (CMOS) See *Complementary Metal-Oxide Semiconductor.*

Complementary Network Services An additional service that can be added to your telephone line. Voice mail is one of these. The service can be provided by the local exchange carrier providing the dial tone or it can be provided by a complimentary network services provider.

Completed Call A call that is connected to its destination. When someone calls a number and the other end is picked up by someone, the call is completed. You would think that a call would be completed when the people were finished talking, but in regard to call routing and switching, that is not the case.

Component Video A video signal transmission and/or reception that consists of separate units for each function. In broadcast television, the video is AM, the audio is FM, and color is PM. They are all transmitted together on a single carrier, but processed and decoded separately.

Composite Signal The whole signal, overhead and payload included. The composite speed of a T1 signal is 1.544 Mb/s.

Computer Telephony Integration (CTI) A wave of products that have been coming out since 1994. CTI applications interface computers with telephone systems, *IVR (Interactive Voice Response)* systems, voice-mail systems, call-accounting systems, and anything else that is telecommunications oriented. A good example of CTI is an OCTEL product that allows users to click and drag voice-mail messages from their phone to any other telephone, they would like to get the message. It also provides diagnostic functions, traffic analysis functions, and administration functions through a *GUI* (*Graphical User Interface*, such as Windows 95).

Concatenation To join two or more ATM channels, ISDN B channels, or T1 DS0 channels together to make a larger single channel that can carry a broader signal. In the T1's case, concatenation is usually called *fractional T1*.

Concentrator Another name for a *multiplexer* in communications.

Condenser An obsolete term for *capacitor*. See *Capacitor*.

Condenser Microphone A microphone that incorporates a capacitor that changes its capacitance as sound waves strike it.

Conditioned Circuit A twisted copper pair within an outside plant network that is modified to carry a digital data signal instead of an analog voice signal. Conditioned circuits have noise-filtering electronics components attached to them instead of load coils.

Conditioning A term that refers to modifying a twisted copper pair in an outside plant network so that it can carry a digital data signal instead of an analog voice signal. Twisted-pair circuits (a circuit is a local loop, which is the pair that connects the central office to the customer) are conditioned by adding noise-filtering electronic components to them.

Conductance The mathematical inverse of resistance. The unit of conductance is Siemons or mhos, which is ohm (the measure of resistance) spelled backwards. To calculate the conductance of a circuit or device, just figure $1/R$ (in ohms).

For example: Convert 500 ohms to a value of conductance.

1/500 ohms = 0.002 siemons.

Conductor An element or chemical that allows electrical current to flow through it easily under the influence of electric forces. The following list of conductors is in order of conductive properties. Sometimes the quality of a conductor is not only viewed by its ability to conduct electricity but its ability to resist corrosion as well.

Conductor

Conduit Conduit is another name for tubing, of which there are many different kinds.

For outside plant applications, 4" PVC conduit is usually used for linking cable vaults with an easy means to add, remove, and upgrade cable. If fiber-optic cable is installed in a 4" PVC conduit, four innerducts of different color are pulled into the PVC conduit first. If conduit is being buried specifically for fiber-optic cable, then a quad-lock is usually used. Quad-lock is four 1" conduits braced together.

For inside plant applications, 4" EMT conduit is normally used. EMT is the standard metal conduit that electricians use. PVC conduit is not allowed in modern buildings because when it burns, it emits very toxic chlorine gas.

Conference Bridge A service where the number of calls or people connected to a conference call are controlled by a single source. Most conference bridge services have everyone that wants to be on a conference call dial a toll-free number, where a conference attendant will answer their call, then connect them through (or bridge them) to the other conference callers.

Conference Call A communications connection where three or more different telephone lines (with different phone numbers or extensions) are connected together.

Confidencer A device that connects to a telephone handset to block out background noise. Some confidencers are in the form of an interchangeable mouthpiece that replaces the existing mouthpiece on your handset. Walker Electronics Company is a well-known manufacturer of confidencers.

Connect Time The duration that a call path through a switch or network is set up. Simply put, how long your phone call lasts.

Connectionless Network A frame type of LAN service that does not use private line or dial-tone service to connect to clients or other hosts. A physical wire connects to the phone company, but the data is sent through the network via address overhead type routing, rather than a physical hard-wired private-line connection or a dial-up line. The telephone company equipment constantly looks for an address from your equipment, then transports the following data packet to that address.

Connection Oriented A protocol model of interconnection that has three phases: connection, transfer of data, and disconnect. Some connection-oriented protocols are X.25, TCP, and a regular telephone call. When studying data protocols, remember that data communication is modeled after voice communication.

Connections per Circuit Hour (CCH) The number of connections or calls completed at a switching point per hour.

Connector A device that is used to attach or plug things together.

Connector, Genderless Sometimes called a *hermaphroditic connector*. A genderless connector developed by IBM that is usually called a *Data Connector*. The Data Connector does not need complementary plugs (male and female) to make a connection, like all other known communications modular connecting systems. The Data Connector is specifically designed and used for switched token-ring backbone applications.

Conservation of Radiance A scientific law that basically says that you cannot amplify or increase light without a light-creating source. So, optical fiber does not make the light brighter as the light travels through it and neither do your sunglasses. It would be cool to have glasses that actually made the night brighter, with no electronics, just the lenses. Conservation of radiance simply states that this is impossible.

Console The large telephone with all the keys and/or buttons on it. It is the traffic-control center for a PBX system. The PBX operator or attendant usually has a console. The two types of consoles designed for two different applications are the hotel PBX operator console and the business PBX operator console.

Contention A type of LAN control scheme used by ethernet, where all the users (PCs) on the network fight for use of the network. The PCs check to see if the network is clear, then transmit. Often more than one PC tries to transmit at once, which causes the data to be garbled (this is called a *collision*). This network-control scheme is called *contention*. I guess you could say that the PCs contend to see which one gets their data on the network first. The best alternative to this type of network is token ring—a completely different method of network control.

Context Keys Buttons on a phone or other device that have a display adjacent to them. They perform different functions, depending on what the display is showing at the time you push that button.

Contiguous Slotting The banding together of two or more adjacent channels in a T1 to get one larger channel. Also called *fractional T1* and *concatenation*.

Control Signal A signal sent as a bit, byte, or tone that prompts a communicating device to do something. When you pick up the handset on your telephone, the switch-hook pops up and makes a 1000-ohm connection across the two telephone wires that go to the phone company. This causes a current flow out of the central office switch. This control signal shows that you are off hook and would like to dial a number. The central office responds to your "off-hook" or current flow signal by sending a dial tone, which prompts you to dial digits. The digits are a control signal that tells the central office switch where to route your call. These are all control signals, and all control signals (regardless of the protocol) are equally as systematic and organized.

Controlled Environment Vault A vault that is designed to have electronics in it. The environment inside the vault must be kept at a certain temperature and humidity.

Controller In telephone equipment, *controller* is another name for CPU. In data communications, such as LANs, a controller controls data transfer between two devices.

Convection Cooling A method that newer telephone electronics uses to cool itself. Rather than having a cooling fan attached to a device to cool it, it is equipped with metal deflectors. The deflectors channel warmer air that is rising out of the top of the equipment, which, in turn, pulls cooler air in through the bottom. With convection-cooled equipment, you still need to have a system that cools and controls the humidity in the room that the equipment is in. This cooling option is available with the Tellabs Titan 5500 Digital cross-connect system, for example.

Cord, Base The cord that goes between your telephone and the wall. It has RJ6x-type plugs on the ends. Base cords are available in 2-conductor (RJ6x2c plugs on the ends), 4-conductor (RJ6x4c plugs on the ends), 6-conductor (RJ6x6c plugs on the ends), and 8-conductor (RJ8x8c plugs on the ends). RJ4x4c means an RJ modular-type plug that is 4-conductor positions wide with four conductors installed.

Cord, Handset The cord that goes between your phone and the handset. Also known as a *curly cord*. It has RJ4x4c (an RJ modular-type plug that is four conductor positions wide, with four conductors installed).

Core Processing Unit Some communications equipment manufacturers call the card or shelf that controls a communications system (e.g., PBX) or portion of the system the *CPU*. This is because they include all of the RAM, subprocessors, buffers, clocking circuitry, and ROM in this part of the system.

Northern Telecom option 81 PBX CPU (Core) Shelf

COS (Class of Service) A type of telephone service or telephone line purchased from a telephone company. See *Class of Service*.

Coulomb A unit of electrostatic charge equal to 6,300,000,000,000, 000,000 electrons or protons (electrons would be a –1C and protons would be a +1C). This unit of electric charge was established by Charles Augustin Coulomb (1736–1806) and is useful because with it we can determine a standard of measurement for electric force, which led to the measurement of electrical charge flow (current), known as the *ampere*. One ampere is equal to one coulomb of charge (the number of electrons) flowing past a point in one second.

Counter Rotating Ring A backbone network architecture that is common to *FDDI (Fiber-Distributed Data Interface)*, *SONET (Synchronous Optical Network)*, *DQDB (Distributed Queue Dual Bus)*, switched token ring, and *CDDI (Copper-Distributed Data Interface)*.

Country Code A code used in international dialing for countries that are not a part of the *North American Number Plan (NANP)*. To dial international long distance from the United States, dial:

011 + county code + city code + number.

For a listing of country codes, see appendix B.

To dial the United States from another country that is a part of the NANP, simply dial the area code the same way you would call long distance to another state. To call the United States from another country that is not a part of the NANP, consult your long-distance company. The United States has a different country code/access code for almost every country that is not a part of the NANP.

Coupling A means of connecting one adjacent circuit to the next. Different types of coupling include: capacitive, inductive, electromagnetic (radio), optical, and direct (hard wire).

Coverage Area The geographical area that is serviced by a cellular or PCS telephone system. Within this area, subscribers can access a cellular or PCS radio signal link and make calls. If the subscriber travels outside of this area, the "no service" or "roam" indicator appears on the phone's display. If the roam indicator is on, the subscriber still has a signal and can make calls, they are just within another cellular company's coverage area and the call will be more expensive. If the "no service" indicator is on, no signal is present and no phone call can be made.

CPE (Customer-Premises Equipment) The equipment that is connected to a phone line. The exact definition is anything beyond the Standard Network Interface, which includes wire, jacks, telephones, answering machines, and any other devices connected to the telephone line.

CPS (Cycles or Characters Per Second) See *Cycle*.

CPU (Central Processing Unit) The device within a computer (or switch or other machine that performs complex tasks) that controls the transfer of the individual instructions from one device connected to its bus (the data or I/O bus) to another, such as ROM, RAM, subcontrollers, decoders, and I/O ports. 2. *Core Processing Unit*. Some communications equipment manufacturers call the card that controls a communications system (e.g., PBX) or portion of the system the *CPU*. This is because they include all of the RAM, subprocessors, buffers, clocking circuitry, and ROM in this part of the system. For a photo, see *Core Processing Unit*.

CR The ASCII control code abbreviation for carriage return. The binary code is 1101000 and the hex is D0.

Craft A reference to nonmanagement RBOC personnel. Many craft personnel are members of the Communications Workers of America (CWA) labor union.

Craft Test Set Also called a *Goat* or *Butt-Set*. A test telephone that is used by technicians to test analog telephone lines (ringing, dial tone, monitor, etc.).

Harris Dracon Craft Test Set "Butt Set" "Goat"

Crimp Tool Tools used to place connectors on the ends of different types of coax and twisted pair.

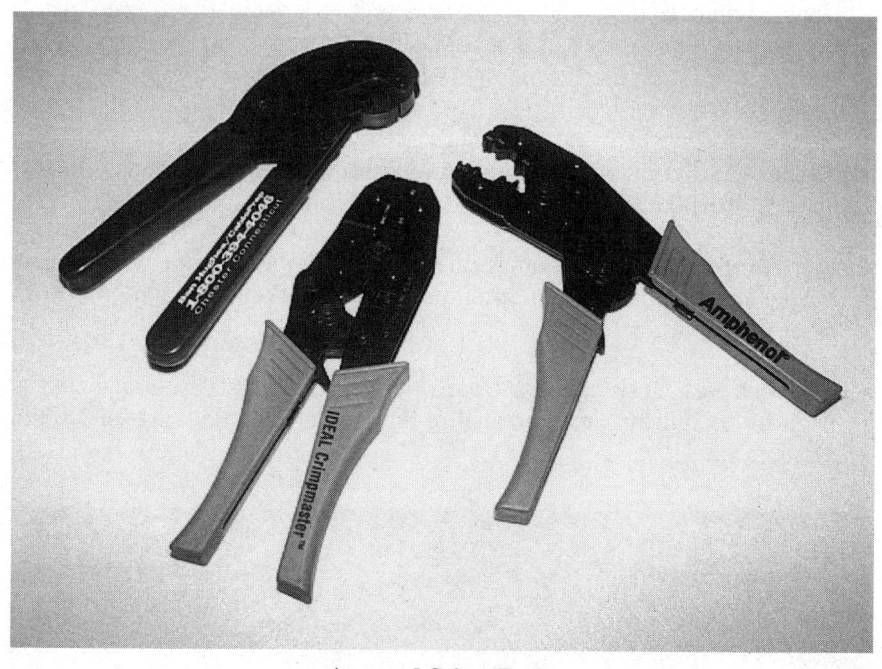

Assorted Crimp Tools

Cross Bar An obsolete telephone switch that was analog and had mechanical relays that connected telephone calls. This is where the term *switch* comes from. Old central office switches contained literally thousands of mechanical switches.

Cross Connect A cross connect is the connection of one circuit path to another via a physical wire. Telephone cable pairs are terminated or "punched down" onto a termination block (usually a 66m150 or an AT&T 110 block) that has extra connections available for each pair so that jumper wires can be easily connected and re-arranged between them.

Cross-Over Cable A connector cable where two or more of the conductors reverse themselves from one end to the other. A null modem cable is a type of cross-over cable.

Cross Talk The two forms of cross talk are inductive cross talk, because of wires touching (or conducting to) each other that shouldn't be and channel seepage because of inaccuracies in multiplexing equipment timing or components. Inductive cross talk is caused by radio or the use of nontwisted-pair wire. Inductive cross talk travels from one device to another via electromagnetic fields generated by different transmissions. A hard cross or physical cross is most commonly caused by water seeping into a telephone cable and conducting the electric signals on pairs in the cable to each other. Channel seepage is usually a situation where the people on one phone call can distantly hear one side (only one person's voice, not both) of another phone call.

CRT (Cathode-Ray Tube) The real name for a TV or monitor screen.

Crystal Oscillator An electronic device that is made from a thin piece of polished quartz crystal. When a periodic voltage is applied to a crystal, it has a piezoelectric reaction. This means that the voltage applied to the crystal distorts it. When the voltage is removed, the crystal physically vibrates. With each vibration of the crystal, a very small AC voltage cycle is produced. The physical size of the crystal determines its oscillating frequency. Crystal oscillators are used because of their reliable timing.

4MHz Crystal

CS (Convergence Sublayer) See *Convergence Sublayer.*

CSDC (Circuit Switched Digital Capability) A 56Kb/s phone line that can carry voice or data. The conditioning equipment is equipped with a digital-to-analog converter and vice versa.

CSMA (Carrier Sense Multiple Access) See *Carrier Sense Multiple Access.*

CSMA/CA (Carrier Sense Multiple Access/Collision Avoidance) See *Carrier Sense Multiple Access/Collision Avoidance*.

CSMA/CD (Carrier Sense Multiple Access/Collision Detection) See *Carrier Sense Multiple Access/Collision Detection.*

CSU (Channel Service Unit) Also called a *CSU/DSU (Channel Service Unit/Data Service Unit)*. A CSU is a hardware device that can come in many shapes and sizes. Rack-mount, shelf-mount, and stand-alone CSUs are available. A CSU/DSU has three main functions. The first function is to act as a demarcation point for a T1 (DS1) service from a local communications company. The second function is to provide line format and line-code conversion (B8ZS to AMI, SF or D4 to ESF, 135 V to 0 V) between the public network and the customer's equipment, if necessary. The third function is to provide maintenance or alarm services and loopback for isolating problems with the T1 line or customer's equipment. Some loop backs can be done remotely, depending on the model CSU/DSU. Some are done with bantam loop plugs.

Rack Mount CSU/DSUs mfg by L XPORT

CTI (Computer Telephony Integration) A wave of products that have been available since 1994. CTI applications interface computers with telephone systems, IVR systems, voice-mail systems, call-accounting systems, and anything else that is telecommunications oriented. A good example of CTI is an OCTEL product that allows users to click and drag voice-mail messages from their phone to any other telephone they would like to get the message. It also provides diagnostic functions, traffic analysis functions, and administration functions through a *GUI* (*Graphical User Interface*, such as Windows 95).

Current The flow of electricity measured in amperes.

Current, Line The average off-hook current of a telephone line is about 35 mA (0.035 A).

Customer-Premises Equipment (CPE) Equipment that is connected to a phone line. The exact definition is anything beyond the standard network interface, which includes wire, jacks, telephones, answering machines, and any other devices connected to the telephone line.

Cut Over The actual changing use of one type of equipment or system to another. If you install a new PBX in your office, the cut over is when you disconnect your old system and begin using a new one.

CVSD (Continuously Variable Slope Delta Modulation) A method of converting analog voice into digital and vise-versa with an on-the-fly variable sampling rate that ranges from 16Kb/s to 64Kb/s, depending on how much bandwidth is available.

CWA (Communications Workers of America) See *Communications Workers of America.*

Cycle One cycle of a waveform's pattern. Cycles are used as a reference to measure the frequency of a waveform or signal. In the following diagram, two waveforms and one cycle of each are singled out. *Cycles* are usually referred to as a number of cycles per unit of time. *Cycles per second* and *hertz* are measurements of the number of cycles you get per second in an analog transmission. *Bits per second* is measurement of how many "square-wave" clock sample sequences are being read from a digital transmission. *Frequency* is a measure of the number of cycles per second (in Hz). One Hz (hertz) is equal to one cycle in one second. Two Hz is two cycles in one second, 1000 Hz is 1000 cycles in one second.

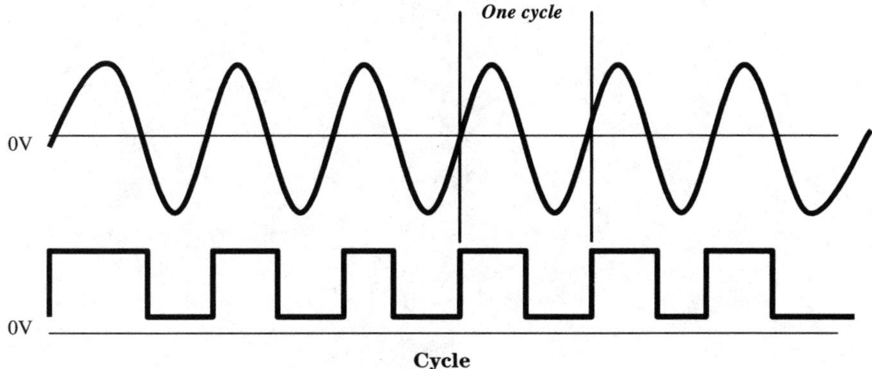

Cycle

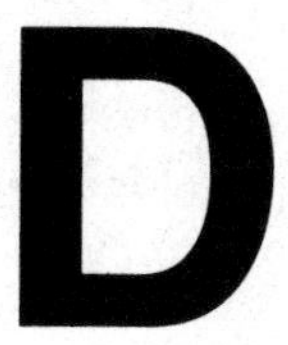

D4 Framing More frequently called the *super frame format* for T1 transmissions. The super frame format consists of 12 standard 193-bit T1 frames. The D4 framing format also incorporates in-band signaling.

DAC (Digital to Analog Converter) A part of a channel bank that encodes analog voice signals into a stream of binary digits. The digital to analog converter or analog to digital converter samples a caller's voice at a rate of 8000 times per second. (The sample rate for a T1 channel is 8000 times per second.) Each sample's voltage level is measured and converted to one of 256 possible sample levels. These levels are from the lowest, 0000000, to the highest, 11111111. The reason for 256 levels is because if you count in binary from 00000000 to 11111111, you end up with 256, the highest number possible with 8 bits. The bits are then transmitted one after another at a high rate of speed to their destination, where the same process happens in reverse. For a diagram, see *Digital to Analog Converter*.

DACS (Digital-Access Cross-Connect System) A DACS is also called a *DCS (Digital Cross-Connect System)*, depending on the manufacturer. A digital cross-connect system is quickly becoming a fundamental part of a local and long-distance company's network. The DACS or DCS is a rack-mountable system that enables any circuit that interfaces with it to be electronically cross-connected from one path to another within the network it is connected to. Circuits that can interface with a digital cross-connect system include DS0, DS1, DS3 (or T3),

STS-1, and SONET OC-1. An incoming circuit can be rerouted by simply making path changes in DACS administrative software. In the following, a T1 circuit coming into the input side (left side) could be cross-connected to exit as one of the channels in a DS3 on the right side. For a diagram of a DACS/DCS, see *Digital Cross-Connect System*.

Daisy Chain A method of connecting devices in a string. The bus topology for Ethernet is an example of a daisy chain. For a diagram of a daisy chain, better known as a *bus topology*. See *Bus Topology*.

DAL (Dedicated Access Line) A private circuit that provides a direct connection (or access) to a long-distance carrier or other communications service, like frame relay or an Internet service provider. Some DALs are a "full-service circuit," which means that if you have a circuit that connects you directly to your Internet service provider, then the only bill you see for that service is from the Internet service provider. The local circuit is in the Internet service provider's name and they pay the phone bill for that service. You, of course, pay a single bill for the entire service. If you are going to get a direct Internet connection, this is the way to go. If it ever stops working, you just call the Internet service provider and they determine where the problem is and fix it.

D Amps *Digital Advanced Mobile Phone Service.*

Dark Current In a photodiode, the dark current is the flow of electricity through the diode when no light is present. Photodiodes are used as light-sensitive switches. When they are exposed to light, they act like a switch and turn on. However, even if no light is present a small amount of electricity still flows through. This is the dark current.

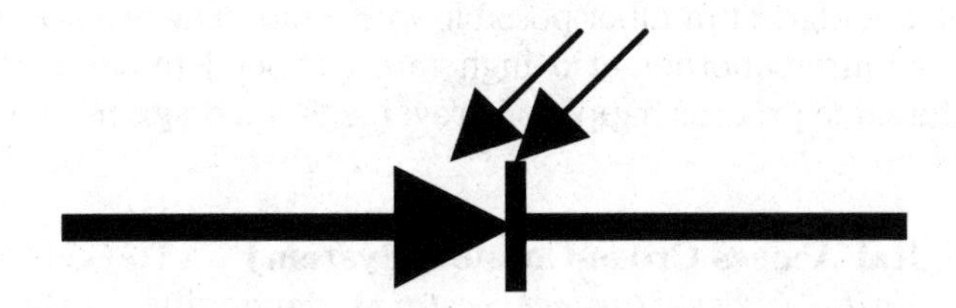

Schematic symbol for a photo diode

Dark Current

Dark Fiber A fiber-optic cable that is installed in a telephone company's outside plant network, but has no electronics connected to it. Sometimes customers like to lease or buy rights to use dark fiber, and connect their own electronics to it in a point-to-point or ring application.

DAT (Digital Audio Tape) A high-fidelity digital storage media for audio applications. DAT tapes are only usable on DAT tape recorder/players. DAT machines record music in the same manner as Compact Disks. For each channel, (left and right) they convert analog sound or music to a 16-bit sample for every 20 microseconds of sound (a rate of about 48,000 samples per second, which is 12 times the accuracy of a T1).

Data In the communications industry, Data is anything that is transmitted or processed digitally. The only thing that is not data anymore is a *POTS (Plain Old Telephone Service)* or analog line. A T1 is a digital circuit. The channels carry voices, computer transmissions, and sometimes video in a digitized data format.

Data Burst A short transmission of data that is not timing sensitive, such as a transmission for a credit-card authorization or an hourly data download. Audio or video is timing-sensitive data.

Data Bus A two-way connecting scheme of 8, 16, 32, or 64 wires or conductors that connect a microprocessor (CPU) to RAM, ROM, and I/O devices. A computing or controlling device needs to have at least three bus systems, a data bus, address bus, and control bus. The data bus provides for transfer of data between the microprocessor, the ROM and RAM, and the I/O devices. The address bus works in only one direction (from the microprocessor to the other devices) and allows the microprocessor to control memory addressing and retrieval. The control bus allows the microprocessor to control data flow and timing for all the various components.

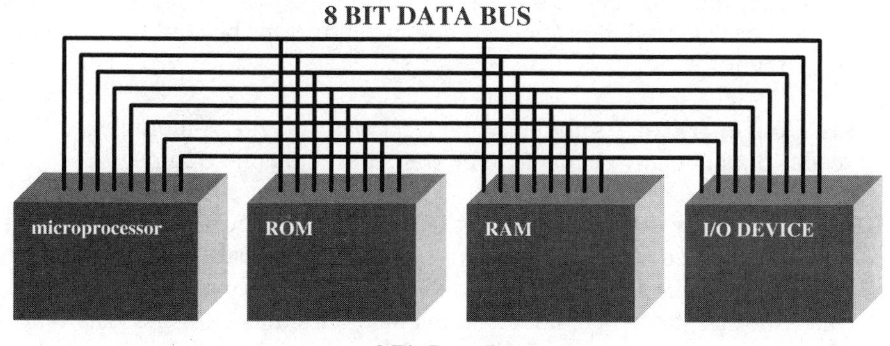

8 Bit Data Bus

Data Circuit Terminating Equipment (DTE, Data Terminating Equipment) DTE equipment receives a communications signal. For a data connection to work between I/O (input output) devices, one needs to be designated the "communications sending" equipment and one the "communications terminating" or DTE. A computer's printer port is a DCE port; a printer is a DTE device. A practical way to classify the two is: DCE is the sender of data and the DTE is the receiver of data.

Data Circuit Transparency A circuit's ability to carry data without any apparent change or restructuring of protocol. In reality, when data is transmitted through the public network, it gets loaded into other protocols, but on the other end of the transmission, the data is received exactly the way it was sent. A clear-channel T1 has data-circuit transparency.

Data Communications To transmit information encoded in binary. Data communications as a whole has many technologies to accomplish this. These techniques are called *protocols*. Some protocols include V.22 1,200 baud modem, ISDN, ethernet, token ring, SONET, switched 56Kb/s, and frame relay.

Data Communications Equipment (DCE) DCE is the equipment that provides the source communications signal. For a data connection to work between I/O (input output) devices, one needs to be designated the "communications sending" equipment and one the "communications terminating" or *DTE (Data Terminating Equipment)*. A computer's printer port is a DCE port; a printer is a DTE device. A practical way to classify the two is: DCE is the sender of data and the DTE is the receiver of data.

Data Compression A method of obtaining higher speeds of data transfer with the same number of bits being transmitted per second. Before the data is transmitted, it is encoded one or more steps beyond the original bit stream.

Data Concentrator Another name for a *multiplexer*. A multiplexer encodes data of many channels to be transmitted on one channel.

Data Connector A genderless connector developed by IBM. This connector is also called a *hemaphroditic connector*. The Data Connector does not need complementary plugs (male and female) to make a connection, like all other known communications modular connecting systems. The Data Connector is specifically designed for switched token-ring backbone networks.

Data Conversion The process of converting data from one protocol to another. Many devices are made to accomplish the transformation of data in one form to another.

Data Extender A device that extends a data signal such as an RS232 or PBX telephone line (or both simultaneously) to another location via a dial-up public telephone line. Data Extender's incorporate their own proprietary data compression protocols, therefore the data extenders on each side of the line must be of the same manufacturer. The advantage of Data Extenders is that they compress two sizeable digital signals (56 kbps/ and higher each) and modulate them to be transported over an inexpensive POTS (Plain Old Telephone Service) line.

Data Integrity A term that refers to how few errors are occurring in the transfer of transmission of data. The lower the error rate, the better the data integrity.

Data Link Control (DLC) The part of a communications protocol that resides in the overhead and provides a user (or the protocol itself) a means to control connect, disconnect, error-correcting, transmission-speed, and other operation-crucial functions.

Data Link Layer A layer in a communications protocol model. In general, the data link layer receives and transmits data over the physical layer media (twisted pair, fiber optic, etc.). The latest model (guideline) for communications protocols is the *OSI (Open Systems Interconnect)*. It is the best model so far because all of the layers or functions work independently of each other. For a diagram of the OSI, SNA, and DNA function layers, see *Open Systems Interconnection*.

Data Packet Switch A device that routes packets of data to another data packet switch until the packet reaches the data packet switch that has access to the address contained in the overhead of the packet. It is also called a *PSE (Packet Switching Exchange)*. Packet switching is a family of protocols for data communications. There are two basic innovations of packet switching. The first is to get away from having to send a huge amount of information at one time and the second is to enable multiple users to utilize the network connection at the same time. When using a modem to transfer a data file, the modem seizes the telephone line, dials a number, and starts sending information. This is very effective, but no one else can use the line, and there is a risk that the call could be disconnected, and all the data would have to be retransmitted (newer modem protocols can pick up where the disconnect occurred, but this is still inconvenient). The other need for improvement is the

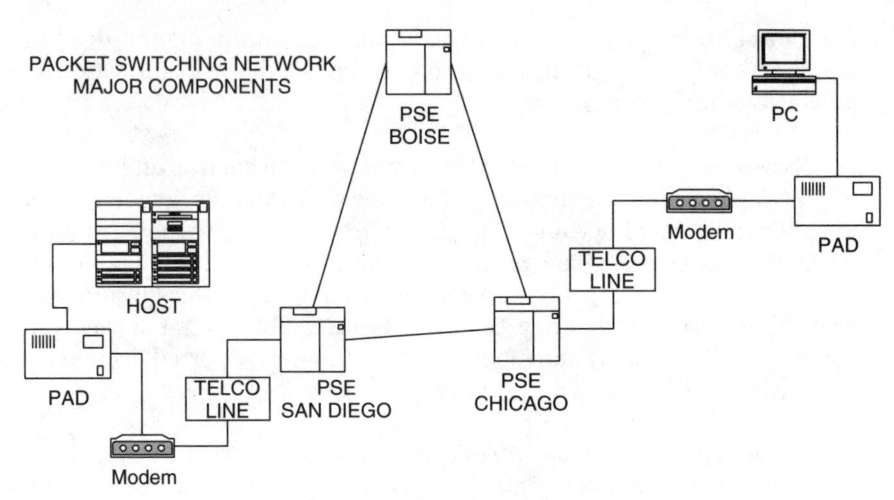

Data Packet Switch

ability for more than one user to use the line at the same time. Packet switching is an excellent remedy for these concerns.

The way that packet switching works is that large amounts of data to be transmitted are broken down into smaller pieces by a PAD (Packet Assembler/Dissembler), which can be a software program in an end-user's computer or a separate piece of hardware on the customer's premises. The PAD then gives each smaller piece of data an address and an identification number relative to the rest of the pieces. The pieces of data with their address, ID number, error-control information, and other HDLC (high-level digital link control) information are called *packets*. The important thing to know here is that each packet has its own address and identity relative to all the other packets of the broken-down file. The packets are then sent to a *PSE (Packet-Switching Exchange)* via a modem that is connected to a packet-switched line that you obtain from your local phone company or packet-switching service provider. The packet-switching exchange then makes copies of the packets, transmits the originals to another PSE that has access to the address contained in the overhead of the packet. A packet can travel through several PSEs before reaching the end PSE. When the packet is safely received by the end PSE, a receipt acknowledgment is sent to the originating PSE, at which time the copy is deleted (or flushed from memory). If any packet is received in other-than-perfect condition, the end PSE sends a request to the originating PSE to retransmit the packet.

Data Service Unit (DSU) Also called a *CSU/DSU (Channel Service Unit/Data Service Unit)*. A DSU is a hardware device that can come in many shapes and sizes. Rack-mount, shelf-mount, and stand-alone DSUs are available. A CSU/DSU has three main functions. The first function is to act as a demarcation point for a T1 (DS1) service from a local communications company. The second function is to provide line format and line-code conversion (B8ZS to AMI, SF or D4 to ESF, 135 V to 0 V) between the public network and the customer-premises equipment, if necessary. The third function is to provide maintenance or alarm services and loop-back for isolating problems with the T1 line or customer's equipment. For a photo, see *CSU/DSU*.

Data Set Ready (DSR) Pin 6 or pin 20 of a DB25 connector wired for the RS232C protocol. This wire is used by a modem or SDI device to send a signal that acknowledges that is ready to receive data.

Data Span Another name for a service purchased from a communications company. It can refer to any digital service, including T1, 56K, ISDN, or any other data-carrying service.

Data Stream A flow of serial bits modulated or sent direct over a transmission media (twisted pair, fiber optic, radio, coax, etc.)

Data Switching Exchange (DSE) Also called *PSE (Packet Switching Exchange)*. A part of a packet switching network that receives packets of data from a PAD (Packet Assembler/Dissembler) via a modem. The PSE makes and holds copies of each packet, then transmits the packets one at a time to the PSE that they are addressed to. The local PSE then discards the copies as the far-end PSE acknowledges the safe receipt of the original.

DATU (Direct-Access Test Unit) Also called *MLT (Mechanized Loop Test)*. Equipment that is either added on or built in to a central office switch. DATU allows a technician to dial the phone number of the DATU or MLT equipment and execute a test for shorts, opens, and grounds remotely. In response to a digital voice, the technician enters a password and a choice of options. The results of the test can be read back to the technician by a digital recording or sent to them via an alpha-numeric pager. DATU units can also send a locating tone on the technicians choice of TIP, RING, or both TIP and RING. The test unit can also short lines and remove the battery voltage for testing purposes.

DB9 A connector used for data-connectivity applications. It has nine pins, and it can be configured for several protocols, including the popular RS-232.

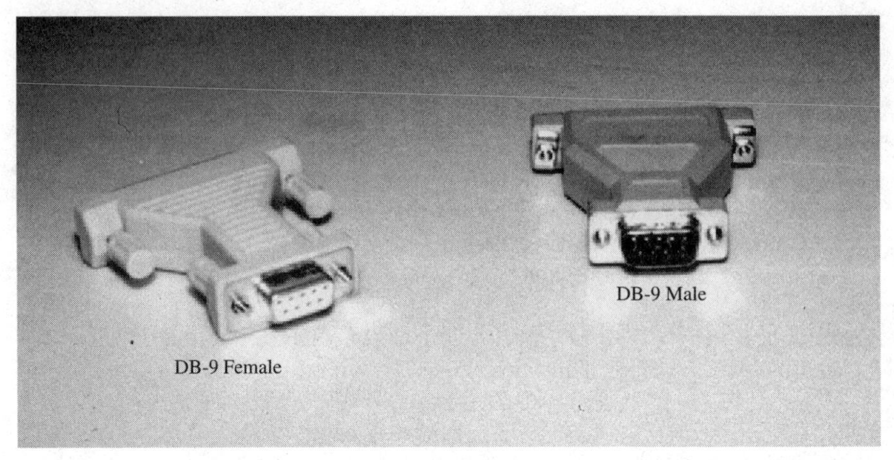

DB9

DB15 A connector used for data-connectivity applications. It has 15 pins and can be configured for several protocols, including the popular RS-232.

DB15

DB25 A connector used for data-connectivity applications. It has 25 pins and can be configured for several protocols, including the popular RS-232.

Db (Decibel) A decibel is 1/10th of a Bel. This is a measurement of increase or decrease of a signal that comes from the ratio of transmitted power to received power. To have a general idea of what a decibel is, remember that negative decibels represent a loss of power. Positive deci-

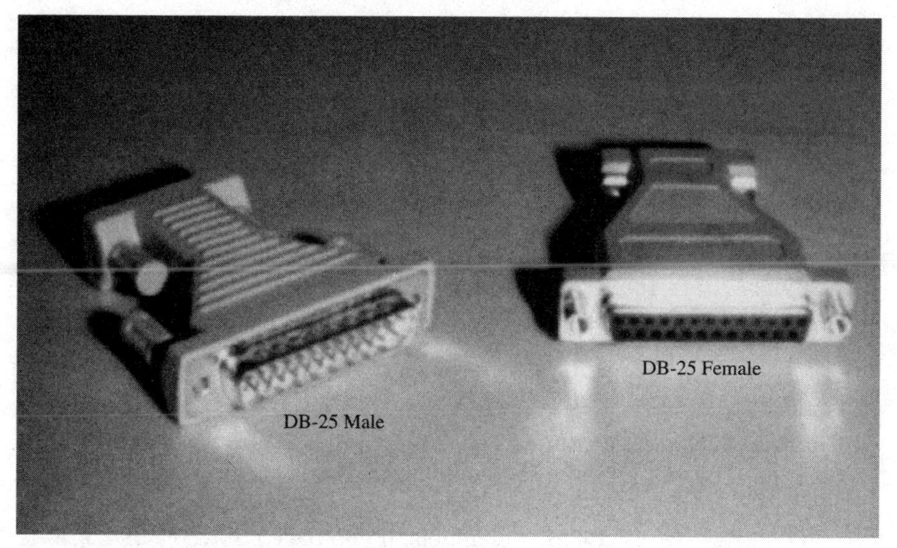

DB25

bels represent an increase in power. If you compare decibels with the way your ear reacts to sound, every –3 decibels would cut the original loudness of the sound in half.

DBA (Dynamic Bandwidth Allocation) The ability of a communications protocol to provide more or less data-transmission speed to different users when the users need it, automatically. Some T1 services are capable of DBA. When only two users are connected to the T1, they would share the entire bandwidth of the T1. If an third user accesses the T1 data link, the full speed or bandwidth of the T1 would be automatically divided among the three users. This automatic bandwidth allocation is efficient because the T1 is being fully utilized, regardless of the number of users connected to it. A T1 without DBA would have to have 23 users (one for each channel, less the 24th signaling channel) to be fully utilized.

D Bank Another term for a *Channel Bank*. A device that demultiplexes (breaks down) a T1 circuit to its 24 channels.

DBm Decibels below 1 milliwatt. This is a measure of power loss with 1 milliwatt as the transmission reference. As a common example if you receive a signal at 1 milliwatt, then you have a loss of 0 dBM. If you receive a signal that is 0.001 milliwatts, then you have a loss of 30 dBM. Some methods of testing analog phone lines include dialing a number that answers and provides a 1-milliwatt reference signal at 1000 Hz. The meter on the line measures the 1000-Hz signal on its end and displays a reading. Most POT's telephone lines are between –20 and –32 DBm.

Dbrn Decibels above reference noise. This is the same method of comparing transmitted and received signals by the log 10 of a ratio, only a reference is premeasured and used as the input (the denominator in the ratio).

DBU Decibels below 1 microwatt. Just dBM at a smaller increment. This is a measure of power loss with 1 microwatt as the transmission reference. As a common example is: If you receive a signal at 1 microwatt, then you have a loss of 0 dBM. If you receive a signal that is 0.001 microwatts, then you have a loss of 30 dBM.

D Channel (Data Channel) The name of an ISDN out-of-band signaling channel. The two kinds of D channels depend on which ISDN circuit you have. If you have a *BRI (Basic Rate Interface)*, the D channel is 16Kb/s and controls two B (Bearer) channels. If you have a *PRI (Primary Rate Interface)*, then the D channel is 64Kb/s and controls 23 B channels. Both D channels carry the same information and perform the same function. A BRI circuit requires one pair (two wires for transmission) and a PRI requires two pairs for transmission. For a diagram that compares the different types of ISDN circuits, see *Integrated Services Digital Network*.

D Connector A 25-pin mini version of the DB25 Connector.

DC The ASCII control code abbreviation for direct control. The binary code is 0001001 and the hex is 11.

DC (Direct Current) DC is current that is induced by a voltage source that does not change direction from positive to negative. DC can fluctuate, and carry an analog signal by varying the DC current and voltage. DC can pulse, it can spike, and it can do many things. The one thing that DC cannot do is change direction. Common sources of DC are batteries, AC power adapters, and power rectifiers.

DCC (Data Communications Channel) An overhead channel in an AT&T SONET ring. It allows the individual nodes to communicate control information to each other.

DCE (Data Communications Equipment) See *Data Communications Equipment.*

DCS (Digital Cross-Connect System A DCS is also called a *DACS (Digital Access Cross Connect System)*, depending on the manufacturer. A digital cross-connect system is quickly becoming a fundamental part of a local and long-distance company's network because of the rapid deployment of broadband transmission equipment (SONET, DS3). Cross connecting broadband services can be cumbersome because larger circuits (OC-1) require coax. The DACS or DCS is a rack-mountable system that enables any circuit that interfaces with it to be electronically cross connected from one path to another within the network it is connected to. Circuits that can interface with a digital cross-connect system include DS0, DS1, DS3 (or T3), STS-1, and SONET OC-1. An incoming circuit can be rerouted by simply making path changes in DACS administrative software. For a diagram and photo of a DCS/DACS system, see *Digital Cross Connect System*.

DCV (Digital Compressed Video) There are several types of DCV. The object of compressed video in general is to transmit an initial picture, then transmit only the parts of the picture that move. A good example is a video phone application, where only a person's mouth and facial features move. Everything else in the video phone picture stays the same.

DE (Discard Eligible) In reference to a committed information rate that a customer has paid for in conjunction with a frame relay circuit, any data sent at a rate that exceeds the committed information rate is discard eligible, which means that it will not be transmitted.

De Facto Standard A standard that has come about because of consumer popularity, not because of formal approval of a standards committee. TCP/IP is a standard protocol for the Internet, but no standards committee has ever formally made it a standard.

Dead Spot A dead spot is an area within a transmitter's range where the radio signal being transmitted cannot be received. Dead spots occur for many different reasons. Sometimes the signal is blocked or reflected, sometimes it is because you are located in a small valley that dips below the radio transmission.

Decibel (dB) A decibel is $\frac{1}{10}$ of a Bel. This is a measurement of increase or decrease of a signal that comes from the ratio of transmitted power to received power. To have a general idea of what a decibel is, remember that negative decibels represent a loss of power. Positive decibels represent an increase in power. If you compare decibels with the way your ear reacts to sound, every –3 decibels would cut the original loudness of the sound in half.

Decimal-to-Binary Conversion For a conversion table of binary to decimal and hexadecimal, see Appendix E.

Decimal-to-Hexadecimal Conversion For a table on decimal-to-hexadecimal conversion, see Appendix E.

Decoder A device that converts a signal or transmission from one protocol to another.

Dedicated Access Reference to a telephone line that is usually provided by an IXC (Inter Exchange Carrier, long-distance company) for exclusive dialing of long distance on their network. Sometimes the customer has the line installed themselves by a *LEC (Local-Exchange Carrier)* and gets billed separately for the dedicated-access line and the long-distance service. Some dedicated-access lines are capable of dialing local calls, but their long-distance service is dedicated to a specific IXC.

Dedicated Channel A channel within a T1 or T3 that is dedicated to a specific customer. Other than that, it is a private line/dedicated circuit.

Dedicated Circuit Also called a *private line*. A private line is a pair of wire or (two pairs of wire for a T1) that runs from your location to a location that you want to be connected to with a dedicated high-speed data connection. Once a private line is installed, it is there all day, every day. There is no dialing on a private line because it does not go through switching circuitry, although it does get re-generated (the data signal on the channel is received and retransmitted). Dedicated lines could be on copper, which they have been very much in the past, but with the explosion of SONET, it is possible to put hundreds of private lines and switched lines on a pair of optical fibers.

Dedicated Line A telephone line from the phone company that is dedicated to one user or device. Most fax machines and modems are on dedicated lines. A dedicated line is not a trunk because a trunk is a line that everyone on the PBX shares. A dedicated line can also be a dedicated circuit (a private data line), but it is less common.

Definity A **PBX (Private Branch Exchange System)** manufactured by Lucent Technologies. For a picture, see *Private Branch Exchange*.

Degaussing Coil Degaussing is to demagnetize. A degaussing coil is simply a long coil of wire that is bent into the shape of a circle. If a CRT (picture tube or monitor tube) becomes magnetized, you will notice an area of discoloration. By waving a degaussing coil around this area, you will demagnetize the screen of the CRT.

Degradation Another term for attenuation. As a signal traverses down a wire, fiber-optic cable, or through the air, it loses power and becomes distorted. This phenomenon is referred to as *attenuation*, *loss*, or *degradation*.

DEL The ASCII control code abbreviation for delete idle. The binary code is 1111111 and the hex is F7.

Delay The time difference between when a signal is sent and when it is received. Equipment is integrated into transmission electronics to make up for this.

Delayed Ring Transfer A PBX and key-system feature that is just like call forwarding, except that before the call is forwarded, it rings a pre-selected number of times. If you are not in your office and someone calls you, you can have the call forwarded to another associate after a certain number of rings. This is delayed ring transfer.

Delta Channel Another name for an *ISDN (Integrated Services Digital Network)* "D" channel, which is the data or control channel of an ISDN line.

Delta Modulation A form of encoding analog signals to digital binary. Instead of sampling a signal and creating an 8-bit binary number, like the standard ADC, delta modulation samples the change (*delta* is the term for change in the physical sciences) of the signal. Delta modulation only looks for two changes in an analog signal, change higher and change lower. This higher or lower signal is sent to a far-end device and the signal

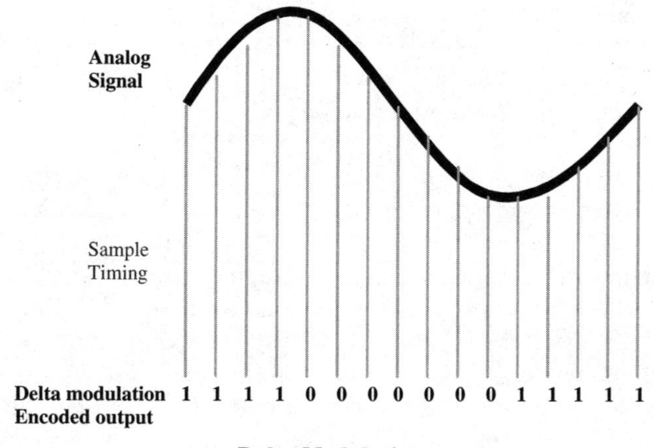

Delta Modulation

is re-created or decoded with these simple higher or lower instructions. Included is a diagram of an analog signal and its delta-code equivalent. Notice that as the signal goes up in voltage, the delta-modulation technique registers a 1 when it goes down in voltage it registers a zero.

Demarc (Dmarc, Demarcation Point) A Dmarc is where the local phone company hands off a telephone circuit. A Dmarc can be in the form of a standard network interface for a residential line, a DSX panel for a T1 or T3, an RJ212X for a business line, or an RJ45 for an ISDN line. The Dmarc separates customer-owned equipment from telephone company-owned equipment. It is also the place where responsibility for the circuit's performance is separated.

Demarcation Point See *Demarc*.

Demodulation The inverse of modulation. When a radio signal is received, the information that was sent over it (audio, video, or data) is still mixed in with the carrier. The process of demodulation removes the carrier signal from the information you want. Some different types of demodulators are AM, FM, and PM.

Demultiplex The inverse of *multiplex*. Multiplexing is the process of encoding two or more digital signals or channels onto one. The reason that channels are multiplexed together in communications is because it saves money. When we use all of the wires in a cable and need more, it costs less to add electronics on the ends of a cable than to install a new one (imagine the expense from LA to NY). A T1 encodes 24 channels into one by using frequency-division multiplexing. In a simpler explanation, a T1 makes it possible to place 24 lines that once needed 24 pairs on only 2 pairs. When a group of signals are multiplexed together, they are all sampled at a high rate of speed, faster than the combined speed of all the channels being multiplexed. For a diagram of the multiplexing process, see *Time Division Multiplexing*.

De Mux Abbreviation for *de-multiplex*. To separate multiplexed channels from one transmission into their original individual channel.

Deregulation The transition of government authority and control away from specific business activities of telephone or cable TV companies. The complete deregulation of the RBOCs will be a gradual process and, depending on the way in which the telecommunications industry evolves, complete deregulation might never happen. The purpose of deregulation is to promote new technology, lower prices, improve service, and create a more abundant supply of telecommunications services. Whether or not this purpose is being fulfilled is still up for question.

Des (Designation Strip) A designation strip, or desi strip is the piece of paper or label that goes under a button on a phone. The type written on the desi strip identifies the feature with the programming for that button.

Desi See *Des*.

Designation Strip See *Des*.

Desk-Top Engineer A person that maintains personal computers and LAN network connectivity. They also maintain front-end services on a PBX switch, which is the connecting, programming, and moving of telephone extensions throughout a network. Desk-top engineers do not usually do switching equipment upgrades or additions and do not do LAN administration. Desk-top engineers are also called *desk-top technicians*.

Desk-Top Technician See *Desk-Top Engineer*.

Detector The circuit inside a radio receiver that detects fluctuations in the modulated carrier (radio signal). The detector simply filters the carrier (radio portion) out of the wanted end signal (audio and/or video). The simplest form of detector is the germanium diode/filter capacitor used in AM receivers.

Dial-By-Name Directory A feature of voice-mail systems that enables a caller to be transferred from the automated attendant to a person (or department) by knowing that person's name, then dialing (spelling) the corresponding letters on the dial pad. These systems are great if you only call people you know. Automated attendants are still regarded as a poor first impression to customers—especially for companies that deal in service or retail.

Dial String A set of instructions that are sent to a device (such as a modem) that is capable of dialing a number on an analog phone line. The instructions in dial strings are the same that you take for granted. They include, go off-hook, wait for dial tone, dial digits, wait for answer, etc.

Dial Tone When you pick up the handset of a telephone that is connected to the phone company or a PBX system, you hear a buzzing hum sound. That sound is dial tone, a signal from the PBX or telephone company central office switch to go ahead and dial your number.

Dial-Tone Delay The time from when you go off-hook and when you receive a dial tone from the host switch.

Dial-Up Line A line that can be dialed into. Some dial-up lines include the *POTS (Plain-Old Telephone Service)* to your house, ISDN, and switched 56 data circuits.

Dial-Up Modem A modem that is intended to be used on the public-switched telephone network. It is connected to a phone line and that phone line has a phone number that people can dial with their modems. These modems are the most common in personal computers. The other type of modem is a short-haul modem, which doesn't dial numbers—it just extends a digital signal (e.g., to the other side of a building for a printer).

DID (Direct Inward Dial) A phone line that comes from the local phone company and connects to your PBX switch. A DID line has a phone number (and DNIS or virtual directory number attached to it) that is targeted to ring directly to a phone on the PBX network without going to a console operator, or anywhere else first. The PBX system usually needs specific DID trunk (incoming line) hardware to make DID lines work.

Dielectric A material that does not conduct electricity. Dielectric materials are used as insulating materials, such as the vinyl coating on copper wires. Good dielectric materials (more frequently called *insulators*) are, glass, ceramic, rubber, and plastic.

Digital A signal that has only two possible levels per cycle, in contrast to analog, which can have an infinite number of possible levels per cycle. The great thing about a digital signal is that is can be regenerated easily. Even though it might pick up noise and RFI as it is transmitted along a wire, when it is regenerated, all the noise is cut out because the regenerating device looks for only two levels of signal to reproduce, 1 and 0. Therefore, all the other stuff, such as white noise and maybe even an unwanted radio station, are not regenerated.

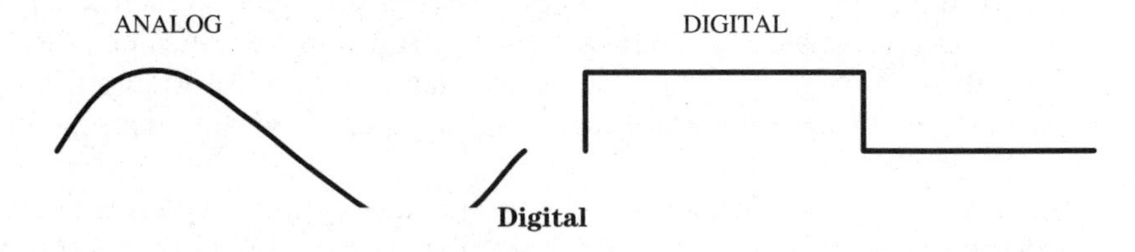

Digital

Digital Announcer A device that stores RANs (Recorded Announcements), and plays them to a specific line/trunk when instructed to do so by an ACD system.

Digital Announcer

Digital Audio Digital audio is really analog audio that is stored in a digital code. It is good to store audio and other information digitally because when the signal is read, decoded, and converted to analog, unwanted noise and interference is filtered out. Other methods of audio storage are subject to unwanted electrical noise that is the electronic nature of the storage media itself. The nature of one of these older alternative technologies, the audio cassette, stores the analog signal directly onto a magnetic tape. The magnetic tape itself has inconsistencies in the metals and other materials used in their manufacture. As the magnetic tape glides against a playback head in a cassette player, it creates its own audio signal which resembles a light hiss sound. In the case of digital audio, the means of storage and encoding being a *Compact Disc (CD)* or a *Digital Audio Tape (DAT)* the imperfections in the storage media are ignored by the electronics that reads the digital code from the DAT or CD. This is because they only read ones and zeros, paying no attention to hisses or popping sounds. Even if the signal is muffled, it is a muffled signal of ones and zeros, which gets decoded just the same.

Digital Compressed Video (DCV) There are several types of DCV. The object of compressed video in general is to transmit an initial picture, then transmit only the parts of the picture that move. A good example is a video phone application, where only a persons mouth and facial features move. Everything else in the video phone picture stay the same.

Digital Cross-Connect System (DCS) A DCS is also called a *DACS (Digital-Access Cross-Connect System)*, depending on the manufacturer. A digital-cross connect system is quickly becoming a fundamental part of a local and long-distance company's network because of the rapid deployment of broadband transmission equipment (SONET, DS3). Cross connecting broadband services can be cumbersome because larger circuits (OC-1) require coax. The DACS or DCS is a rack-mountable system that enables any circuit that interfaces with it to be electronically cross connected from one path to another within the network it is connected to. Circuits that can interface with a digital cross-connect system include DS0, DS1, DS3 (or T3), STS-1, and SONET OC-1. An incoming circuit can be rerouted by simply making path changes in DACS administrative software. In the DACS shown below, a T1 circuit coming into the input side (left side) could be cross-connected to exit as one of the channels in a DS3 on the right side.

Digital Frequency Modulation Another term for *Frequency-Shift Keying*.

Digital Line Protection *Digital line protection* refers to the protection of modems from the higher line voltages of digital lines. Lines connected to a digital line interface card on a PBX system are wired and look the same as normal phone lines from the phone company, but they operate at a higher voltage. Digital service lines from the phone company (T1) lines are also a higher voltage, about –135 Vdc. Modems are not for digital lines, they are for 52-V analog lines. If a modem is connected to a digital line by mistake, it could be destroyed. Digital line protection is a feature designed into modems that protects them from a mistaken connection to a digital line.

Digital Loop Back A feature of transmission equipment that allows a user to reroute a signal back to the source instead of into the termination or end equipment. By doing this, the user can see if the signal going into the equipment is good or bad. If the signal loops back and is good, but the signal is bad coming out of the equipment on the far end when the loop back is removed, then the trouble is most likely in the end equipment. In many modems and digital service units, the loop back can be controlled remotely.

Digital Microwave Digital microwave has become a very economical way to bypass construction costs of broadband private line services. Many *CAPs (Competitive-Access Providers)* have access to microwave

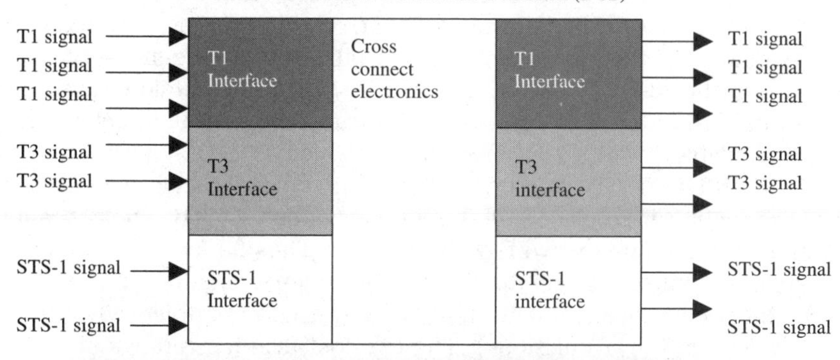

Digital Cross Connect System (DCS)

radio resources, such as licensing, equipment, and installation. Digital microwave is also called an *eyeball shot*, *38 gig*, or is just referred to as *radio*. Most of the microwave being installed for private-line service today is in the 33-GHz to 39-GHz frequency range. These microwave units use an FM-FSK over two sidebands for transmitting at full duplex. They are available in T1, DS3, and STS-1 (which is a DS3 formatted for SONET). The 38-GHz microwave has a range that depends on the size of the antenna (dish) placed on the outdoor radio unit. The choices in antenna size are one or two feet in diameter. The one-foot antenna has a maximum range of one to three miles, depending on the regional weather conditions (rainfall, snow, and especially fog drastically attenuate microwave transmissions). The two foot dish has a range of two to seven miles, also dependent on the weather in the region. For a diagram of a microwave application, see *Terrestrial Microwave*.

Digital Service Cross-Connect (DSX) A reference to a digital service termination/patch panel that allows DS1 and DS3 circuits to be monitored by test equipment, such as a TTC Tberd or T-ACE. DSX panels are usually terminated via wire-wrap. For a photo of wire-wrap terminals, see *Wire-Wrap*.

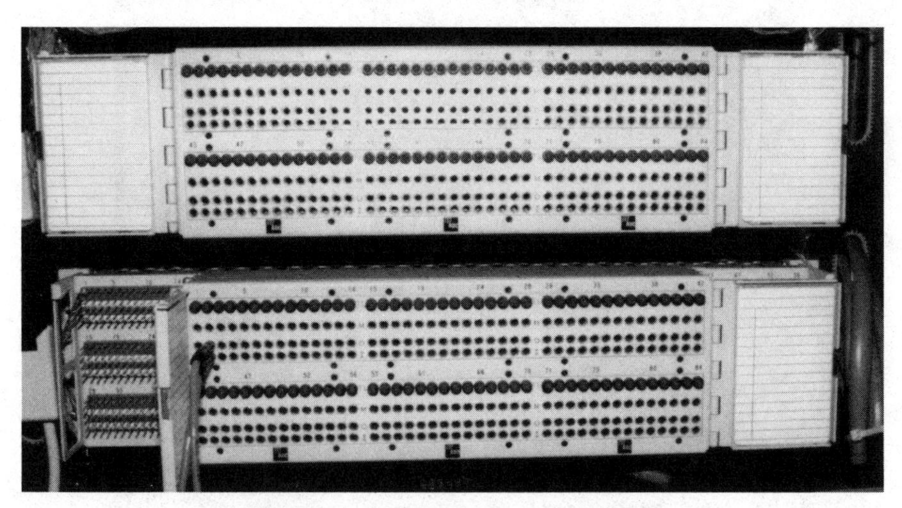

DSX Cross Connect Panels Mfg. by ADC

Digital Subscriber Line Another name for an *ISDN BRI (Basic Rate Interface)*.

Digital-to-Analog Converter (DAC) A part of a channel bank that performs the function of encoding analog voice signals into a stream of binary digits. The analog-to-digital converter samples a caller's voice at a rate of 8000 times per second. Each sample's voltage level is measured and converted to one of 256 possible sample levels. The DAC converts all of the digital numbers back into an audio signal. See also *Analog-to-Digital Converter*.

Digital Versatile Disk (DVD) A newer version of the common 650-MB CD (Compact Disc). The DVD is capable of storing 17 GB. DVD players can play newer DVD discs containing audio and video information, as well as your old audio-format CDs.

DIN Connector A screw-on type connector that is installed on coaxial cable in RF/microwave applications. DIN connectors have better intermodulation suppression and power-handling capabilities than N-type and other coax connectors.

RG 8 Coax with DIN Connector

Diode An electronic semiconductor device that simply put, only conducts electricity in one direction. Whether or not the device conducts is dependent on which direction the device is "biased." Diodes (or rectifiers) are used to change alternating current (AC) to direct current (DC). If a more positive voltage is applied to the anode lead of the diode, then the diode simply acts like a wire. If the more positive voltage is applied to the cathode lead, then it acts as if there is no connection. The illustration is of the schematic symbols of the first diode, which was a vacuum tube, and a modern solid state silicon diode. Below those figures is an illustration of a pair of diodes converting AC to DC.

DIP (Dual Inline Pin) A way a component is physically made. A DIP component has two rows of pins (pins are a means to solder the component

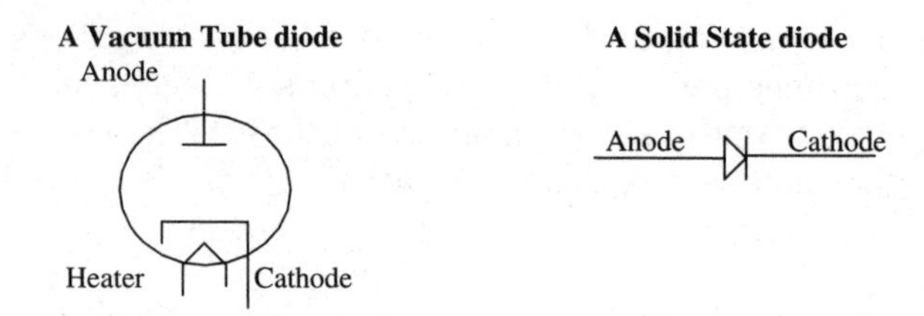

A simple one diode rectifier circuit, with a filter capacitor to eliminate DC fluctuations.

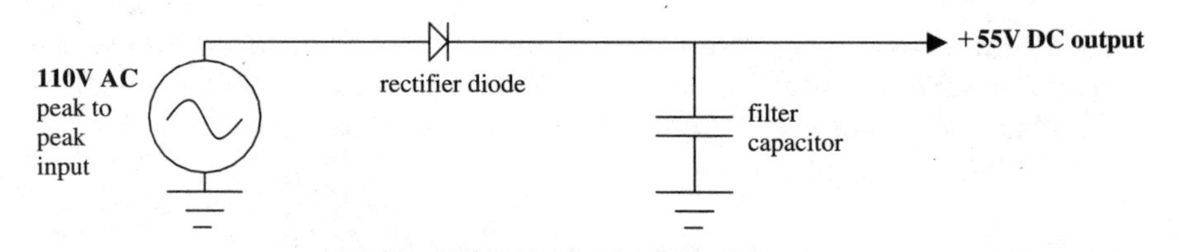

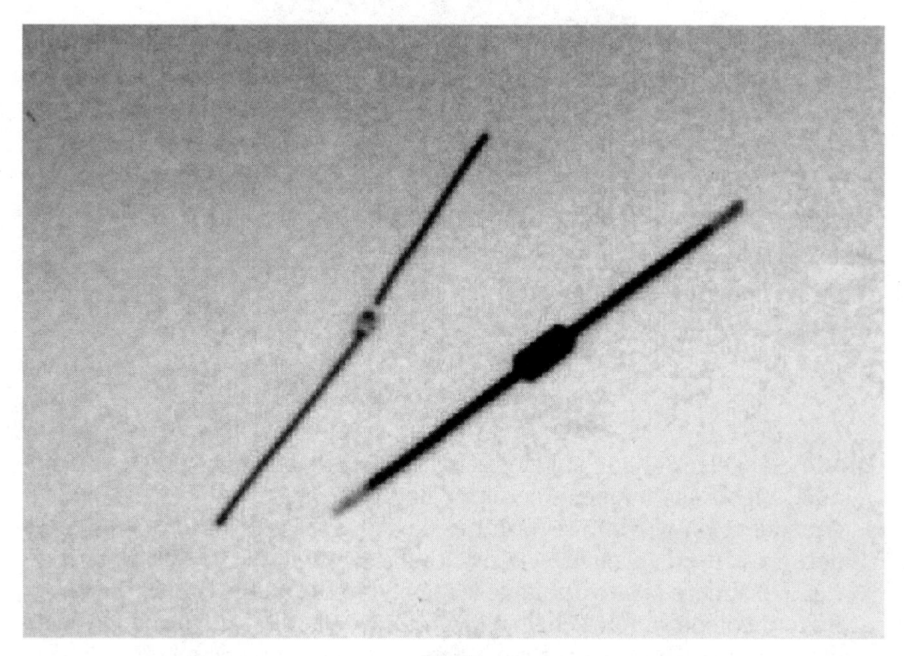

Diode

into a circuit). Many components are "packaged" this way. Some of them are DIP 8-segment display, DIP integrated circuit, and of course, DIP switches.

DIP Switch A very small manual switch that comes in a DIP package. DIP is an abbreviation for *Dual-Inline Pin*, it is the way a component is physically made. See *DIP* for more information.

Dipole A type of balanced antenna with two elements that are fed by transmission line. Dipoles are mildly directional. A common dipole antenna is the "rabbit-ear" style, is used with TVs.

Direct Current (DC) DC is electrical current that is induced by a voltage source that does not change direction from positive to negative. DC can fluctuate, carry an analog signal by varying the DC current and voltage. DC can pulse, it can spike, it can do many things. The one thing that DC cannot do is change direction. Common sources of DC are batteries, AC power adapters, and power rectifiers.

Direct-Inward Dial (DID) A phone line that comes from the local phone company and connects to your PBX switch. A DID line has a phone number that has a DNIS or virtual directory number attached to it and is targeted to ring directly to a phone on the PBX network without going to a console operator or anywhere else first. The PBX system usually needs specific DID trunk (incoming line) hardware to make DID lines work.

Direct Inward System Access (DISA) A feature of PBX phone systems where a user can dial a number that terminates into the PBX and either get another dial tone (with which to make long-distance calls) or to access their voice-mail system.

Direct Outward Dial (DOD) A feature of a PBX system that allows telephone stations to access outside dial tone or not access outside dial tone. If you pick up a phone on a PBX system and dial "9" for an outside dial tone, you might hear a siren sound instead. If you do, that means that the particular phone you are dialing on does not have DOD enabled. In most PBX systems, even though DOD is not enabled, an emergency "911" call will still go through. It is common to place special instructions on the phone to explain how to make an emergency call.

Direct Station Selection (DSS) A device that can be added to a PBX telephone set that has additional buttons on its face so that a user can see what extensions are in use (off hook) and which are free. When a call comes in to an answering agent, he/she can look at the direct station-selection module attached to their phone and see whether the desired person is on their phone or not. Calls can be made and transferred to the extensions appearing on the DSS by pressing their associated button.

Directional Antenna An antenna that is sensitive to the direction of the received or transmitted signal. Rabbit ears (dipole antennas) and "dish" type antennas are directional. Dish-type antennas are not only directional, but they focus a received transmission to the element (some call it the *stinger* because it is raised to the front of the dish). They also focus transmitted signals that are bounced into the dish and out to their destination.

Directional Coupler A device that is engineered into a microwave antenna system that allows transmit and receive signals to be used on the same antenna. The device accomplishes this by differentiating the powerful transmit signal from the weak receive signal.

Dirty Power Power that comes directly from the power company. The power in our homes is "dirty" power because it is not a pure 120V AC. As it travels cross country on power lines, it collects all kinds of EMI of all frequencies. Dirty power is also subject to unpredictable outages. Some devices that are used to clean up dirty power are UPS systems and surge suppressors.

DISA (Direct Inward System Access) A feature of PBX phone systems where a user can dial a number that terminates into the PBX and either get another dial tone (to make long-distance calls on) or access their voice-mail system.

Discard Eligible In reference to a committed information rate that a customer has paid for in conjunction with a frame-relay circuit, any data sent at a rate that exceeds the committed information rate is discard eligible, which means it will not be transmitted.

Disco (Disconnect) Many telephone and cable companies call their orders to disconnect a service "disco orders."

Disconnect Supervision The ability of a PBX switch to recognize the disconnecting of the far end of a call. Keep in mind that in early switch days, people accessed trunks by picking up a phone and released them when the call was over. Machines do not know when the call is over because they are not the ones having the conversation. When you call someone, have a normal conversation, then hang-up, you assume that the PBX system disconnected the path from your phone to the trunk and disconnected the link between that same trunk and the central office. Without disconnect supervision, the PBX does not know when to release (hang up) a central-office trunk. Without disconnect supervision, your trunks will soon all be busy, but no one will be on the phone!

Dish In telecommunications, a *dish* refers to a parabolic dish antenna. It has this name because its shape is a parabolic curve, so all radians from a single point are reflected into one direction. For a diagram, see *Parabolic Dish Antenna*.

Disk Drive A hardware device that is manipulated by a software program called a *disk operating system*. Some disk drives have a disk built in to them and some have interchangeable disks. The function of a disk drive is to store, read and write memory from the disks that are made for them. Different disk drives are capable of storing different amounts of data (measured in bytes). A 3.5" floppy disk is capable of storing up to 1.44 MB (1.44 million bytes). A CD-ROM disk is capable of storing up to 650 MB. Another kind of disk is a hard disk, which is built-in to a hard-disk drive. These disks are not capable of being interchanged, but they can hold many times more memory. A common hard-disk drive in a personal computer can store more than 2 GB (2,000,000,000 bytes) of information.

Disk Operating System (DOS) A software program that manipulates a disk drive. The DOS contains the information that tells the disk drive how to format, and read and write to disks. MS-DOS (Microsoft disk operating system) is probably the most common known disk operating system to PC users. Without a DOS program loaded onto your computer, it is virtually useless.

Display One of the several communications output interfaces to a user. Displays enable a computer or controlling device to communicate with a user visually. Displays are available in the form of monitors, LCD screens, and light emitting diodes. Other output devices that computers and controlling devices use to communicate with their users are printers, speakers, and lights (such as alarm indicators).

Distance-Sensitive Pricing The pricing of communications services based on the distance between the two points connected by the phone company or service provider. Long-distance companies use a method of figuring the distance between two cities "airline mileage" (See *Airline Mileage* for more information). The price of the service is then based on the mileage that is calculated. The more mileage, the higher the price. Local exchange companies that provide a service across town figure the price by the number of central offices the line passes through. Each office requires a channel termination (also called *chanterm*) of the line as it enters and leaves each central office. Physically, a channel termination is a cross-connect from a channel of one transmission device to another.

Distinctive Ringing A feature of a PBX system that allows telephone sets to ring differently. This is a very nice feature if you would like outside calls to ring differently than internal calls from co-workers. This helps a user to know if they should say "hello" or "Emergency service, may I help you?" Distinctive ringing is also used in offices where many phones are in close proximity to each other. When one phone rings, all the people in the office know who's phone it is by tone, pattern, and pitch of the ring.

Distortion Any change to a signal's original waveform, except size (amplitude). Changing the size of a waveform is amplification (for larger) or attenuation (for smaller). The most common form of signal distortion is "clipping," which you can hear when the volume on a cheap stereo is turned up to high.

Distributed Queue Dual Bus (DQDB) IEEE 802.6 Standard. A broadband protocol that is full duplex and implemented on fiber optic. The Distributed Queue Dual Bus is an architecture that is made of two serial busses, called *a* and *b*, which carry data transmissions in opposite directions simultaneously (hence full duplex). The busses can be implemented in a straight line or in a ring. If a fiber is cut for some reason, the node equipment reconfigures itself to accommodate for the disconnection. When this happens, the DQDB is divided into two networks and the individual nodes adjacent to the cut automatically restructure themselves as head ends.

Distribution Cable Cable that connects a PBX switch or telephone company central office to it's customers. It is the cable system of outside plant. Distribution cable usually has two parts, an F1 (facility 1) and an F2 (facility 2). The F1 goes from the central office to an access point (AP) or cross connect point where it is then cross-connected to F2 pairs. The telephone cable that runs along neighborhood streets is usually F2 cable and the cables that run along main roads are usually F1 cables. For a picture of a cross box, see *Access Point*.

Distribution Frame This is also called a *Main Distribution Frame (MDF)*. It is the place where all the wire, fiber optic, or coax for a network is terminated. The distribution frame is usually placed as close to the central office switch or PBX as possible.

Dithering A variable error in a *GPS (Global Positioning System)* latitude and longitude signal. The error is purposely integrated into the positioning system to prevent anyone other than the government from having absolutely precise positioning information. A typical dithering error in a civilian-purchased GPS unit is about ±0 to 50 feet.

Divestiture The break up of AT&T by the federal government due a business practice was considered to be of monopolistic nature, effective

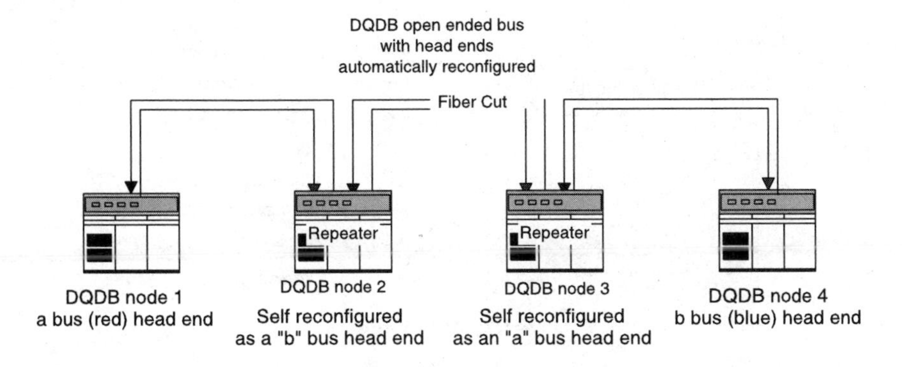

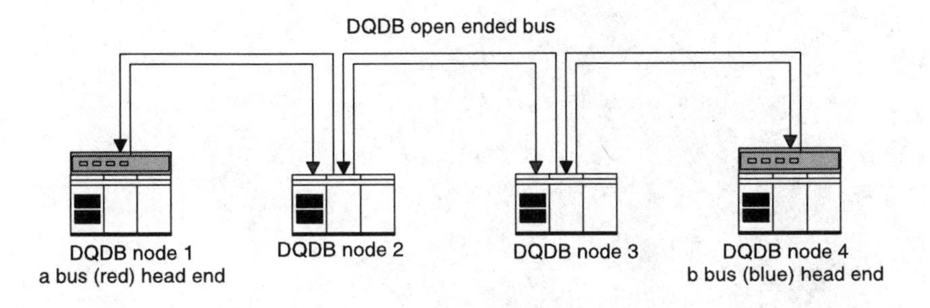

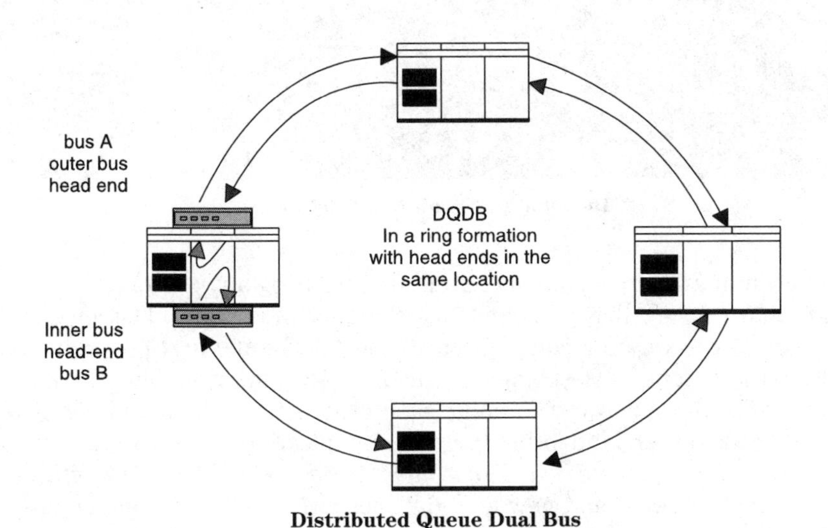

Distributed Queue Dual Bus

December 30, 1983. AT&T and its 22 Bell operating companies were separated. The 22 Bell companies were combined into seven regional Bell operating companies (RBOCS). AT&T was legally limited to the long-distance business (although they were allowed to be in the computer business), and the seven RBOCs were limited to the local telephone business. Both AT&T and the seven RBOCS were restricted from manufacturing

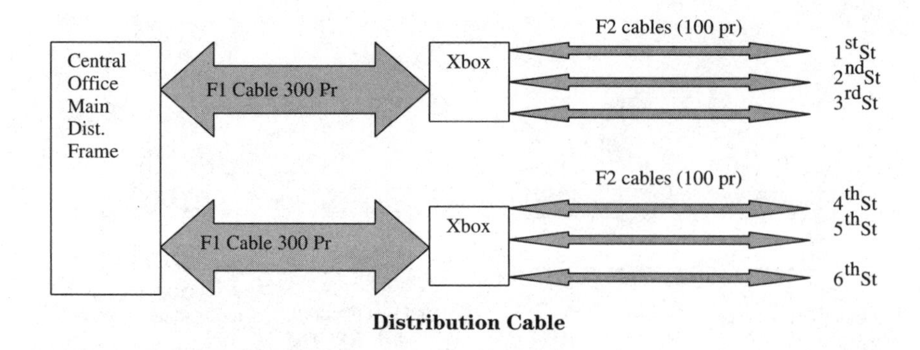

Distribution Cable

Distribution Frame 66M150 Blocks

telecommunications equipment and from sharing any customer or market information. This judgment was made by Judge Harold Greene, and it paved the way for competition in the telecommunications industry. The stipulations on the companies involved are now changing. AT&T and the RBOCS are allowed to enter each other's businesses and compete with each other, and other new communications companies (called *CLECs*, *Competitive Local-Exchange Carriers*, and *CAPs*, *Competitive-Access Providers*, and new long-distance companies) are building networks and offering communications services.

DLC (Digital Loop Carrier) Equipment that is used to provide two dial tones over a single twisted copper pair. If a customer wants to have an additional telephone line installed and no more twisted pairs are left to feed their area, the telephone company can install a DLC,

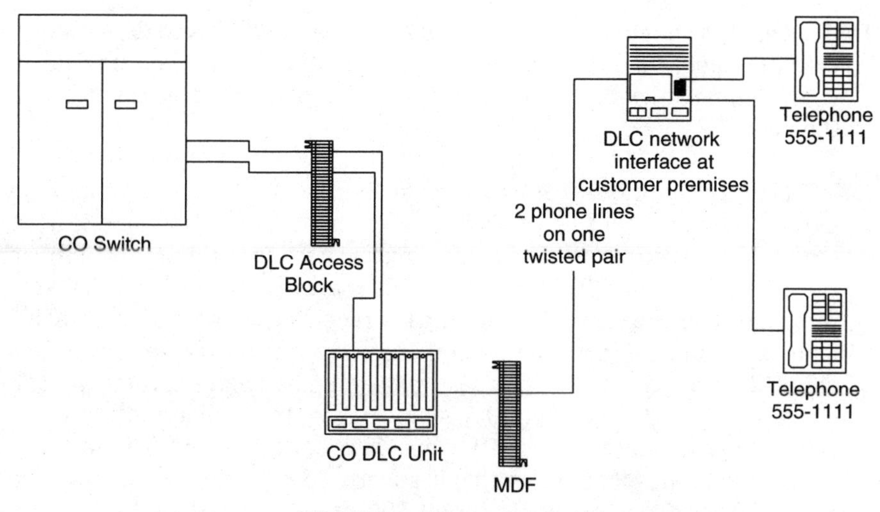

DLC (Digital Loop Carrier)

which will make two phone lines work on one pair. The DLC has two parts. The first part is a central office DLC unit, which is rack mountable. The two phone lines (the original one and the new one) are cross-connected into this unit. The second part is a modified network interface that contains electronics. It receives the two phone services transmitted from the central office DLC unit over one twisted pair and provides a hand-off to the customer's wire as two separate twisted pairs. The DLC accomplishes this by taking the two analog lines and digitally multiplexing them onto one pair. DLC works great for voice applications, but it can have abnormal effects on fax machines and modems.

DLCI (Data Link Connection Identifier) A group of bits in a frame relay connection's overhead that identifies the virtual circuit or channel the data is being sent to its address on.

DLE ASCII control-code abbreviation for data link escape. The binary code is 0000001, the Hex is 01.

Dmarc (Demarcation Point) A Dmarc is where the local phone company hands-off a telephone circuit. A Dmarc can be in the form of a standard network interface for a residential line, a DSX panel for a T1 or T3, an RJ212X for a business line, or an RJ45 for an ISDN line. The Dmarc separates customer-owned equipment from telephone company-owned equipment. It is also the place where responsibility for the circuit's performance is separated.

DND (Do Not Disturb) A feature of PBX telephone sets to disallow any calls or pages while the feature is activated. The feature is usually activated and deactivated by pushing the "do not disturb" (DND) button on the phone.

DNIS (Dialed Number Identification Service) A service from your phone company that is similar to Automatic Number Identification (caller ID), except, instead of providing the caller's number, the number (or a four-five digit DID type routing DN is provided) the caller dialed is provided. This is useful when a company receiving the call has several incoming numbers. If certain numbers dialed by customers through a specific DID trunk determine how the call should be handled, then an *ACD (Automatic Call Distributor)* system can use those digits to route the call to a certain extension. If 10 trunks have four different 800 numbers ring to them, the 800 number that is advertised in Spain can be routed to Spanish-speaking personnel and an 800 number advertised in France can be routed to French-speaking personnel.

Documentation Information regarding a network that is updated to allow others that are involved in managing a network do so in an efficient and timely manner. Managing a network without the proper documentation on the components and the way in which the components are connected can be very cumbersome, unreliable, and costly.

DOD (Direct Outward Dial) A feature of a PBX system that allows telephone stations to access outside dial tone or not access outside dial tone. If you pick up a phone on a PBX system and dial "9" for an outside dial tone, you might hear a siren sound instead. If you do, the particular phone you are dialing on is not DOD enabled. In most PBX systems (even though DOD is not enabled), an emergency "911" call will still go through. It is common to place special instructions on the phone to explain how to make an emergency call.

Doghouse The closure that contains cellular/PCS transmission equipment. Doghouses can come with heater/air-conditioner units (environmental control) and are about the size of a small doghouse. Near each cellular/PCS antenna is a small building, called a *hut*. Inside the hut is where the doghouse is located.

Dongle (Dongle Key) A device for protecting copyrights on computer software that looks very much like a DB25 gender-changer/adapter. Inside the dongle is usually an encoded ROM circuit with a user-rights serial number burned into it. If the dongle is not plugged into the printer port of the PC that the software is loaded on, the software does not work.

Dongle

Do Not Disturb See *DND*.

DOS (Disk Operating System) See *Disk Operating System*.

Drag Line A string or rope pulled into a conduit for making future wire or cable installation easier.

Drift When a carrier frequency changes unintentionally because of a transmitter problem. Drift can occur because of a temperature change. Drift can also be caused by bad connections, or defective components. Crystal oscillators are the most drift-reliable circuits. Frequency drift can also be caused by temperature changes in the atmosphere, because of the diffraction of the radio signal as it travels through different densities of air. Looking down a road toward the horizon on a hot day you might notice that the road and other objects look like they are wet or wavy. This is a visual example of atmospheric diffraction.

Drive Ring A ring with a nail attached to it, used to fasten or hold drop wire to telephone poles or sides of buildings.

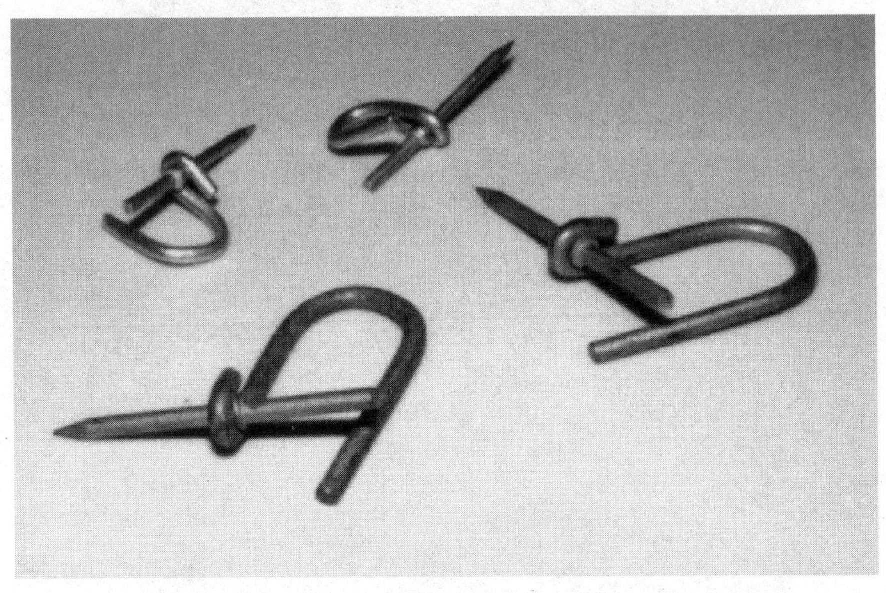

Drive Rings

Drop Another term for *service wire*. The service wire is the aerial or underground wire that runs from your home or office to the terminal in your back or front yard. Abbreviations are *ASW (Aerial Service Wire)* and *BSW (buried service wire)*.

Drop Cable Another term for *service wire*. See *Drop*.

Drop Clamp A device that is used to attach an aerial service wire to a "J hook" or "Ram's horn," which is attached to a building or pole.

Drop Reel A reel that is used to transport and distribute drop wire during installation.

Drop Wire Another term for *service wire*. See *Drop*.

Dry T1 A T1 that without the –135V DC battery voltage. A CSU/DSU has the ability to convert a wet T1 to a dry one. The T1 circuit is transmitted with a –135V DC voltage on the public telephone network to power repeaters and other conditioning equipment. A dry T1 is also called a *DS0*.

Drop Wire Reel

DS (Digital Service) The prefix for digital service circuits. A comparison of the DS-level circuits and other carriers is in the following table.

DS0 (Digital Service Level Zero) 64Kb/s. Equivalent of one voice (or one analog POTS) line. A DS0 is the basic building block of which all the other DS services are comprised. A DS0 can come in two flavors, in-band signaled and out of band signaled. The in-band signaled DS0 is best suited for carrying voice applications. It has a 56Kb/s user bandwidth and an 8Kb/s channel is bit-robbed from the total 64Kb/s DS0 bandwidth. The signaling channel is for carrying dialed digits, of hook and dial-tone signals for a central office. With the in-band signaling format, 24 DS0s can be carried on a T1. The other flavor is an out-of-band signaled DS0, which has a bandwidth of 64Kb/s. This format is best suited for data transmissions. You can get 23 out-of-band signaled DS0s on a T1. The 24th channel is used for the signaling of the other 23.

Bandwidth	Equivalent DS0	Equivalent DS1	Equivalent DS3	Comments:
64Kb/s	1	*	*	one phone line
1.544Mb/s	24	1	*	popular service
3.152Mb/s	48	2	*	equipment
2.048Mb/s	32	1	*	European
6.312Mb/s	96	4	*	equipment
8.448Mb/s	96	4	*	European
44.736Mb/s	672	28	1	popular service
34.368Mb/s	512	16	1	European
139.264Mb/s	2016	80	6	long haul radio
51.84Mb/s	**672**	**28**	**1**	**electrical OC1**
51.84Mb/s	**672**	**28**	**1**	SONET
255.520Mb/s	**2,016**	**84**	**3**	SONET
622.080Mb/s	**8,064**	**336**	**12**	SONET
2.488Gb/s	**32,256**	**1,344**	**48**	SONET
4.976Gb/s	**64,512**	**2,688**	**96**	SONET

DS (Digital Service)

DS1 (Digital Service Level 1) 1.544 Mb/s. Another name for a T1. The specific difference between a DS1 and a T1 is that the T1 is on copper and comes with a –135-V battery voltage, and the DS1 is a dry circuit, on copper or fiber-optic lines, with no battery voltage. Other than that, they are the same. A DS1 has a total bandwidth or transmission speed of 1.544 Mb/s. The 1.544 Mb/s is divided into 24 64Kb/s channels. A DS1 (T1) is available in several different packages that offer different line formats and framing formats. The package that a customer requests from a phone company depends on what they want to use the DS1 for and what kind of equipment they have. Telecommunications customers use DS1 circuits as private lines to connect data devices from one geographical place to another or to transport large amounts of dial tone to the premises. DS1 circuits are also used to connect directly to a long-distance company for broadband WAN service. Telecommunications companies also use DS1 (they are T1 circuits within their own network) circuits to provide more telephone service where a shortage of twisted pairs is available (see SLC96). For a diagram see *T1*.

DS1circuit/line types and applications

Line format/coding	framing format	signaling	Application
AMI	SF/D4	in-band	24 voice/modem channels
AMI	ESF	in-band	24 voice/modem channels
AMI	ESF	out-of-band	23 voice/modem or digital/data channels
B8ZS	SF/D4	in-band	24 voice/modem channels
B8ZS	ESF	in-band	24 voice/modem channels
B8ZS	ESF	out-of-band	23 voice/modem or digital/data channels

DS1 (Digital Service Level 1)

DS1c (Digital Service Level 1) A digital signal that combines two DS1 channels. The aggregate frequency is 3.152Mb/s, which contains 3.088Mb/s of payload, (two DS1s) and 64Kb/s (one DS0) of overhead for transmission control.

DS2 (Digital Service Level 2) 6.312Mb/s. A DS2 is four DS1 channels multiplexed together within a DS3 multiplexer. A DS2 is not available to customers. It is just a step in the creation of a DS3.

DS3 (Digital Service Level 3) 44.736Mb/s. A DS3 is a circuit that is provided to customers by telephone companies. It is a transport for 28 T1 circuits, which adds up to 672 DS0 circuits (voice channels). Telecommunications customers use DS3 circuits as private lines to connect data devices from one geographical place to another or to transport large amounts of dial tone to the premises. DS3 circuits are also used to connect directly to a long-distance company for broadband WAN service. Telecommunications companies also use DS3 circuits to provide more telephone service where a shortage of twisted pairs is in their cable plant. Sometimes it is less expensive for a telephone company to install the DS3 electronics in areas, rather than long feeds of large twisted copper-pair cables. DS4 (Digital Service Level 4) 274Mb/s. A DS4 is a transport for six DS3 circuits. Its capacity in DS1 circuits is 168. The capacity in DS0 circuits is 4032.

DSR (Data Set Ready) Pin 6 or pin 20 of a DB25 connector wired for the RS232C protocol. This is the wire that a modem or SDI device uses to send a signal that acknowledges that it is ready to receive data.

DSS (Digital Switched Service) A service offered by telephone companies where a telephone line is switched while still in it's digital form. Many telephone lines that leave a customer's premises via T1 are converted to analog when they reach the telephone company central office. DSS lines are run directly into the central office telephone switch in digital form (64 Kbp/s per line).

DSS (Direct Station Selection) A device that can be added to a PBX telephone set that has additional buttons on its face so that a user can see what extensions are in use (off hook) and which are free. When a call comes in to an answering agent, he or she can look at the direct station selection module attached to their phone and see whether the desired person is on their phone or not. Calls can be made and transferred to the extensions appearing on the DSS by pressing the associated button.

DSU (Data Service Unit, CSU/DSU, Channel Service Unit/Data Service Unit) A DSU is a hardware device that is available in many shapes and sizes. Rack-mount, shelf-mount, and stand-alone DSUs are available. A CSU/DSU has three main functions. The first function is to act as a demarcation point for a T1 (DS1) service from a local communications company. The second function is to provide line format and line-code conversion (B8ZS to AMI, SF or D4 to ESF, 135 V to 0 V) between the public-network and the customer-premises equipment, if necessary. The third function is to provide maintenance or alarm services and loop-back for isolating problems with the T1 line or customer's equipment. For a photo, see *CSU/DSU*.

DSVD (Digital Simultaneous Voice and Data) A modem and software combination that allow a voice and data to be sent over the same connection, via your computer.

DSX (Digital Service Cross-Connect) A reference to a digital-service termination/patch panel that allows DS1 and DS3 circuits to be monitored by test equipment, such as a TTC Tberd or T-ACE. DSX panels are usually terminated via wire-wrap. For a photo of wire-wrap terminals, see *Wire-Wrap*. For a photo of a DSX panel, see *Digital Service Cross-Connect Panel*.

DTE (Data Terminating Equipment) DTE is equipment that receives a communications signal. For a data connection to work between I/O (input/output) devices, one needs to be designated the "communications-sending" equipment and one the "communications-terminating" or DTE (data-terminating equipment). A computer's printer port is a DCE port, a printer is a DTE device. A practical way to classify the two is: DCE is the sender of data and the DTE is the receiver of data.

DTMF (Dual-Tone Multiple Frequency) The tones that you hear when you dial a single-line push-button phone. The tones are a mixture of two frequencies. For a diagram, see *Dual-Tone Multiple Frequency*.

DTMF Cut Through A feature of voice-response systems, voice-mail systems, and auto attendants to hear the digits that you dial and play a RAN (recorded announcement) at the same time. This feature reduces the frustration level for people who hate to listen to voice-message systems because when the listener makes their choice (pushes a key) the selection is executed and the RAN stops immediately.

DTR (Data Terminal Ready) A light of the front of a modem or data-communications device that indicates that it is ready to receive a handshake signal from another communications device.

DTU (Digital Test Unit) Some DTUs are stand-alone devices, and some are add-ons for integration into telecommunications equipment.

Dual Homing Another term for *alternate routing*, *redundancy*, or *self-healing* that is used in telecommunications networking.

Dual Ring of Trees A network topology that uses a dual ring topology as a backbone for other ring or star networks.

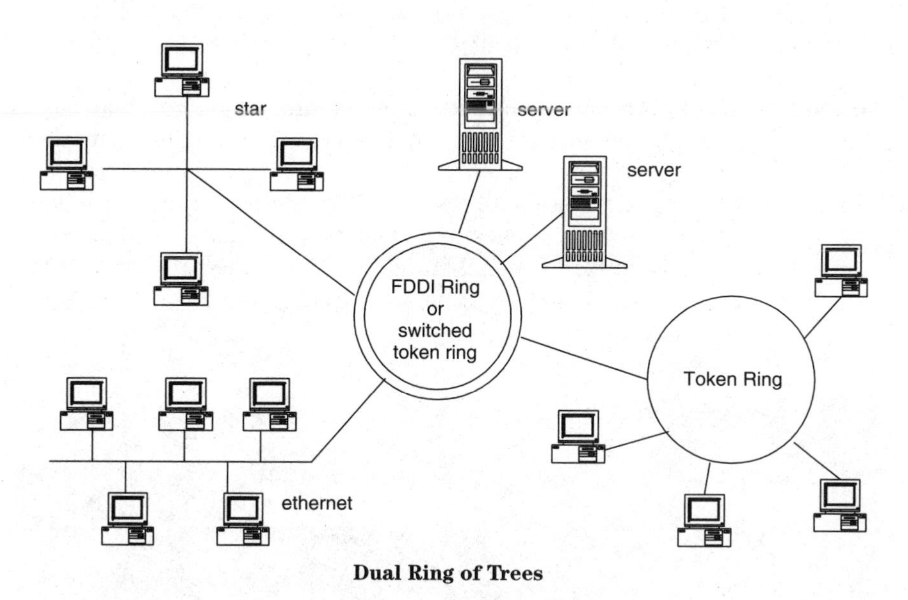

Dual Ring of Trees

Dual-Tone Multiple Frequency (DTMF) The tones that you hear when you dial a single-line push-button phone. The tones are a mixture of two

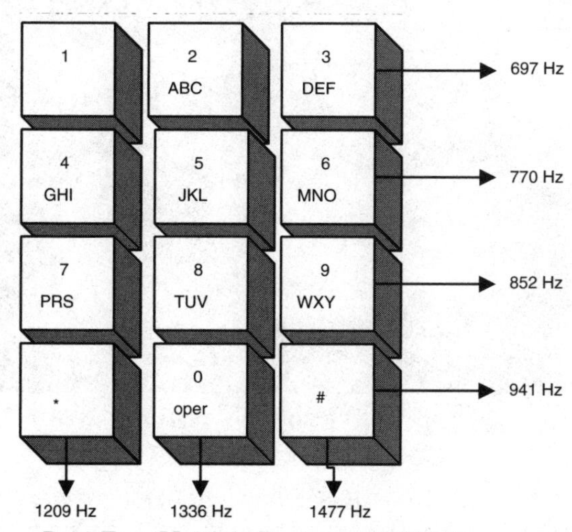

Dual-Tone Multiple Frequency (DTMF)

frequencies. The frequencies are connected according to the diagram provided.

Duct A conduit or pipe that runs from one location to another, within a building, connecting buildings, or connecting cable vaults. It is a good practice to install a duct, then pull cable into it, rather than directly mounting or burying the cable.

Dumb Terminal An I/O (input/output) communications terminal. They are called "dumb" because they do not do any processing of information, they just display it. Dumb terminals are used to display input/output information from ACD systems, PBX switches, SONET transport equipment, or other interface applications. Dumb terminals are not used as much anymore because "terminal-emulation" programs are available for personal computers. Examples of popular dumb terminals are the Wyse 50 and VT100.

WYSE 50 Dumb Terminal

Dummy Load A device that is connected to electronic output equipment, such as radio transmitters, power supplies, and even stereo systems to test them under full strain. One dummy load for a home stereo system, for example, is two large 8-ohm resistors, one for each channel (left and right) in place of the speakers. The resistors allow a full volume-range (power output) test without having to endure the sound.

Duplex Jack A jack or connecting block with two jacks on its face.

Duplex Transmission (Full Duplex) A communications protocol that has the ability to send and receive at the same time. A DS1 is a full-duplex protocol that carries other protocols. The alternative to full duplex is half duplex, where two communications devices take turns sharing a line. Humans speak half duplex, because it is to hard to have a meaningful conversation while both people are talking at the same time. CB radios and "Walkie Talkies" are also half duplex.

DVD (Digital Versatile Disc) See *Digital Versatile Disc*.

Dynamic Bandwidth Allocation The ability of a communications protocol to change the amount of bandwidth allotted to each user while in use.

Dynamic Load Balancing A feature of *ACD (Automatic Call Distributor)* systems to evenly distribute incoming calls to agents.

Dynamic Memory (DRAM) RAM (random access memory) that holds its data as long as the power is on. The other popular RAM memory is static memory, which is slower, but holds its data when the power is interrupted. See also *NVRAM* and *RAM*.

E

E1 A European standard that is equivalent to an American T1. The E1 and T1 are not completely the same. They both use 64Kb/s channels, but the T1 has 24 and the E1 has 32. The following table compares E1 and T1. The European standards are used in all countries, except the United States, Canada, Japan, and Singapore.

	Total Bandwidth	total number of 64Kb/s Channels	Number of Channels used for Out of Band Signaling
T1	1.544 Mb/s	24	1
E1	2.048Mb/s	32	2

E1

E2 A European standard that carries four E1 circuits. It does the same thing in Europe that a T2 does in North America. A comparison of E2 and T2 circuits follows. The European standards are used in all countries, except the United States, Canada, Japan, and Singapore.

	Total Bandwidth	total number of 64Kb/s Channels	Equivalent E1/T1 Carried
T2	6.312 Mb/s	96	4
E2	8.448 Mb/s	128	4

E2

E3 A European version of a T3. An E3 has a smaller bandwidth and carries fewer sub-channels in comparison to a T3. The European standards are used in all countries, except the United States, Canada, Japan, and Singapore.

	Total Bandwidth	total number of 64Kb/s Channels	Equivelent E1/T1 Carried
T3	44.736 Mb/s	672	28
E3	34.368 Mb/s	512	16

E3

E911 (Enhanced 911) Enhanced 911 service uses an *ANI (Automatic Number Identification)* signal that comes in with the call and cross references it to a database of addresses and displays the result on a *PSAP (Public Safety Answering Point)* agent's monitor screen. Standard 911 uses the ANI only. E911 is useful for the times that the caller is not emotionally capable or knowledgeable enough of the area to provide a correct address in a timely manner.

E&M (Ear and Mouth) A type of telephone line or trunk. What makes E&M trunks different than ground-start and loop-start trunks is that E&M trunks use four wires (two pair) instead of two (one pair). One pair is used for transmit (mouth) and the other is used for receive (ear). There are also six-wire (three pair) E&M trunks that use the third pair to control the other two. Older PBX systems used E&M signaling/trunking to connect to the telephones within its office. E&M signaling is an out-of date standard, and it has recently become a discontinued service from USWest Communications, and other RBOCs (Regional Bell Operating Companies).

Earth Ground The electrical potential of the earth (0 V). To maintain a good earth ground (a metallic rod) is driven into the ground (the length of a standard grounding rod can vary, depending on the geographical location). Any wire connected to that rod is "grounded." The power company installs a rod like this when they connect power to your home. The telephone and cable TV companies wire their network interfaces (lightning protection) to the power company's earth-ground rod. The alternative to earth ground is a "floating ground." A floating ground is simply a reference point that is not "earth grounded." The negative terminal of your car battery is a floating ground and any home appliance that has a two-prong electrical plug is also a floating ground (in newer homes that are wired correctly).

ECH (Enhanced Call Handling) A reference to devices that handle telephone calls, such as voice-mail systems, integrated voice-response systems and *ACD (Automatic Call Distribution)* systems.

ECP (Enhanced Call Processing) A reference to an Octel (a voice-mail system manufacturer now owned by Lucent Technologies) voice-mail feature that allows callers to route themselves to their destination via a prompt/response system of messages.

Eddy Current Eddy current is the electrical current produced in the core (the central piece of metal that coils of wire are wrapped around) of a transformer. As the transformer's core magnetizes and demagnetizes in conjunction with the AC electricity flowing through the coils wrapped around it, magnetic fields are created. These magnetic fields cause electric current to flow in the core. The core heats up because of the current flow, and this heat is considered an inefficiency. This inefficiency is called *eddy current loss.*

EDO RAM (Extended Data-Out Random-Access Memory) A Dynamic RAM technology that reduces access times by as much as 10%.

EEPROM (Electrically Erasable Programmable Read-Only Memory) A type of *EPROM (Erasable Programmable Read-Only Memory)*. A microchip that contains circuitry that is capable of storing binary instructions, then being erased or reset. Read-only memory is where instructions that tell a CPU how to work are stored. The different types of EPROM devices include *EEPROM (Electrically Erasable Programmable Read-Only Memory)* and *UVEPROM Ultra-Violet Erasable Programmable Read-Only Memory)*, which can be erased by ultraviolet light exposure.

EIA Standards (Electronic Industries Association Standards) EIA standards are available from the EIA's headquarters in Washington, DC.

Electrolysis The use of electricity to change the properties of chemicals or electroplating. Many wires and conductors are electroplated with different metals to increase conductivity. This is the opposite of creating electricity with the use of chemicals (batteries).

Electromagnetic Interference (EMI) Interference caused by a radio signal or other magnetic field, inducing itself onto a medium (twisted/non-twisted pair wire) or device (telephone or other electronics). The world we live in is full of radio waves that are emitted from electric appliances, such as blenders, automobile engines, transmitters, and even fluorescent lights. Even though we take preventative measures to avoid picking up these unwanted signals, they sometimes find their way into places where they are not wanted. Electromagnetic interference is usually caused by one of two things. The first is when a wire connected to a device acts as an antenna and picks up the EMI, which is then passed on to the electronics inside the device and is amplified. The second is when an electronic component inside a device acts as an antenna because of poor design, poor shielding, or because the component is defective.

Electromotive Force (EMF) Another name for *voltage*, which is the origination of the designator used in Ohm's Law formulas. In the formula for calculating voltage, E represents voltage (in volts), R represents resistance (in ohms), and I represents current (in amps). $E = I \times R$

Electronic Switching System (ESS) A family of telecommunications switches manufactured by Lucent Technologies. The 5ESS is a common central office switch used by RBOCs.

Electronic Warfare A professional field in the armed forces that specializes in the science of disabling communication and control equipment with the use of EMI, EMP, and by creating "ghost" or deceptive radar images.

Electrostatic Discharge (ESD) Static electricity. ESD became a big deal when computer and electronics manufacturing companies started using *CMOS (Complementary Metal-Oxide Semiconductor)* electronic components in the devices that they make. Many microchips contain CMOS transistors called *MOSFETs (Metal-Oxide Semiconductor Field-Effect Transistors)*. CMOS components are in every PC made today. They are also used to make the circuitry for LCD watches, telecommunications equipment, home electronics, and many others. The advantage of CMOS components is that they use less power than other components (such as *TTL, Transistor/Transistor Logic*). This is why you can have a tiny battery power your wrist watch or calculator for months. However, the disadvantage of CMOS is that it is extremely sensitive to static electricity. Whenever handling CMOS components, be sure that your body is grounded (to drain off any static that might be on your body). CMOS components can be damaged by static fields, such as one that is created when you brush your hair. Even if the static doesn't arc out into a component, its field can still damage or weaken it. If CMOS components are weakened by static, they usually fail unpredictably in the future.

E Link (Extended Link) An *SS7 (Signaling System 7)* signaling connection between a signaling-end point translator and a signal-transfer point. SS7 is the protocol that controls call transfers between central offices in North America.

EM ASCII control code abbreviation for *End of Medium.* The binary code is 1001001, the Hex is 91.

E-Mail (Electronic Mail) A software program that you can load onto a computer network that allows the users on the network to write each other notes and send copies of documents. Lotus Notes and CCMail are two examples of this software.

EMI (Electromagnetic Interference) See *Electromagnetic Interference.*

Emission A reference to electromagnetic waves (this includes heat, radio, and light) radiating from a source. For example, the sun emits ultraviolet radiation and radio stations emit electromagnetic signals.

EMF (Electromotive Force) See *Electromotive Force.*

EMT (Electrical Metal Tubing) The metal tubing that electricians use to encase electrical wire. EMT is also used to provide a path into and throughout buildings in telecommunications applications. EMT is 2 to 4 inches in diameter and protects the communications cable (copper or fiber) from being cut easily.

Emulation The use of a PC to act or communicate as a dumb I/O terminal. To use a PC in this application (such as to plug into a microwave link and boost its power), it must be equipped with terminal-emulation software. The microwave-link device has its own microprocessor and only needs a device to communicate with, that device is usually a VT100 terminal. Terminal-emulation software allows your PC to "look like" a VT100 terminal to the microwave radio equipment.

End Device/Instrument A telephone, fax machine, modem, terminal adapter, PBX system, computer, or anything else that terminates a communications link.

End Office (EO) The telephone company central office that serves or connects to the end user/customer. The telephone line in your home connects to an EO. In other telephony applications, a line might connect directly to a long-distance company's switch, bypassing the EO. This is called a *Dedicated Access Line (DAL).*

End to End A reference to the ability of a circuit to communicate/signal from one end user to another without altering service. Regular *POTS (Plain Old Telephone Service)* telephone lines are end-to-end signaled communication lines. After the circuit is established, you can still dial digits into a voice-mail system to reach an extension.

End User The customer, the one that uses or consumes a product or device.

Enet Abbreviation for Ethernet.

Engineer Furnish and Install (EF&I) A way to purchase something. If you would like to install a SONET ring in your campus environment and you ask Northern Telecom for pricing, they will offer you the option of just buying the equipment or buying the equipment, and having them engineer and install it. If you can afford to buy equipment that has the option of EF&I with the purchase, I recommend that you spend the extra money. It saves in the long and the short run.

Enhanced 911 Enhanced 911 service uses an ANI signal that comes in with the call and cross references it to a database of addresses. It displays the result on a *PSAP (Public Safety Answering Point)* agent's monitor screen. Standard 911 uses the ANI only. E911 is useful for the times the caller is not emotionally capable or knowledgeable enough of the area to provide a correct address in a timely manner.

Enhanced DNIS A step above standard *DNIS (Dialed Number Identification Service)*. Enhanced DNIS comes with *ANI (Automatic Number Identification)*, better known as *caller ID*.

ENQ The ASCII control-code abbreviation for *enquiry*. The binary code is 0101000 and hex is 50.

Envelope 1. Reference to the modulated carrier signal in a radio transmission. 2. A data block in a packet transmission network that contains addressing or other data in binary form.

Environment Electronic equipment usually has specified requirements for the environment in which it is located. The requirements are usually listed in the literature that comes with the equipment. Common environmental requirements are: –35 degrees F to 85 degrees F, 20% to 60% humidity, and dedicated 120V AC power.

EO (End Office) See *End Office*.

EOT The ASCII control-code abbreviation for *end of transmission*. The binary code is 0100000 and hex is 40.

EPP (Enhanced Parallel Port) A parallel port on a computer or other data device that allows data to be transferred to a device connected to it at twice the rate of a regular parallel port. The EPP acts as an extension of a device internal bus system. They are commonly used to interface detachable disk or tape drives.

EPROM (Erasable Programmable Read-Only Memory A microchip that contains circuitry capable of storing binary instructions, then being erased or reset. Read-only memory is where instructions that tell a CPU how to work are stored. The different types of EPROM devices include *EEPROM (Electrically Erasable Programmable Read-Only Memory)*, and *UVEPROM (Ultra-Violet Erasable Programmable Read-Only Memory)*.

EQ (Equalization, Equalizer) To adjust the tone or sound of a circuit by diminishing or augmenting specific frequency bands. The tone control on a radio is a type of equalizer. A radio transmitter might have a tendency to amplify low-end signals, such as the sound of a bass guitar or drums better than high-end signals, such as the sound of a voice or cymbals. An equalizer can be used to reduce or increase the amplification of either end of the broadcast for an even and accurate reproduction of the input.

Equalization See *EQ*.

Equipment Cabinet There are many types of equipment cabinets, but the most common is 7 feet high by 24 to 26 inches wide with a 22- or 19-inch wide mounting rack built into it. Quality equipment cabinets are equipped with blower fans to circulate air through them and have a locking door. Some have clear plastic doors so you can view the alarm/status lights on the front of your equipment. Any time equipment uses a fan to keep it cool, you should place it in the cleanest, dust-free environment possible. If you have a dusty environment, the fans will blow dust into your equipment, which will build up and act as an insulating blanket. This will cause overheating and failure of electronic components.

Erasable Programmable Read-Only Memory (EPROM) See *EPROM*.

Erlang A one-hour unit of telephone traffic. This can be one phone call that lasts for one hour, or two phone calls that last for 30 minutes each, etc. Erlangs consist of *CCS (Centum Call Seconds)*.

ERP (Effective Radiated Power) The actual power in watts radiated from a transmitter's antenna. A typical FM radio station has an ERP of 15,000 watts (15 kw).

Error Checking The methods used by modems and other transmission equipment to detect errors in the data received in a transmission. Common error-checking methods are *VRC (Vertical Redundancy Checking)*, *LRC (Longitudinal Redundancy Checking)*, and *CRC (Cyclic Redundancy Checking)*. The most accurate error-checking method used in modems today is called CRC (cyclic redundancy checking). CRC is designed to check a frame-type protocol.

Here is a simplified version of CRC logic. Imagine that a binary number is read off (transmitted) to you. You hear the ones and zeros, and as you hear them you write them down (remember them). Then, you are given another number to divide the first number by (*FRC, Fame Check Sequence*), then another number (algorithm) that should match up as the answer. If the answers match, then there is a 99.99995% chance that you heard the first number correctly and there are no errors. Drawn out in a very simplified form, CRC error-checking logic looks something like this:

`Bit block to be transmitted:`	`11001000` (this is equal to 200 in decimal).
`FRC frame:` (added to the end)	`1010` (this is equal to 10 in decimal).
`Algorithm:` (the answer)	`10100` (this is equal to 20 in decimal).

So, the receiving equipment divides the block transmitted by the FRC frame and should get the algorithm. If the numbers don't match, then it requests a retransmission.

Error Rate Some transmission test equipment (TTC Tberd), some computer software (Novell), and some network adapters (SMC ethernet) have the separate ability to transmit a known group of packets over a network, then see how accurately they are returned. When they are returned, the equipment divides the total number of packets by the number of packets that have errors. The end number is a percentage, which is the error rate.

ESC The ASCII control code abbreviation for escape. The binary code is 1011001 and the hex is B2.

ESD (Electrostatic Discharge) See *Electrostatic Discharge.*

ESF (Extended Superframe Format) A type of T1 Line. An additional innovation of the D4/SF (Superframe) format. In a T1 circuit, each channel is sampled (which is an 8-bit DAC sample) and the bits are sent on down the line. If you take 8 bits and multiply it by 24, you get 192. If you add one "timing" bit to the end of the 192-bit chain, you get 193. A Superframe is 12 of these 193-bit frames, chained together. This allows each of the 12 framing bits (193rd bit) in each of the 193-bit sequences to mean something other than a timing signal. Different meanings include signaling, such as dial tone, digits dialed, and off-hook or busy. Then, in-band signaling comes into play, in which ESF can be configured as either in-band or out-of-band. Within the 24 channels, the least-significant bit (the one that will have the least effect on the accuracy of the DAC conversion) of the 6th and 12th samples of each frame are used for additional signaling, control, and maintenance. This is called *bit robbing*, and it is the reason why in-band signaled T1 lines have only 56Kb/s of bandwidth, as opposed to "non-bit-robbed" (clear channel) T1 lines, which have 64Kb/s of bandwidth.

ESS (Electronic Switching System) A family of telecommunications switches manufactured by Lucent Technologies. The 5ESS is a common central-office switch used by RBOCs.

Essential Service A term used by the FCC and PUC regarding telephone service. It is now regarded that a telephone line and a telephone is equally as important as heat or electricity for a household. Under different utility regulations, telephone service is offered to low-income households at a reduced rate. Reduced rates are almost always granted if the household is the residence of a person with a life-threatening health condition. In this case, the telephone line is needed for 911 service and the ability to call a doctor. If you know of an elderly person that has a very low income, you might be able to help them by calling the local phone company and requesting a form for reduced-rate service.

ETB The ASCII control-code abbreviation for end-of-transmission block. The binary code is 0111001 and hex is 71.

Ethernet A family of *LAN (Local Area Network)* protocols. Ethernet is one of the oldest communication protocols for personal computers. When a LAN is mentioned, two things should immediately come to mind. Physical topology and the protocol the LAN uses to manage communications between devices. Ethernet can be implemented in a bus or star physical topology. The alternative family of LAN protocols is the token-passing type, which are configured as a ring topology. See *Token Ring*.

In an Ethernet LAN, computers are given a means to communicate with each other called a protocol. A protocol is a set of rules and instructions for communicating. Within the protocol is a "logical topology." Even

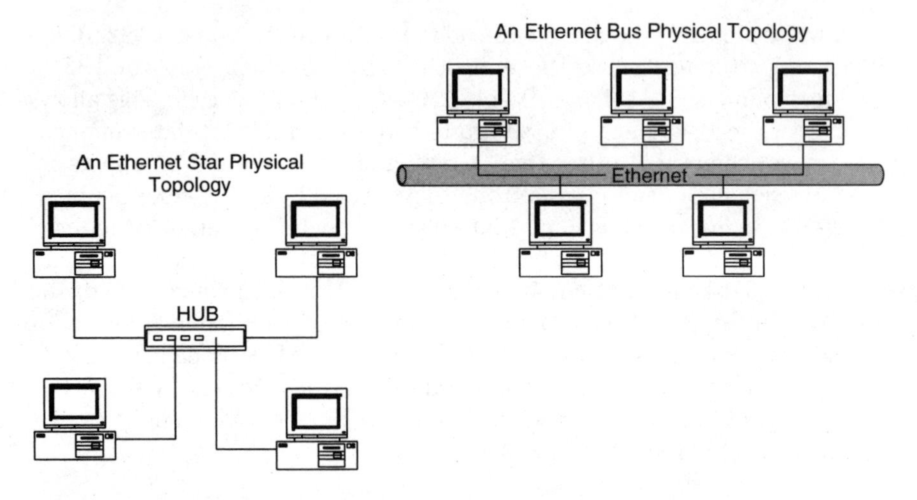

ETHERNET TYPES

PROTOCOL	PHYSICAL TOPOLOGY	WIRING USED
10 BASE 2	BUS	RG 58 COAX (50 ohm)
10 BASE 5	BUS	RG 8 COAX (50 ohm)
10 BASE T	STAR	CAT 4 or 5 UTP/STP*
100 BASE T	STAR	CAT 5 UTP/STP*

* unshielded twisted pair / shielded twisted pair

Ethernet

though a network can be connected as a star, it can still look like a bus to the communications equipment because all of the computers/devices are connected to the same wire (in the star diagram, the hub is a device that connects all the wires together). The way Ethernet works is similar to the way people talk in a group. Instead of using wire to carry the binary coded information as Ethernet does, people use air to carry sound information. When there is a silence, then one of the persons in the group is able to speak. When the persons speaks, they might say "Johnny, do you know the answer for 5+5?" Even though all the people in the group hear this message, they know it is for Johnny because the message was "addressed" to him. So, only Johnny will respond "10." Then, imagine as Dawn and Vicki acknowledge a silence and try to speak at the same time. This is confusing and no one understands the information. Ethernet has the same problem and it is called a *collision*. Collision is the disadvantage of Ethernet. Because of the possibility of collisions (which happen very frequently), Ethernet is called a "contention based" protocol because all of the connected devices are contending for use of the network. Manufacturers have come out with new ways to avoid collisions, called

CSMA/CD and *CSMA/CA*. Ethernet has many different types of wiring to connect devices, and many different *NICs (Network Interface Cards)* to select from that need to be installed in each computer or device on the network. The following is a list of Ethernet protocols and the type of wiring used for each.

Ethernet Switch An Ethernet protocol that gives each computer or device connected to the network its own channel to communicate with. In plain old Ethernet, all of the devices communicate on the same wire (channel). In Switched Ethernet, the single wire is multiplexed into a variable number of channels—one for each device that is communicating. How fast each channel is depends on the number of devices communicating at any one time. Collision is eliminated in Switched Ethernet, which was the largest inefficiency with the previous Ethernet protocols (CSMA/CD and CSMA/CA). This newer protocol transfers data at a much faster rate because of the newfound efficiency. Switched Ethernet requires its own special *NICs (Network Interface Cards)* and other hardware. It also requires its own software drivers.

ETX The ASCII control code abbreviation for end of text. The binary code is 0011000 and the hex is 30.

Even Parity A method of bit-stream checking. Parity is used in error correction. The number of logic "ones" is counted in a bit stream. There is "odd parity" and "even parity." Which is used depends on if you like odd or even numbers, or if the modem you are trying to connect with likes odd or even numbers. Parity is a part of error-checking protocol. It is simply the part of the protocol where the two devices are told if they are counting odd number bits or even number bits. In odd parity, if the number of ones is an odd number, then a parity bit is set to "one" at the end of the bit stream. This is *odd parity* because the parity bit is set to one when the number of "ones" is odd. In *even parity*, the parity bit is set to "one" when the number of "one" bits is even.

Exa See *ExaByte*.

ExaByte Exa is a unit of scientific/engineering notation. 1×10^{18} power. Or you could think of Exa as 18 zeros on the end of a number. Five ExaBytes (5EB) is equal to 5,000,000,000,000,000,000 bytes.

Exchange The area that a single central office services. Soon, when number portability is fully implemented, an exchange will not be associated with a central office. It will be associated with an area and the legal regulations imposed on communications companies in that area. Currently, each central office is assigned a group of numbers that it can use. The numbers are the first three digits (not including the area code). The numbers (801)-355-xxxx, (801)-237-xxxx, and (801)-575-xxxx are assigned to the USWest Salt Lake City, Utah main central office.

Exchange Area An area served by multiple communications companies and multiple central offices. Each exchange area has its own legal regulations regarding how companies can compete (or price and package their services) and what services they are required to provide.

Expansion Slots Space allocated in a KSU for the addition of circuit cards in the future. Additional circuit cards needed might be trunk interface cards or station/extension interface cards.

Extended Superframe Format (ESF) See *ESF*.

Extension A telephone or equipment connection on a *PBX (Private Branch Exchange)* or key system. Extensions can be from two to seven digits long. An extension is sometimes referred to as a *DN (Directory Number)*. An extension can be an electronic keyphone if it is connected to a digital PBX interface or it can be a modem, fax machine, or analog phone (like the ones made for home use) if it is connected to an analog interface.

External Modem Also called a *stand-alone modem*. A modem that comes in its own package (case) and comes with a cable that plugs into a COM port/serial port on a computer or data device. External modems are popular for dial-up remote-access administration for PBX switches. The alternative to external modems are internal modems, which are popular in PCs and come in the form of circuit cards that plug into the PC's motherboard. Both do the same job equally as well, but most computers only have two COM ports. If you use an internal modem, then you can use your two COM ports for other applications, you don't have to have another power outlet, and internal modems are usually less money.

Eyeball Shot Another name for a terrestrial microwave link. The link is made by two radio transceivers equipped with parabolic dish antennas pointed directly at each other. Radio can carry point-to-point transmissions of many bandwidths, including DS1, DS2, DS3, STS1, and OC1. Their range can vary, depending on the size of the antenna (dish), weather in the region, and the amount of power emitted. Including all of the previous factors, a link can range from 0 to 50 miles.

F Connector A connector for coax cable. The standard for cable TV is 75-ohm coax with an F-type connector on the end. For other types of coax connectors, see *Coax*.

F Connector

Face Plate 1. Some telephone jacks come with separate face plates that snap on to the front of a telephone jack. The snap-on/snap-off design makes it easier to access the wire connections, but this style of jack is usually a poor choice for a long-term application. 2. Some telephone sets have face plates that fit over the buttons and display of the phone. Instead of buying the phone that is the color of your choice, you simply buy a universal phone, then buy the face plates with your choice of color.

Facilities A term that refers to physical equipment, such as cable, switches, and transport/carrier equipment, that is commissioned into the field for service. Switches and other equipment are considered inside facilities and telecommunications cabling and its cross boxes, vaults, and splices are considered outside facilities.

Facilities-Based Carrier A telephone company that has its own switches and communications facilities, unlike a reseller.

FACS (Facilities) A portion of a phone company that tracks the use of facilities (cable pairs and central-office switch ports). If you are a telephone network technician and you find that a pair in a cable (cable pair) is bad, you call FACS and notify them that the pair is unusable. This way, they don't think that they have an extra pair for future service.

Facsimile A fax machine. A machine that can dial a telephone number connected to another fax machine and generate a copy of a document fed into it on the far end.

Fan Out A multiplexer. The device that breaks a DS1 or DS3 service down into the size that a customer wants. A fan out on a DS3 line breaks the 28 DS1 channels out for a customer and a fan out on a DS1 line breaks the DS1 into 24 DS0 channels.

FAP (Fuse Alarm Panel) A power-distribution panel that is installed at the top of a relay rack. All equipment in the rack is wired to the panel for power. Each device has its own fuse within the panel to protect the rectifier from an "over-current" condition if a device fails or a wire shorts. If any of the fuses blow, an alarm indicator is displayed. For a photo, see *Fuse Alarm Panel*.

Farad The standard unit of capacitance. A capacitor is an electronic device with two special properties. It only allows alternating current to pass through it, and it can store an electric charge. One of the many applications of capacitors is to filter AC out of DC power supplies and rectifiers. This is done by placing a capacitor from the DC output to ground. The capacitor appears as an easier path to voltage fluctuations and RFI, and an impossible path to direct current (DC). Physically, a capacitor is two plates of metal separated by an insulator (mylar is common). The physical size of a 1-F capacitor would be two sheets of tin foil the size of a football field, insulated (or separated) by a thin sheet of mylar. The farad is a huge unit of capacitance, so most capacitors are measured in microfarads (μF). For a schematic symbol of a capacitor and a photo, see *Capacitor*.

FAT See *File Allocation Table*.

FAX See Facsimile.

Fax Jack A device that connects to a phone line and has two jacks on the other end, one for a fax machine and one for a telephone. These devices are for users that want to use only one phone line for faxes and voice calls. When a call arrives, the fax jack answers the line immediately and waits for a mechanical tone from another fax machine. If it does not hear the tone, it assumes that the call is a person and not a fax machine. It then rings the phone on its other end. If the fax jack would have "heard" a tone, it would have connected the line to the fax machine plugged into its other end. The only bad thing about fax jacks or other line-sharing devices is that when they seize the line immediately, they block caller-ID (ANI) signals.

Fax on Demand A feature of voice-response systems that enable a caller to listen to a recorded message that gives them a selection of information that they can receive via a fax. After making a selection, the caller is then prompted by another recording to enter the number that they would like the information faxed to.

Fax Switch See *Fax Jack*.

FCC (Federal Communications Commission) An organization of the federal government that was set up by the Federal Communications Act of 1934. The FCC works in conjunction with the 50 state Public Service Commission bodies and Congress. It has the legal authority to regulate the following three areas of communications. Communications being defined as radio, video, telephone, and satellite communications within the United States. 1. Regulate who is permitted to manufacture and sell telecommunications equipment and service. 2. Regulate the price of interstate long distance. 3. Determine the electrical standards for telecommunications, such as operating frequency of transmitting devices.

FCC Tariff A ruling on a type of communications service. A tariff defines a service and the price that certain companies are allowed to charge. Tariffs usually restrict RBOCs and AT&T from being competitive by forcing them to sell service at higher prices than the companies wish. If the FCC did not impose these tariffs on the communications giants, it would be impossible for new smaller companies to become established and compete. AT&T, USWest, PAC BELL, NYNEX, etc., would simply drop their rates so low that the other companies would be driven out of business or be driven to being bought out by one of the larger companies.

FCS (Frame Check Sequence) A part of Cyclic Redundancy Checking error correction in data transmissions. See *Error Checking* for more information.

FDDI (Fiber-Distributed Data Interface) LAN backbone protocol that requires its own fiber-optic cabling, NIC cards, and software to configure them. FDDI is a 100Mb/s protocol that came about when Ethernet and token ring were in their 10Mb/s and 16Mb/s infanthood. Since then, Ethernet has developed 100 Base-T, which is a 100Mb/s protocol, but it still has only a maximum 40Mb/s throughput (actual transfer speed). Token Ring has now developed into Switched Token Ring, which has a maximum throughput of 80Mb/s. The maximum throughput of FDDI is a true 100Mb/s, and its self-healing ring capability is only matched by Switched Token Ring and exceeded by SONET. FDDI is a great LAN backbone architecture, but it is twice as expensive as its Ethernet and Token Ring competition, and it requires fiber-optic cabling. If your backbone architecture will carry important data that cannot have any downtime, FDDI and Switched Token Ring are your options. If you have crucial data flow and long distances (A ring of more than a 2-mile circumference) or an environment where EMI is abundant (such as a factory with large electrical equipment), then FDDI on fiber optic will be your prime backbone architecture. A FDDI ring has a maximum circumference of 62 miles (100 km), but a repeater or node must be spaced every 1.25 miles (2 km). Up to 1000 nodes can be placed on a FDDI ring. For a diagram of a FDDI application, see *Fiber Distributed Data Interface.*

FDM (Frequency-Division Multiplexing) Multiplexing is the process of encoding two or more digital signals or channels onto one. Channels are multiplexed together in communications to save money. When we use all of the wires in a cable and need more, it costs less to add electronics on the ends of a cable than to install a new one (imagine the expense from LA to NY). A T1 encodes 24 channels into one by using time-division multiplexing. Many radio stations are put in the same airspace by using frequency-division multiplexing.

FDMA (Frequency-Division Multiple Access) Another name for *frequency-division multiplexing* that the cellular telephone industry uses.

FDX (Full Duplex) A full-duplex line or communications path is able to communicate both directions, transmit and receive, at the same time. A T1 is a full-duplex line, with one pair used for transmit and the other used for receive. Full duplex can be accomplished on one pair of wires by using two multiplexed channels, one for receive, and one for transmit. There are two other types of transmissions. One is half duplex, where transmit and receive are sent one at a time. A CB radio works in half-duplex mode: one person talks while another listens, and vice versa. The other type of transmission is simplex, where communication is one way only. An FM radio station or TV broadcast is simplex.

Feature Buttons Buttons on PBX telephone sets that activate features such as hands free, call forward, do not disturb, transfer, and speed dial. Feature buttons can be changed or customized to a user's liking. For example, if John likes to have the top feature button as "hands free" and the bottom button as "transfer," then the buttons can be programmed for those functions. If Sally likes the bottom button to be her sister's "speed call" button, then the feature button on the bottom of the phone can be programmed that way.

Feature Cartridge A cartridge containing ROM or RAM that permits a phone system (PBX) to use different features. I always thought that if you buy the system, why not put all the software features on it, and make it simple. If you would like to have the features, you need additional software, and software is expensive. Feature cartridges give customers the option of buying what they want.

Feature Code If you want to use a feature on a PBX system, but don't have a button on your phone that performs that feature, you can use a feature button in combination with a code. If you would like to transfer a call on a phone system, but you don't have a transfer button on your phone, you can transfer the call by pushing the feature button and then entering a code (e.g., 52) on the dial pad. Feature codes are also used in the public telephone network. If you are a USWest customer in Utah, you can activate a feature that calls the party that just called you (if you did not get to the ringing phone in time to answer it). The feature code for this service is *69. Just pick up the phone and dial *69. There is a charge for this service just like there is a charge for directory assistance.

Feature Group In PBX systems programming individual telephone extensions can be very tedious if you have 300 of them on your system. Some PBX manufacturers build a feature into their programming called a *feature group*. You can assign features to a feature group, such as: transfer on button 1; speed dial on buttons 5, 6, 7, 8, 9; do not disturb on button 2; voice mail on button 3; etc. Then you can assign extensions to the feature group. Each extension assigned to a feature group will have the features of that feature group. This is much easier than programming every button on all 300 phones.

Feature Phone A reference to a PBX telephone set that is capable of select features, such as do not disturb, call forwarding, speed dial, transfer, etc.

Federal Communications Commission (FCC) See FCC.

Federal Telecom Standards Commission An organization established in 1973 to assist in the development in telecommunications interface standards. Federal Standards begin with FEDSTD.

Federal Universal Service Fee A Federal tax placed on telecommunications services provided by telephone companies.

Feedback The reintroduction of an amplifier's output signal back to its input. If you have been at a public gathering or speech and heard the microphone make a loud squeal sound through the loudspeakers, you have heard a type of feedback. This type is caused by the sound from the speakers finding its way back into the microphone and the amplifier.

FEP (Front-End Processor) A communications "front-end" device that can be loaded with a "firewall" to prevent unwanted users from accessing the communications network. An FEP can also perform routing, and differentiate between different communications protocols, depending on the software that runs on it. For a diagram of an FEP, see *Front-End Processor*.

FER (Frame Error Rate) Some transmission test equipment (TTC Tberd), some computer software (Novell), and some network adapters (SMC Ethernet) have the separate ability to transmit a known group of packets over a network and then see how accurately they are returned. When they are returned, the equipment divides the total number of packets (or frames) by the number of packets that have errors. The end number is a percentage, which is the error rate.

Ferric Oxide A compound that is sometimes used as a coating on magnetic tapes. It (and some other compounds) can be magnetized.

Ferrule 1. A part of a fiber-optic connector that holds the connector ends in alignment when they are connected together. 2. A metal ring that is sometimes found on power cords. It helps to reduce RFI passing from the power company into the device using the cord.

FET (Field-Effect Transistor) The two different varieties of transistors, bipolar and field-effect, are designed to manipulate electricity flowing through them in different ways. Bipolar transistors are current-controlled devices and field-effect transistors are voltage-controlled devices. The advantage of field-effect transistors is that because they are voltage controlled, they can switch from one to zero and draw hardly any current. Current is what drains batteries and field-effect transistors help make batteries last a long time. Bipolar transistors are composed of different types of silicon stacked on top of each other. Field-effect transistors are composed of one piece of silicon, with a different type of silicon added to the sides. Field-effect transistors are available in different types: Junction, MOS (Metal-Oxide Semiconductor), IS (Ion-Sensitive), DE (Depletion-Enhancement) MOSFET, and E (Enhancement-Only) MOSFET.

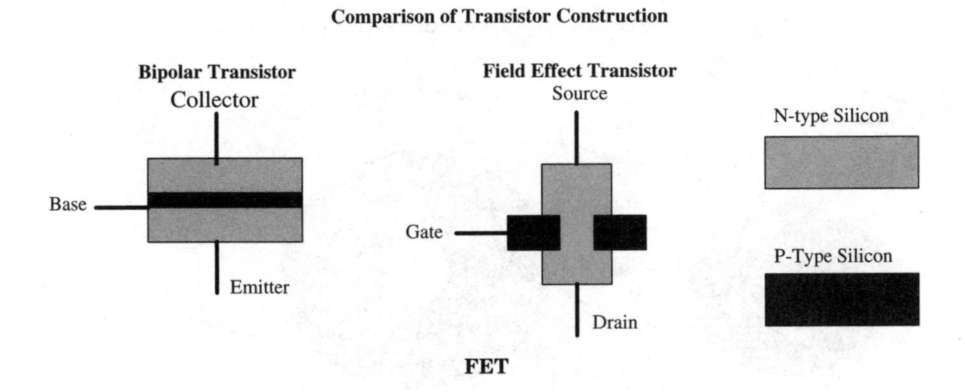

FET

FF ASCII control code abbreviation for form feed. Binary code is 1100000 Hex is C0.

FFDI (Fast Fiber Data Interface) PlusNet Phoenix AZ.

Fiber-Distributed Data Interface (FDDI) See *FDDI.*

Fiber Optic A thin strand of tiny layers of glass that have different refractive properties. The layers of material that have different refractive properties enable the thin strand to channel light through it by bending the rays of light. The light travels through the core of a fiber, and is bent back toward the core when it enters the cladding. Fiber-optic cable can be: multi mode or single mode. Multi-mode fiber optic has a larger core

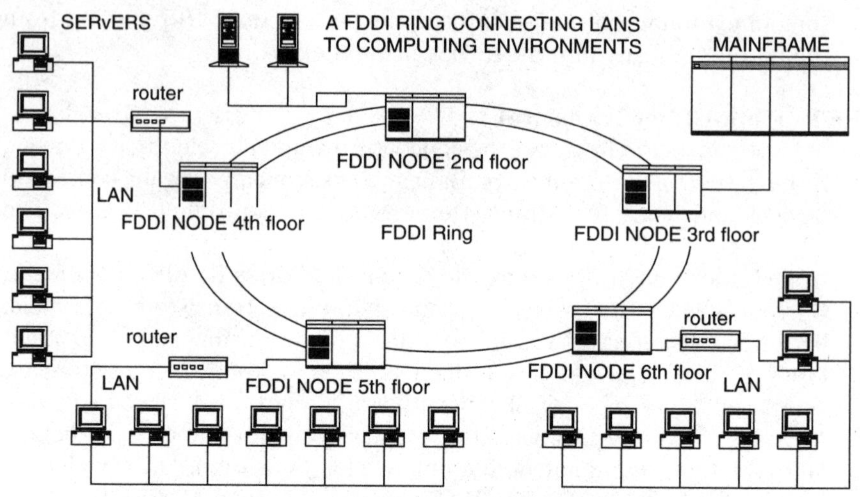

Fiber Distributed Data Interface

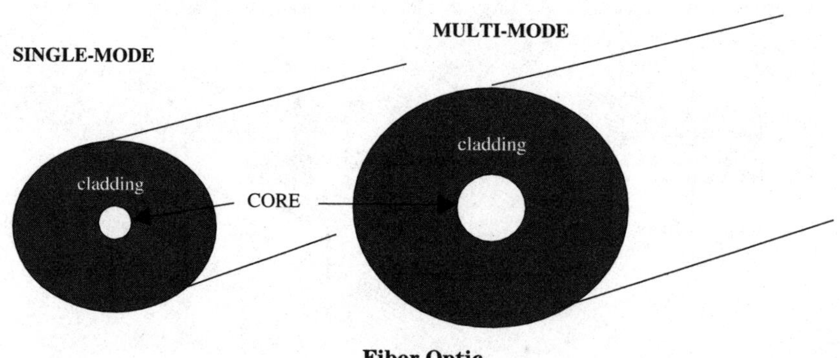

Fiber Optic

than single mode. Single mode is better for transmitting long distances, and multi mode is better for transmitting multiple colors of light (or sending more than one signal on a single fiber). Single-mode fiber is much more widely used in telecommunications than multi-mode cable. SONET is a fiber-optic based protocol standard that uses single-mode, graded-index fiber optic.

Fiber-Optic Attenuator A small device with two connectors, one on each side. A fiber-optic attenuator works like your sunglasses, it reduces the level of light passing through it, just as sunglasses reduces the level of light entering your eyes so that you can see more effectively.

Fiber-Optic Buffer The plastic coating on individual fibers. The color of the buffer colors distinguishes fibers from each other. The 12 different colors for buffers are shown in Appendix F.

Fiber-Optic Color Code See Appendix F.

Fiber-Optic Connector The three main different types of fiber-optic connectors are: SC, ST, and FT.

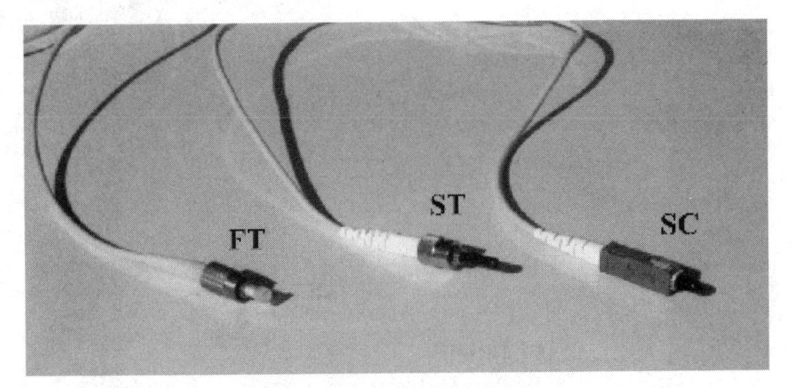

Fiber Optic Connectors

Fiber-Optic Distribution Panel Also called a *fiber-optic distribution bay*, this unit is a fiber-optic termination device and organizer. It also

Fiber Optic Distribution Panel Mfg by ADC SC Connector Type

houses fiber-optic splice trays, where the connector plugs (called *pigtails*) are spliced to the ends of fiber-optic cables.

Fiber Remote A solution made by Northern Telecom that extends IPE (intelligent peripheral equipment, Northern Telecom's name for PBX telephone station equipment) a distance of up to five miles.

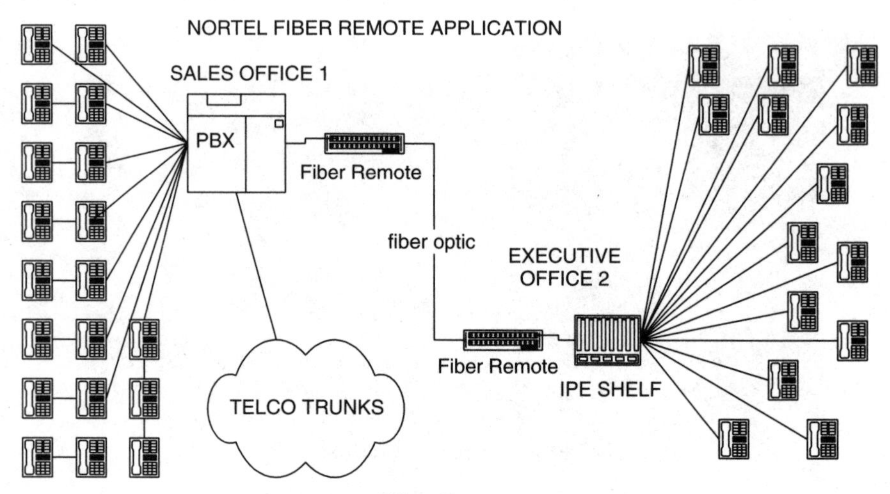

Fiber Remote

FIFO (First In First Out) This means that the first bit into a memory (for temporary storage) is the first out when it is retrieved. In some memory-handling schemes, the last bit in is the first out (LIFO). You can portray LIFO by imagining that you have a box. As you stack books into the box, and then remove them one by one, the first book you put in is the last one out. If you fill the box in the same manner and then turn it over and open the bottom to remove the books, you have a FIFO input and retrieval scheme.

Fifth Generation A reference to the Lucent 5E family of electronic telephone service switching systems, which is far more advanced and flexible than the 4E and previous switches.

Filament 1. A part of an electronic vacuum tube. The filament is often called a *heater*. The filament in a tube heats up and causes electrons to be emitted from the cathode. 2. The part of an incandescent light bulb that heats and lights. Filaments in light bulbs are often made of Tungsten. The light bulb no longer lights when the tungsten filament breaks.

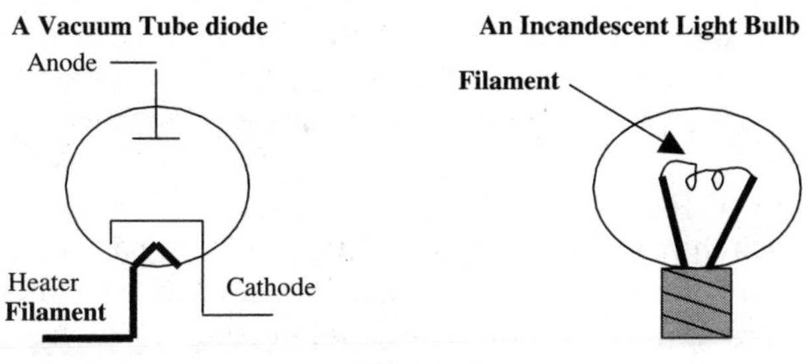

Filament

File Server A computer in a client-server environment that stores files or data. All the information regarding customer account transactions could go to a *file server*. When a PC on a network wants to run a report about customer accounts, the program will retrieve the data from the file server. The other type of server is an *application server*. This means that the program (application is another word for computer program) that runs the report is located on (or in) the server. The PC (client) requests the application server to calculate reports, then the server downloads the reports to the PC or a printer.

Fish Tape A tool for pulling wire through walls, conduit, or anywhere else that wire needs to go. It is coil of thin and stiff metal tape encased in plastic. It has an opening where the stiff wire tape can be pulled out and extended through a conduit (or other place) where the wire needs to be installed. The wire is then attached to the end of the fish tape so that it can be pulled through. As the person holding the fish tape is pulling the wire through the conduit, they re-coil it back into the plastic case. This tool is necessary for electricians and telecommunication wire installers.

FITC (Fiber To The Curb) Some day, telecommunications companies will have fiber-optic networks that run through cities and neighborhoods that terminate at pedestals that are now homes for twisted-pair telephone wire. From the terminal (which is usually located near a curb or in the backyard within a public easement) to the house or building will be twisted pair and/or coax, which will have the capability to provide data, voice, and cable-TV services. Having one company provide all these services on the same network is only available in special test sites in the United States.

FITH (Fiber To The House) After telecommunications companies establish *FITC (Fiber To The Curb)* networks, the next step will be fiber to the house. Fiber to the house will enable one telecommunications company to provide data, voice, and cable-TV services that are all integrated. You will be able to order a movie from your computer that will play on your

TV when you want it to start. You will probably even be able to pause the movie and start it again at your discretion. Data and Internet connections will be much faster, and, of course, you will still be able to make phone calls.

Flash A form of telecommunications signaling. To send a flash signal, press the switch-hook of a telephone briefly. If you have call waiting on your telephone line and another call comes in (you hear the beep), you briefly push in the switch hook on your telephone to switch to your other call. When you want to revert back to the original call, briefly press the switch hook on your phone again. Some telephones have a flash button on them, which is a more convenient and less cumbersome way to send a flash signal than flipping the telephone's switch hook.

Flexible Ringing Also called *distinctive ringing*. A feature of a PBX system that allows telephone sets to ring differently. This is a very nice feature if you would like outside calls to ring differently than internal calls from co-workers. This helps a user to know if they should say "hello" or "Emergency service, may I help you?" Distinctive ringing is also used in offices where many phones are close to each other. When one phone rings, all the people in the office know who's phone it is by tone, pattern, and pitch of the ring.

Flip Flop The individual devices that dynamic computer or data-device memories are made of. One flip-flop can store one bit of information. The RS-flip flop is the main component of a 555 timer (an electronics industry-standard circuit). A flip flop is a member of the logic component family. It is comprised of two transistors that turn on and off inversely to each other and four bias resistors. On the schematic symbol, *S* represents for set and *R* represents reset. Output Q is positive when a positive pulse finds its way to the S lead of the RS flip flop; the Q output stays positive and "remembers" the pulse. The Q output is reset to low when the R lead receives a positive pulse.

SCHEMATIC SYMBOL FOR AN RS-FLIP FLOP

Bit pulse input — S Q — Bit storage

Positive pulse reset — R $\overline{Q}$ — for 555 timer use (discharge)

Flip Flop

Floppy Disk A magnetic disk used for storing digital data. A floppy disk is literally a floppy plastic disk, coated with iron oxide (or another magnetic compound) inside of a plastic case. A 3.5" floppy disk is capable of storing up to 1.44 megabytes (1,440,000 bytes) of data.

Flush Jack A telephone or data-connection jack that is mounted in a wall in the same manner as an electrical outlet. Flush jacks can have as many as six connections/plugs on its face. AMP, Siecor, Leviton, Lucent, and Amphenol are all companies that manufacture a wide variety of flush jacks and other connectivity solutions.

Flux A material that is used in the center of solder rolls. Some solders have acid cores, which cleans the connection to be soldered. If the connection is dirty, then the solder will not adhere to the metal.

FM (Frequency Modulation) See *Frequency Modulation.*

FNC (Federal Networking Council) The council that coordinates the communications networking among Federal agencies, such as NASA and the Department of Defense.

FOD (Fax on Demand) A feature of voice-response systems that enables a caller to listen to a recorded message that gives them a selection of information that they can receive via a fax. After making a selection, the caller is then prompted by another recording to enter the number that they would like the information faxed to.

Football Another name for an aerial service wire splice. For a photo, see *Aerial Service Wire Splice.*

Footprint 1. The path that a satellite makes across the Earth's surface. If you drew a line of the satellite's path on the face of the Earth, that line would be its footprint. 2. A reference to a cabinet or the space a device takes up. If a sales person tells you that you can have a bunch of new features in the same footprint, then you don't have to change the cabinet, you might only need to change some software and a card or two.

Forced Account Code Billing A service offered by long-distance companies and a feature of PBX systems. The way the feature works is that every time an individual wants to make a long-distance call they must input (or dial) an account code (and sometimes a password) after they dial a 1 (1 is the first number dialed in long-distance calls). After they dial the account code and/or password, they hear a confirmation tone, then they continue by dialing the long-distance phone number. When the telephone bill comes, each call on each account code is itemized. This is a convenient way to keep track of who is making which long-distance phone calls, to where, and when. Then, you will have the ability to go to the person and ask why.

Foreign Exchange Service A telephone number that is served by another exchange. If you are on the south side of town, that area of town is probably serviced by a different central office exchange than the north side of town. Let's say that your business moves to the north side of town. The telephone company notifies you that you will need to change your phone number because you are moving to a different exchange area. (Each central office exchange has its own number plan for the first three of seven digits). You cannot change your phone number because you have paid a fortune to advertise it. You don't want your clients to be hassled by getting an intercept message saying that "The number you have dialed xxx-xxxx has been changed. The new number is xxx-xxxx". So, the telephone company offers you a service where your telephone lines are forwarded from your old office (south) to your new office. This service is called *foreign exchange service*. Your lines feed from a different (or foreign) central office than the office you are actually served by. Your customers then call you on the same number and never know that you moved, and they don't have to listen to a recording. You will probably only need to make your advertised number's foreign exchange (abbreviated *FX*) lines or trunks. The rest can be regular phone lines that are serviced from your exchange area. Soon, FX lines will be a thing of the past. Because of the Telecommunications Act of 1996, all phone numbers within an area code must be portable from one exchange or service provider to another. This is called *number portability*. This means that in the previous scenario, you could have had your numbers moved to any exchange within the area code and even changed phone companies without having to change your phone number.

FORTRAN (Formula Translation) A high-level computer programming language.

Forward A feature of PBX systems and a service offered by local, long-distance, and cellular telephone companies. The way that the Forward feature (also called *Call Forwarding*) works is if you know you are not going to be at your phone, you can make all your calls ring at a different number.

FPDL (Foreign Data Processor Link) Rockwell's version of a serial data interface from a PBX switch to a PC or other I/O equipment. Northern Telecom calls FPDLs *MSDLs (Meridian Serial Data Link)*. It is usually an additional card that plugs into a PBX system's common equipment-expansion area, which is equipped with an RS-232 communications port.

Fractional T1 A way to configure a T1 (DS1) service so that two or more of the DS0 circuits are joined together to make a larger data communications channel. If a full T1 is too big and expensive, but a 56K line is too small, then a fractional T1 might offer a good mid-range solution.

Fractional T3 A private line service where 4, 8, 12, or 16 T1 circuits (multiples of DS2) are combined together for a private data communications line.

FRAD (Frame Relay Access Device) A device located on the customer's premises that acts as the Network Interface for frame relay services.

Fragment An incomplete data frame.

Frame One complete sampling and conversion cycle of a multiplexed data transmission. A frame is sometimes confused with a packet. A packet is an envelope that contains data and an address that the data is sent to. A packet contains data to be transmitted, error-correcting information for the data in the packet, an address, timing information, and other bits of data, depending on the protocol that the packet was formed under. A frame is a momentary picture of a multiplexed data transmission, containing bits of data, or samples from each channel.

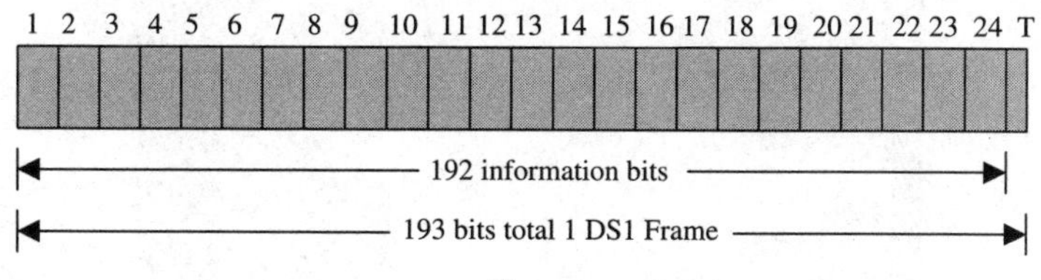

Frame

Frame-Check Sequence (FCS) A part of Cyclic Redundancy Checking error correction in data transmissions. See *Error Checking* for more information.

Frame Rate (FPS) Frames per second. The number of frames transferred in a second.

Frame Relay A service from telecommunication companies, as well as a protocol for transmitting data over MAN or WAN networks at rates as low as 64Kb/s to as high as 2.4 Gb/s. Frame relay is a packet-switching technology. As data is sent from a computer to a frame-relay interface, the frame relay equipment breaks the data stream down into manageable pieces. Then it assigns each piece of the data stream a reference number (e.g., 437 of 7435 packets), error-detection information, and gives it an address. The address is an identification number for the device that the packet is going to. When the packets are sent into the frame relay network, which is all over the world, each packet-switching center looks at the address and sends it to a center that has access to that address. This enables packets to take separate routes through the frame-relay network. Frame-relay services come with a *CIR (Committed Information Rate)*. CIR is the rate in Kb/s or Mb/s that a communications company will guarantee over a frame-relay circuit that they provide to their customer. If you purchase a frame-relay circuit, there is a place on the order agreement that you state the rate of information that you want to transmit. The choices are 16Kb/s, 32Kb/s, 48Kb/s, and 64Kb/s. If you enter 32Kb/s for your committed information rate, you will pay more for your service than the 16Kb/s CIR. You have the capability to transmit and are permit-

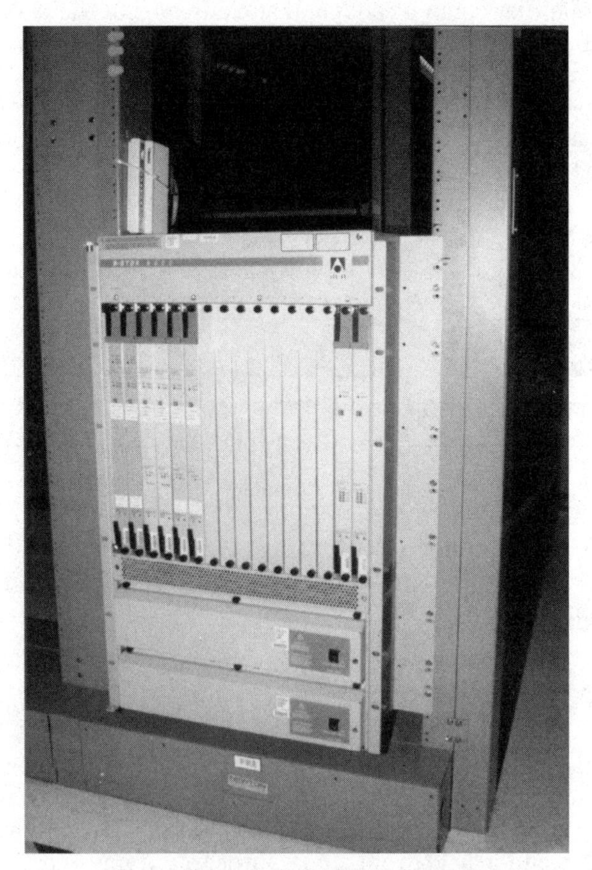

Frame Relay Switch Mfg by Ascend

ted to transmit at rates up to the full 64Kb/s, unless the frame-relay network becomes congested, then only your committed rate of 32Kb/s will get through. This is similar to purchasing the use of lanes on the freeway. If the freeway has very few cars travelling on it, then there is no sense in buying the right for multiple lanes because no one is using them anyway. However, if the freeway is congested, then you are getting your money's worth with your lane rights. You never know how congested a frame-relay network will be, or when and where bottlenecks will occur in the network.

Framing Bit A bit added to a multiplexed data stream that separates each complete cycle of sampling, DAC conversion through all the channels being multiplexed. A framing bit can also be called a *timing bit*. See *Frame* for more details.

Franchise A right (license) to provide telecommunications service to a community, county, or city. Telephone companies and cable TV companies must have a franchise agreement before they sell their services. Copies of franchise agreements are public documents and you can get one from your local PSC office.

Free-Space Communications A reference to radio communications, such as terrestrial microwave, satellite, and cellular.

Frequency *Frequency* is a measure of the number of cycles per second, and its unit is the hertz (Hz). One hertz is equal to one cycle in one second. Two Hz is two cycles in one second, 1000 Hz is 1000 cycles in one second *(CPS)*. Cycles are used as a reference to measure the frequency of a waveform or signal. Cycles are usually referred to as a number of cycles per unit of time. CPS and Hz are measurements of the number of cycles you get per second in an analog transmission. BPS is measurement of how many "square wave" clock sample sequences are being read from a digital transmission. For a diagram of one cycle in a waveform, see *Cycle*.

Frequency Band A range of frequencies that a certain class of radio communications operates within. For example, FM radio operates within 87.9 MHz and 108 MHz, which falls within the VHF frequency band. For a general list of frequency ranges, see *Band, Frequency*.

Frequency-Division Multiple Access Another name for frequency division multiplexing that the cellular telephone industry uses.

Frequency-Division Multiplexing (FDM) See *FDM*.

Frequency Hopping A method of radio transmission where the carrier frequency changes at fixed intervals.

Frequency Modulation (FM) See *FM*.

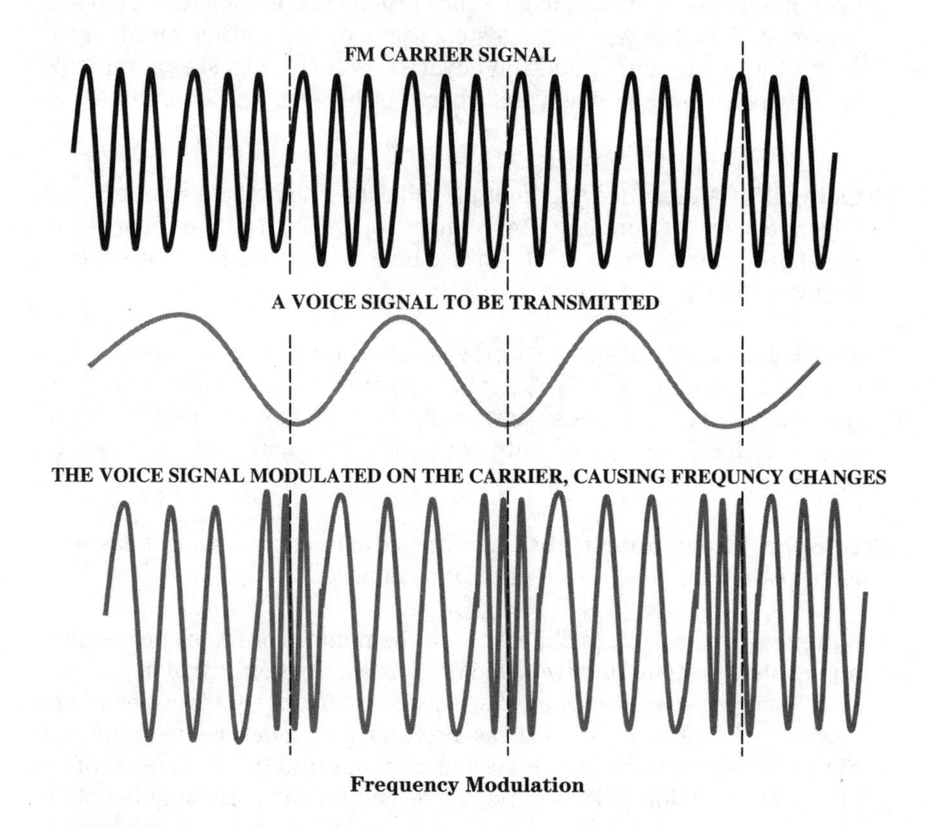

Frequency Modulation

Frequency Response A range of frequencies. An excellent home stereo system might have an output frequency response of 20 Hz to 20 kHz (the range of human hearing). A telephone line has a frequency response of 500 Hz to 3500 Hz (the range of human voice).

Frequency-Shift Keying (FSK) A method of binary signal modulation. FSK is the way that modems send bits over the telephone lines. Each bit is converted to a frequency. A 0 is represented by a slower frequency, and a 1 is represented by a higher frequency. Morse Code is a method of binary keying, where all signals have two states (binary), long and short. FSK is the same technology, only two different frequencies are used, instead of two lengths of beeps or lines.

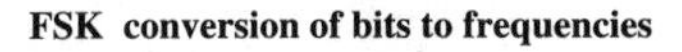

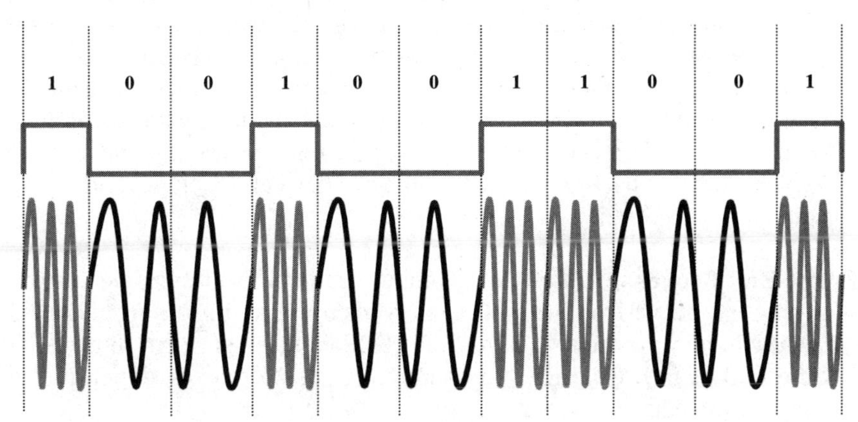

Frequency Shift Keying

Fresnel Loss A loss of signal in a fiber optic because a splice or a crack in the glass of the fiber. The splice or crack causes a type of reflection, called a fresnel reflection. This causes the light to scatter in different directions, rather than "focus" its way down the fiber.

Fresnel Reflection 1. A reflection of light because of a splice or a crack in a fiber optic. 2. A reflection of a terrestrial radio signal off of an object (such as a building, hill, or body of water) from its fresnel zone. Fresnel reflections can cause a reduction in signal strength or a total loss of the signal altogether.

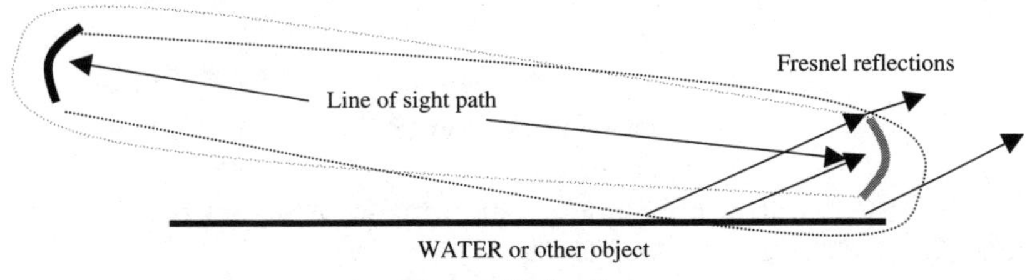

Fresnel Reflection

Fresnel Zone The area within a terrestrial microwave dispersion pattern, but outside of the line-of-sight path. The fresnel zone should be clear of any obstructions. Obstructions in the fresnel zone can cause fresnel reflections.

Friction Electricity Another name for *static electricity*. The opposite of static electricity is *dynamic electricity*, like that from a power outlet.

Static electricity consists of free electrons (electrons that are not a part of the valence shell of an element). The free electrons have no means to flow to a positively charged media, so they build up and form a charge or a voltage. The voltage is not measurable because when a measuring device is introduced to the static field of electrons, it provides a path for them to flow. When the electrons start flowing, they are no longer static. Typical static or frictional charges on a dry day can reach beyond 30,000 volts.

Front-End Processor (FEP) A communications "front end" device that can be loaded with a "firewall" to prevent unwanted users from accessing the communications network. An FEP can also perform routing and differentiate between different communications protocols, depending on the software that runs on it.

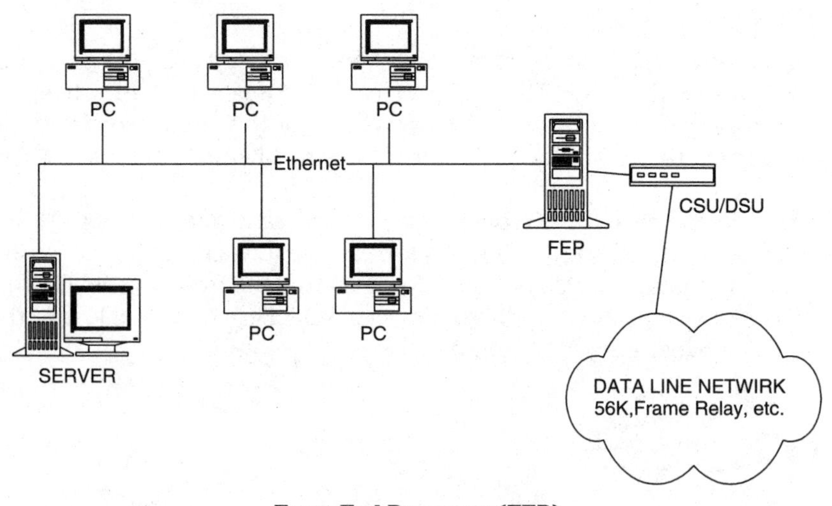

Front-End Processor (FEP)

FS The ASCII control code abbreviation for form separator. The binary code is 1100001 and the hex is C1.

FSK (Frequency-Shift Keying) See *Frequency-Shift Keying*.

FT Connector A metallic, screw-on fiber-optic connector. For a photo, see *Fiber-Optic Connector*.

FTP (File-Transfer Protocol) A type of packetized communications protocol. Many file-transfer protocols divide large files into smaller packets of data and send them to another device across a communications media. The idea of a FTP is that it is unlikely that a large file sent over a public network in one giant data stream will complete. If there are any errors, the transmission will have to start over. This is a waste of telecommunications dollars. FTP protocols break the larger file into smaller pieces, then gives each piece a reference (e.g., packet number 425 of 750) so that it can be identified from the rest when it is re-assembled on the far end. If a packet is corrupted, the FTP will request that the packet be retransmitted. It is much easier to retransmit a packet than an entire file.

FTTC (Fiber To The Curb, FITC) See *FITC*.

Full Duplex See *FDX*.

Functional Entity An entity, if it is physical or logical, that performs a task. An example of a functional entity is the data link layer of the OSI protocol. It is not necessarily a thing, it is just a phase in a process.

Functional Signaling Signaling of an ISDN line or other communication where the signaling of the circuit is performed in a manner that the user understands as well as the machines that make the communications work. An ISDN signal that is functional is a display reading that says "incoming call from John Doe," the ISDN could just give a ring or some other notification, but the call-signaling information bits in the ISDN circuit are actually decoded and passed on to the user.

Fuse A current-sensitive protection device. A fuse is designed into equipment so that if a component within that equipment should fail, it will blow the fuse. When a component fails, it draws excessive current. Excessive current causes excessive heat, which causes fire and ruins other components (not to mention danger to people). Always replace a fuse with the correct size. Replacing a blown fuse with one that has a higher amperage rating could damage the equipment.

Fuse Alarm Panel (FAP) A power-distribution panel that is installed at the top of a relay rack. All equipment in the rack is wired to the panel for

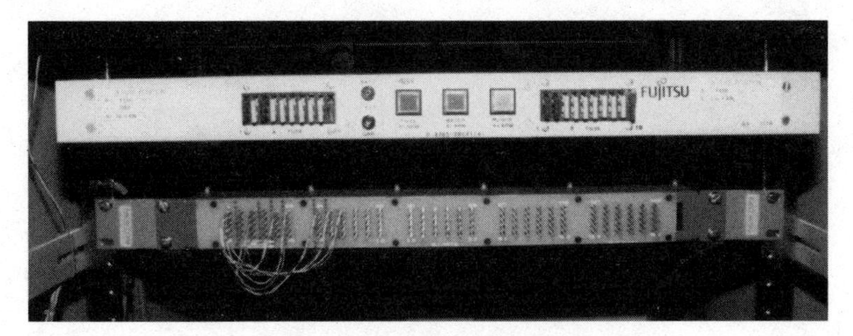

Fuse Alarm Panel (FAP) Mfg by Fujitsu

power. Each device has its own fuse within the panel to protect the rectifier from an over-current condition if the device fails or a wire is shorted. If any of the fuses blow, an alarm indicator is displayed.

Fusion Splicing A method of splicing fiber-optic cable. Fusion splices have less fresnel refraction when they are complete and, therefore, less loss than mechanical splices. The way a fusion splicer works is after the two fiber ends to be spliced are inserted into the splice housing, the splicer cleaves the ends, butts them together and thermally fuses the ends together. Fujikura is a popular manufacturer of fusion-splicing equipment.

FX Line See *Foreign Exchange Service.*

FXO (Foreign Exchange Office) A telephone company central office that is of another exchange, which means that it has a different numbering plan. See *Foreign Exchange Service.*

FX Trunk See *Foreign Exchange Service.*

G

Gaff What telecommunications and power company personnel wear to climb wooden telephone and power poles. The official name for these devices are *lineman's climbers*. They are also called *climbers*, *hooks*, and *spurs*. They consist of a steel shank that has straps on it so that it can be strapped to a person's leg. On the inside of the shank is a spike that is used to stab into the pole. For a photo, see *Climbers*.

Gallium Arsenide A new innovation in semiconductor technology. Gallium arsenide is more tolerant of heat than silicon, and it also has the special ability to convert light into electricity and vise versa.

GAN (Global-Area Network) A network classification. Other network classifications are: *LAN (local-area network)*, *MAN (metropolitan-area network)*, and *WAN (wide-area network)*. American Express utilizes a GAN to provide financial services to its customers worldwide.

Gas Carbon A device the telephone companies use for lightning protection. When telephones were first used, there was a big problem with lightning striking telephone wires and subsequently electrocuting people or burning homes down. Lightning protectors give lightning an easier path to ground compared to IW or a person. The early lighting protectors were made of carbon. When they were hit by lightning, they would short to ground, then the phone line would be out of service until a telephone technician came and replaced them. The new lightning protectors are made with a gas. When the lightning hits them, they temporarily short to ground, then re-enable the phone line automatically. This innovation greatly reduced the number of bad phone lines that a telephone company would have after a thunderstorm. Gas carbons have no carbon in them, they are just called that because the old lightning protectors were made of carbon. The new gas lightning protectors are the same shape and size as the old ones, so they can easily fit into older network interfaces. For photos of lightning protectors, see *Lightning Protector* and *2 Line Network Interface*.

Gate An electronic logic device. The three different functional types of logic gates are: the AND gate, the OR gate, and the inverter (NOT circuit). Gates are reactive devices. For a certain input, they react and produce an output. These small, simple circuits and other circuit devices (such as latches and flip-flops) make up all microprocessors and control devices. Thousands of logic gates and other components are used to make a single microprocessor or control chip.

Gate Array A circuit, usually on one microchip, that contains many gates. The gates are connected together to perform a function, such as decoding.

Gateway A gateway is a demarcation point for different networks. International telecommunications are done through gateway central offices. Gateway central offices (class 1 central offices) connect communications to other countries. The gateway does the translation from T1 to E1, T3 to E3, and vise-versa. A gateway can also be a translating device on a smaller network as well. A smaller gateway would connect two different LAN networks of different protocols (e.g., Token Ring to Ethernet).

Gateway Cities The United States has five International Telecommunications gateway cities. Any international call is routed through one of these cities. They are: New Orleans, New York, Washington DC, Miami, and San Francisco.

Gauge (AWG, American Wire Gauge) A measurement standard for copper wire. The gauge rating is the thickness of a solid copper wire. The larger the gauge, the smaller the wire (go figure). Most telephone wire ranges from 19 AWG to 26 AWG. Cat 3 is commonly 24 AWG. The electrical wire in your home is probably 12 AWG.

Gender Bender See *Gender Changer*.

Gender Changer A small device used to mate two plug ends of the same type.

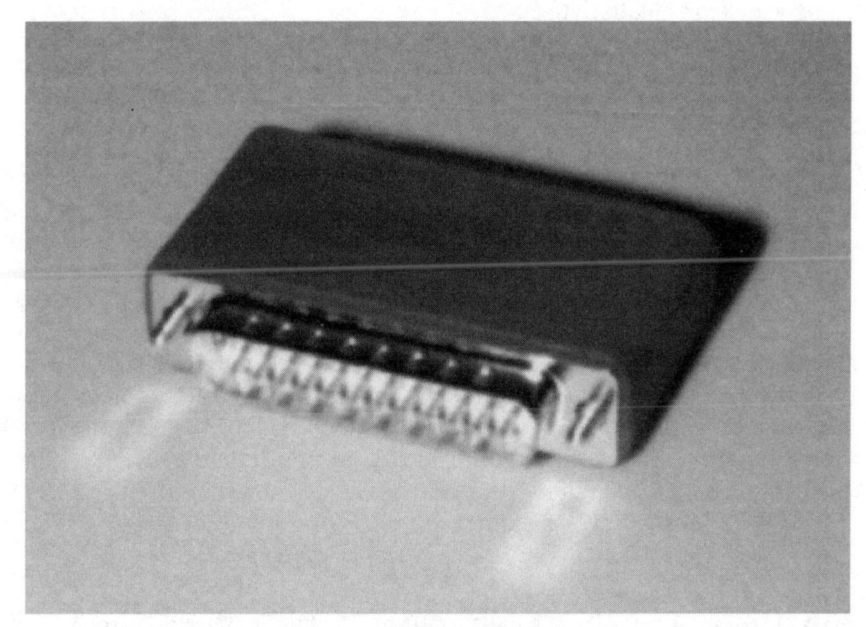

Gender Changer

Genderless Connector See *Data Connector*.

Gender Mender See *Gender Changer*.

GFCI (Ground Fault Circuit Interrupt) A type of circuit breaker. In newer homes, GFCI electrical outlets are required within a certain distance of a sink or bathtub. Unlike a normal circuit breaker that simply disconnects power when a current to AC common is exceeded (e.g., 15 A is common), a GFCI breaks the circuit when current to ground (instead of AC common) is detected. The regular common circuit breakers are located in the circuit breaker box in your home. The GFCI breakers are located inside of the electrical outlet itself.

GHz (Gigahertz) *Giga* means billion. 38 GHz means 38 billion Hz or 38,000,000,000 Hz.

Giga (G) Engineering notation for billion. One gigabyte is equal to 1 billion bytes. Hard disk-drive memories with gigabytes of capacity are now becoming very affordable.

Gigabit One billion bits.

Gigabyte One billion bytes.

GII (Global Information Infrastructure) Standards are still being set for GII by the ANSI at this writing.

Global-Positioning System (GPS) A system developed by the U.S. Department of Defense. The GPS system uses geostationary satellites to triangulate the position of a GPS receiver located on the face of the earth. Location is given in latitude, longitude, and altitude. The GPS receiver can even calculate bearing (direction) and speed. GPS receivers are available to the general public in many electronics magazines. You can even buy a GPS at your local Radio Shack. The receivers that are sold on the civilian market do not decode the dithering of the satellite signal and don't work indoors. *Dithering* is a random error that is introduced to the carrier frequency so that the GPS system would be slightly inaccurate. The deviance in location and altitude is anywhere from ±50 feet. This is so hostile entities cannot use the Department of Defense's location tool against the United States with known accuracy.

GNE (Gateway Network Element) An interface on SONET-node equipment that enables it to interface and exchange network data with other SONET nodes located within a network of SONET rings.

Goat Another name for a *craft test set*. For a photo and explanation, see *Craft Test Set*.

GPS (Global-Positioning System) See *Global-Positioning System*.

Grade-1 Cable A type of twisted-pair cable that is designed for PBX, telephone, RS-232, and other low-speed (1Mb/s or less) data applications.

Grade-2 Cable A type of twisted-pair cable that is designed for transmissions up to 4Mb/s. It is good for older IBM 3270 applications, IBM PC networks, and ISDN.

Grade-3 Cable A type of twisted-pair cable that is designed for older LAN networks like 10BASE-T and 802.5 Token Ring.

Grade-4 Cable A type of twisted-pair cable designed for Ethernet speeds up to 10 Mb/s. See *CAT 4* and *CAT 5*.

Grade-5 Cable A type of twisted-pair cable designed for speeds up to 100 Mb/s. Used on IBM Token-Ring networks. This is not CAT 5 cable. Grade 5 cable is two twisted pairs of stranded copper.

Graded-Index Fiber The core of graded-index fiber optic is made of many layers of glass, consisting of many refractive indexes that cause the light to gradually bend as it approaches the outside of the fiber. Graded-index fiber (like stepped-index fiber) is available in multi-mode or single mode, and it is the more expensive. The alternative to graded-index fiber is stepped-index fiber, which has a core made of glass, consisting of one refractive index.

LIGHT AS IT TRAVERSES THROUGH THE CORE OF A FIBER OPTIC

Stepped index fiber

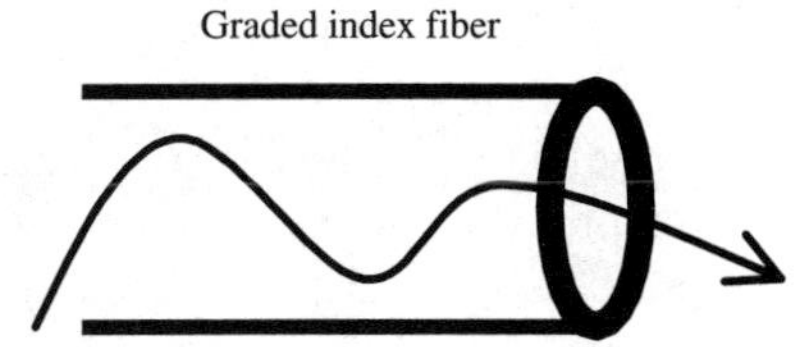

Graded Index Fiber

Graphic Equalizer To adjust the tone or sound of a circuit by diminishing or augmenting specific frequency bands. The tone control on a radio is a type of equalizer. A radio transmitter might have a tendency to amplify low-end signals, such as the sound of a bass guitar or drums better than high-end signals, such as the sound of a voice or cymbals. An equalizer can be used to reduce or increase the amplification of either end of the broadcast for an even and accurate reproduction of the input.

Gray, Elisha The possible true inventor of the telephone. Alexander Graham Bell beat him to the patent by a few hours.

Greenwich Mean Time (GMT) Also known as *zulu time*. This now-obsolete term was named after the very accurate clock standards in Greenwich, England. The clocks are incredibly accurate because of the cesium timing reference standard, which is the time-keeping element in "bits clocks" (timing devices used in central office nodes to synchronize SONET equipment. Greenwich Mean Time is now known as *Universal Time Coordinated (UTC)*.

Grid The part of an electronic vacuum tube that the input signal is fed to. The grid manipulates current flow between the filament and the plate (anode).

Ground Earth Ground. The electrical potential of the earth is 0 V. To maintain a good earth ground, a metallic rod four to six feet long is driven into the ground. Any wire connected to that rod is "grounded." The power com-

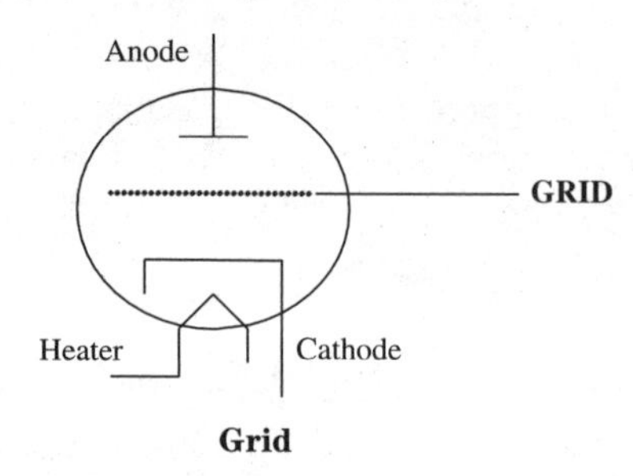

Grid

pany installs a rod like this when they connect power to your home. The telephone and cable-TV companies wire their network interfaces (lightning protection) to the power company's ground rod. The alternative to earth ground is a *floating ground*. A floating ground is simply a reference point that is not "earth grounded." The negative terminal of your car battery is a floating ground and any home appliance that has a two-prong electrical plug is also a floating ground (in newer homes that are wired correctly).

Ground Clamp A clamp or strap that is used to make a secure connection to a water pipe or grounding rod. The ground clamp then provides a way to connect a wire to earth ground.

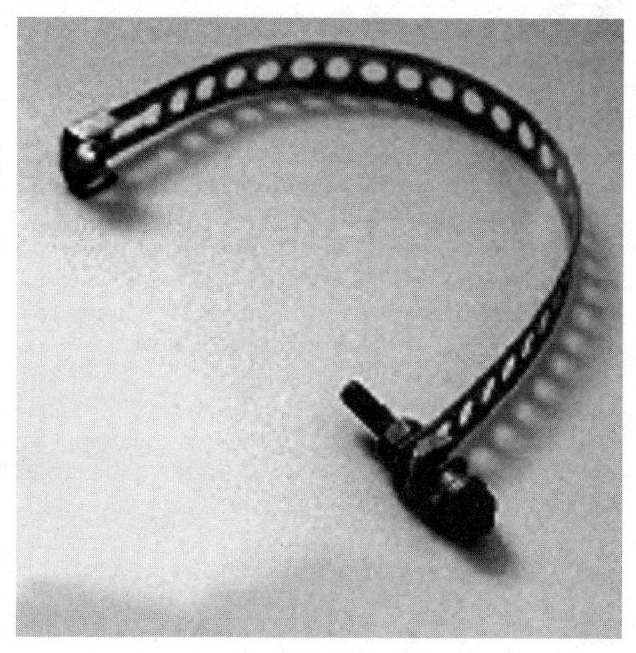

Grounded Clamp

Ground Fault A trouble on a telephone line that is caused by one of the twisted-pair phone wires coming into contact with electrical ground. A ground fault causes a hum on a telephone line. A ground fault can occur in many places on a telephone line. Ground faults are commonly caused by a portion of the phone line being wet or by a bad lightning protector.

Grounding Field An array of grounding rods placed in the ground and connected together around an antenna site or central-office site. The idea of a grounding field is to provide the best possible earth ground for electronic equipment.

Ground Start Trunk A phone line that uses a ground instead of a short (loop-start trunks use a short between tip and ring) to signal the central office for a dial tone. Some PBX telephone systems require the use of ground-start lines (trunks). Most newer PBX systems use loop-start (individual or on a T1) trunks or ISDN PRI trunks.

GS The ASCII control code abbreviation for group separator. The binary code is 1101001 and the hex is D1.

GS Trunk (Ground-Start Trunk) See *Ground-Start Trunks*.

GTE (General Telephone and Electronics) An independent telephone company (independent of the RBOCs). GTE owns and operates smaller local telephone companies across the United States.

GTP (General Telemetry Processor) A device manufactured and implemented to receive and process telecommunications equipment alarming protocols, such as *TBOS (Telemetry Bit-Oriented Serial)*.

Guard Band A frequency band separator between radio channels. A guard band prevents multiple or nearby radio stations on the dial from being received simultaneously.

Guest Mailbox A temporary voice mailbox set up in a voice-mail system. It can be attached to a phone or it can be a "virtual mailbox," where the user simply has a phone number that other people can call to leave messages.

Guy Hook A hook that is bolted to power/communications poles and used to attach guy wires.

Guy Thimble A device used to attach a guy wire strand to a bolt, which, in turn, attached to an anchor in the ground.

Guy Wire A steel cable that provides lateral support for a vertical structure, such as a telephone pole or radio tower/antenna.

Guy Hook

Guy Thimble

Half Duplex Two-way communications, one direction at a time. CB radio is an example of half-duplex operations; two people take turns transmitting and receiving. The two other types of transmissions are *full duplex* and *simplex*. A full-duplex line or communications path is able to communicate both directions, transmit and receive at the same time. A T1 is a full-duplex line, with one pair used for transmit and the other used for receive. Full duplex can be accomplished on one pair of wires by using two multiplexed channels, one for receive, and one for transmit. Simplex is one-way communication only. An FM radio station or TV broadcast is simplex.

Half-Wave Antenna An antenna that is one half the wavelength of the frequency that is designed to receive or transmit. For example, the wavelength of an antenna for a 38-GHz microwave signal is equal to:

$$\textit{Wavelength} = \frac{\textit{Speed of light} \text{ (in meters/second)}}{\textit{Frequency} \text{ (in CPS)}}$$

$$= \frac{300{,}000{,}000}{38{,}000{,}000{,}000} = 0.0079 \text{ meters}$$

$$.0079 \text{ Meters} = 7.9 \text{ mm.}$$

A full-wave antenna for 38 GHz would be 7.9 mm. A half wave would be half of that, 3.9 mm. It's a small antenna because it is for a small wavelength. That's why they call it *microwave*.

Hand Hole A small cable vault that is basically big enough to get your hands into. It is used for outside cable plant splices. Hand holes are most common in the fiber-optic realm of outside plant. A typical hand hole is about two feet wide, three or four feet long, and two feet deep.

Hand Hole

Hand Off To connect a phone call or other service from one telephone company to another. Hand-offs usually happen in a place called a *co-location*.

Handset The device attached to a telephone that you hold to your head during a telephone call. It has a speaker and a microphone.

Hands Free Also called *speaker phone*. Hands free is a feature of PBX telephones and 2500 (single-line standard) telephones. It allows the user to talk on the telephone as if it were an intercom, not using a handset. Speaker phones have a microphone and a speaker built into the telephone set itself.

Hand Shake The initial connection set-up part of a protocol for modems. When you dial out on a modem, often you can hear a beep after the modem on the far end picks-up. This is a *handshaking signal.* After the handshaking signal the modems go through the handshaking process, which is to exchange information specific to what speed they will transmit and what error-detection protocol will be used. After the handshake is complete, data transmission begins.

H PAD (Host Packet Assembler/Dissembler) A PAD that is specifically located on the host end of a communications link. Even though PADs are the same on each end, sometimes technicians refer to them in a specific manner. The H PAD is a device that is located at the host end of a virtual communications link in a frame protocol environment that reassembles and disassembles large files of data. The HPAD also adds and removes address, envelope, and HDLC information.

Hardware Physical electronic equipment. Most electronic equipment needs software, which are the programs and instructions that tell it how to run.

Hardwire To be physically connected by wire, either with cross connects or customized cables. The alternative to hardwiring equipment is to use modular equipment, such as a modular jack. A modular jack is equipped with a plug so that devices can be easily attached and detached. Old jacks, which can still be found in old homes, are *hard wired*, which means that the telephone cord had to be permanently affixed to the terminals inside the jack with screws. The same went for nonmodular or hard-wired telephones. If you wanted to have a longer cord, you couldn't buy one at the store and just plug it in. You had to call the phone company and they would send a telephone technician out to install a longer cord for you.

Harmonic A frequency that is a multiple of a lower frequency. For example, 6000 Hz is a harmonic of 3000 Hz. 12,000 Hz and 15,000 Hz are also harmonics of 3000 Hz. Multiply 3000 by any positive integer and that is a harmonic frequency.

Harmonic Distortion The tendency of a circuit to amplify and pass harmonics of an input signal. Feedback is an example of harmonic distortion. Feedback is the squealing sound you often hear when a person approaches a microphone at a public speech. See also *Feedback*.

Harmonica Adapter An adapter that converts a 25-pair cable plug into 12 four-conductor RJ11 plugs or 24 two-conductor RJ11 plugs. Harmonica adapters are frequently used as an alternative connectivity (as opposed to hardwired 66M150 blocks) on temporary (and sometimes an inexpensive permanent) installations of key or PBX telephone systems. For a photo, see *Modular Adapter* and *258A Adapter*.

HDLC (High-Level Data-Link Control) A framing format for packet and frame-relay protocols.

HDSL (High Bit-Rate Digital Subscriber Line) A type of T1. An HDSL T1 can be transmitted up to 20,000 feet on twisted-pair copper wire before it needs to be repeated/regenerated. The smaller the wire (the larger the gauge), the better HDSL works. 24 gauge is optimal. 19 gauge reduces the distance that HDSL can transmit because of the increased inductance and capacitance of the larger wire.

HDTV (High-Definition Television) HDTV was approved by the FCC in 1997. It will take a while to implement because any TV station or CATV company that wants to broadcast HDTV will have to install HDTV transmission equipment, which is different from the broadcast-standard equipment currently in use. People that want to view HDTV will have to buy a new TV that is of the HDTV standard. If you would like to record programs on HDTV, you will need to get a new VCR too. When HDTV equipment goes on sale to the public, the prices of HDTV sets is expected to start at about $1300.00.

HDX (Half Duplex) Half duplex is two-way communications, one direction at a time. CB radio is an example of half-duplex operations; two people take turns transmitting and receiving. The two other types of transmissions are full duplex and simplex. A full-duplex line or communications path is able to communicate both directions, transmit and receive at the same time. A T1 is a full-duplex line, with one pair used for transmit and the other used for receive. Full duplex can be accomplished on one pair of wires by using two multiplexed channels, one for receive and one for transmit. Simplex is one-way communication only. An FM radio station or TV broadcast is simplex.

Head End Where cable-TV signal processing takes place. Located at the head end is the array of satellite dishes that the cable-TV company uses to pick-up their programming transmissions. They re-channelize the stations (e.g., put WGN on channel 8 and HBO on channel 14), add in local commercials and local programming. When all the signal processing is done, the cable-TV company then sends the broadband (a multi-channel signal) TV signal down its coax cables so that it can be distributed to subscribers.

Held Order A telephone or cable-TV service that cannot be installed because of a shortage of equipment. The shortage could be because the equipment is not installed yet (such as telephone cables in a new neighborhood) or the available facilities have run-out (i.e., all the pairs in a particular area are used, or the central-office switch has utilized all of its line interfaces).

Henry The unit of inductance. Inductance is also referred to as reactance. An inductor, or coil of wire, is a reactive device. Reactance is the resistance that a component gives to an AC or fluctuating DC current. The two components that cause reactance: inductors (coils) and capacitors. The difference between resistance and reactance is that resistance is always the same, regardless of the voltage amplitude or frequency applied to the resistive device. The reactance of a component changes along with frequency changes, or the speed at which an AC current changes direction. The higher the frequency, the higher the reactance or resistance to that frequency. The reason that coils of wire cause reactance is that as electricity flows through them, they force the electricity to create a magnetic field every time it changes direction. A perfect inductor has zero reactance to a DC current, and has a specific reactance or resistance to every AC current. Each coil or inductor has a value in henrys. The higher the number of henrys, the more it will resist AC or fluctuating DC. Coils are used to filter out ("choke" out) DC fluctuations in power supplies. They are also used to help tune in radio or other frequencies.

Reactance is also caused by other electronic conditions where it is not useful. All wire and electronic components possess a small amount of reactive properties (e.g., the reason twisted-pair wire causes attenuation of signal strength is because of the inductance of the copper wire and the capacitance of the two wires being next to each other.

Hermaphroditic Connector A genderless connector developed by IBM that is usually called a "Data Connector". The Data Connector does not need complementary plugs (male and female) to make a connection, like all other known communications modular connecting systems. The Data Connector is specifically designed and used for switched token ring backbone applications.

Hertz (Hz) A measure of the number of cycles per second in a waveform. One Hz, or Hertz, is equal to one cycle in one second. Two Hz is two cycles in one second, 1000 Hz is 1000 cycles in one second, or Cycles per Second (CPS). Cycles are used as a reference to measure the frequency of a waveform or signal. Below is a diagram of two waveforms and one cycle of each singled out. Cycles are usually referred to as a number of cycles per unit of time. Cycles per second and Hertz are measurements of the number of cycles you get per second in an analog transmission. Bits per second is measurement of how many "square wave" clock sample sequences are being read from a digital transmission. For a diagram of one cycle (hertz), see *Cycle*.

Heterodyne To mix a radio-frequency signal with an audio or other signal to be carried in a transmission.

Hexadecimal A number system based on 16 numbers, instead of 10, like the one we count with in our every day lives. Hexadecimal is a convenient shortcut to inputting 16-/ bit binary numbers during machine-language programming. Hexadecimal counts as follows: 0, 1, 2, 3, 4, 5, 6, 7, 8, 9, A, B, C, D, E, and F. For a conversion table of hexadecimal, binary, and decimal numbers, see *Binary-to-Decimal Conversion*.

Hexadecimal-to-Binary Conversion For a conversion table for binary (base 2), hexadecimal, and decimal (base 10) numbers, see *Binary-to-Decimal Conversion*.

Hexadecimal-to-Decimal Conversion For a conversion table for binary (base 2), hexadecimal, and decimal (base 10) numbers see *Binary-to-Decimal Conversion*.

HF (Hands Free) Also called speaker phone. Hands free is a feature of PBX telephones and 2500 (single-line standard) telephones. It allows the user to talk on the telephone as if it were an intercom (not using a handset). Speaker phones have a microphone and a speaker built into the telephone set itself.

HFU (Hands-Free Unit) In NEC Dterm (Digital Terminal) PBX phones, the telephone does not have speaker phone ability until a daughter board, called an *HFU*, is installed inside it.

Hi-Fi (High Fidelity) An attempt a reproduction of an audio signal that is as close as possible to the original using the most advanced technology possible within a price range.

High and Dry A test result of direct-access test units attached to central office switching equipment that means a twisted pair is clear of shorts, grounds, and equipment, including telephones and sometimes load coils (depending on how the DATU is configured.

High-Definition TV (HDTV) See *HDTV*.

High Fidelity (Hi-Fi) See *Hi-Fi*.

High-Level Data-Link Control (HDLC) A framing format for packet and frame relay protocols.

High-Level Language A computer programming language that interfaces meaningful instructions (to humans) to lower-level programming languages, such as machine language. FORTRAN, COBOL, C, BASIC, and SAS are high-level programming languages.

High-Pass Filter An electronic device that eliminates frequencies below a specified frequency. The two categories of frequency filters are: active and passive. Active filters use active devices that require power, such as transistors and op amps to amplify the desired signal and attenuate the undesired signal. Passive filters are made with components that do not require external power, such as capacitors and inductors. Capacitors and inductors have reactive properties that cause them to resist or pass an AC signal.

HIVR (Host Interactive Voice Response) A telecommunications and data-processing technology that interfaces a person to information held in a computer by using a phone line. If you have ever called your bank and entered your account number, a password, and a prompt so that a computerized voice can read back your bank account balance, then you have used HIVR.

Hold Recall A feature of PBX telephone systems and key systems that makes calls put on hold ring back to the person who put them on hold. The hold-recall time interval can be set to a time specified by the system administrator (e.g., 15, 30, 45, or 60 seconds).

Hollow Pipeline A term that is usually used to describe a private line out-of-band signaled (CCC, Clear Coded Channel) DS1. There is no timing, no framing, no error correction. You input your bit stream on one end and out they come out the other end in the same order. There is only a maximum speed you can transmit. For a DS1, it is 1.536Mb/s. This is 1.544 Mb/s less the framing overhead of 8 Kb/s.

Home Run A telephone or data communications wiring scheme that means the wire that is installed runs from a jack to a point where it can be cross-connected or terminated to DCE equipment or a telephone NI. Each station (PC or telephone must have its own dedicated wire. CAT 5 wiring and PBX wiring must be installed in a home run manner. Token ring is installed station to station.

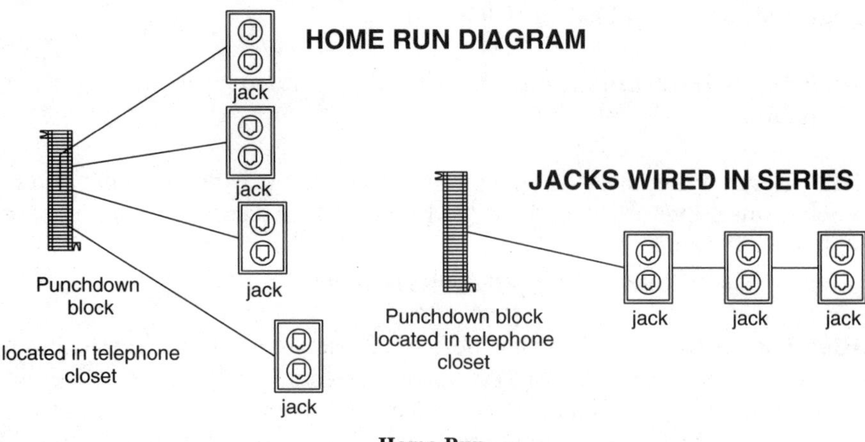

Home Run

Hook Flash Hook flash or "Flash" is a form of telecommunications signaling. To send a flash signal, press the switch-hook of a telephone briefly. If you have call waiting on your telephone line and another call comes in (you hear the beep), you briefly push in the switch hook on your telephone to switch to your other call. When you want to revert back to the original call, you briefly press the switch hook on your phone again. Some telephones have a flash button on them, which is a more convenient and less cumbersome way to send a flash signal than flipping the telephone's switch hook.

Hook Switch (Switch Hook) The switch that is pressed when you hang up a telephone handset.

Hooks Also known as *climbers*, *lineman's climbers*, *spurs*, and *gaffs*. Climbers for use on power poles have a shorter blade on the shank than climbers made for tree climbing (lumber/trimming applications). For a photo, see *Climbers*.

Hop Count The number of devices, such as bridges, routers, and hubs between a communicating device and its destination.

Horizontal Blanking As the beam inside a picture tube or monitor scans across the screen, it must be turned off while it retraces back to the starting point of the next line that it is going to "paint." The turning off of the beam is called *blanking*.

Horizontal Output The power amplifier that amplifies the horizontal output sync signal in a TV or monitor. The output is fed into a deflection yolk, which creates the magnetic fields that control tracing of the CRT

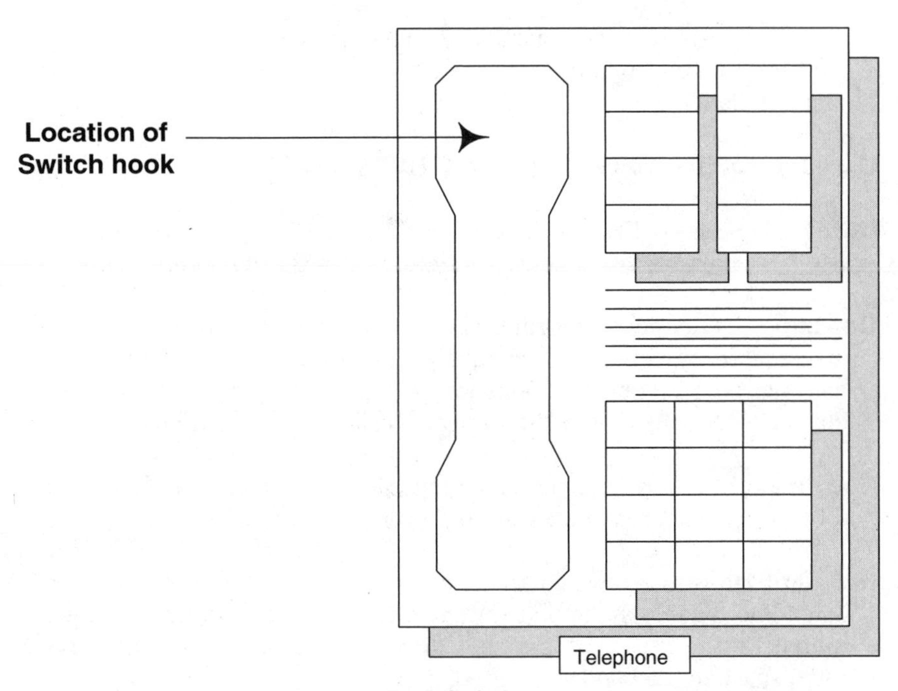

Hook Switch

beam in a sideways direction. There is also a vertical output amplifier, which does the same for the up and down tracing of the CRT beam. The horizontal output frequency for a standard TV is 15.73425 kHz. When you turn on some TVs, you can faintly hear the very high pitched "dog whistle" sound of the horizontal output circuitry.

Horizontal Polarization The pointing of a microwave dish antenna so that the transmission dispersion is in a sideways, or horizontal pattern. The headlights on cars are polarized in a horizontal manner, so the light dispersion is spread across the horizontal surface of the road. The other kind of polarization is vertical, where the transmission dispersion is in an up-and-down pattern. The two antennas or dishes employed in a point to point application need to be polarized the same way.

Horsepower In electronics, it is sometimes useful to compare wattage power to horsepower to get an idea of what a watt of power really is capable of. One horsepower is equal to 746 W. Some electric motors are stamped with horsepower rating.

Host Computer Usually a reference to a mainframe computer (although many hosts are actually servers) that controls the storage and retrieval of data in extremely large databases. Host computers can simply retrieve data for another computer (PC or Server) or do calculations and summary on data itself.

Host Interactive Voice Response (HIVR) See *HIVR*.

Hotel Console A PBX console specially designed for hotel front-desk use. Consoles are available in two types: business and hotel.

Hot Line A telephone that rings another phone with no dialing. Hot lines are created by using devices called *hot-shot dialers*, which automatically dial a phone number. Telephones connected to hot-shot dialers usually have no dial pad. Hot lines are also called *ring-down circuits*.

Hot Pluggable A reference to a circuit card which can be installed or removed without turning off system power.

Hot-Shot Dialer A device used to create a hot line or ring-down circuit. A hot line is a telephone that rings another telephone with no dialing required by the user. The hot-shot dialer automatically dials the phone number when the handset is lifted.

HSDL (High-Speed Digital Subscriber Line) An older term for ASDL. This is a service in the making to provide video to the home over twisted-pair telephone lines. Its current line format is T1 AMI, 16Kb/s to the CO, (for control to change the channel) and 1.528Mb/s to your TV. The twisted pairs are incorporated with adaptive digital filtering to help correct attenuation and noise. The format will probably change before it is offered widely. Rumors say that a 6Mb/s bandwidth is the future of ADSL.

HT The ASCII control code abbreviation for horizontal tab. The binary code is 1001000 and the hex is 90.

Hub A device used to broadcast data information over many lines. Simply stated, a hub makes a star wiring configuration look like a bus configuration to all the devices connected to it. Hubs are utilized in ethernet networks.

Hunt A telephone line feature. If you have several phone lines that are answered by a group of people, the telephone company can make those several phone lines work together. If a call comes in to a line that is associated with other phone lines in a hunt group, the call will rotate from

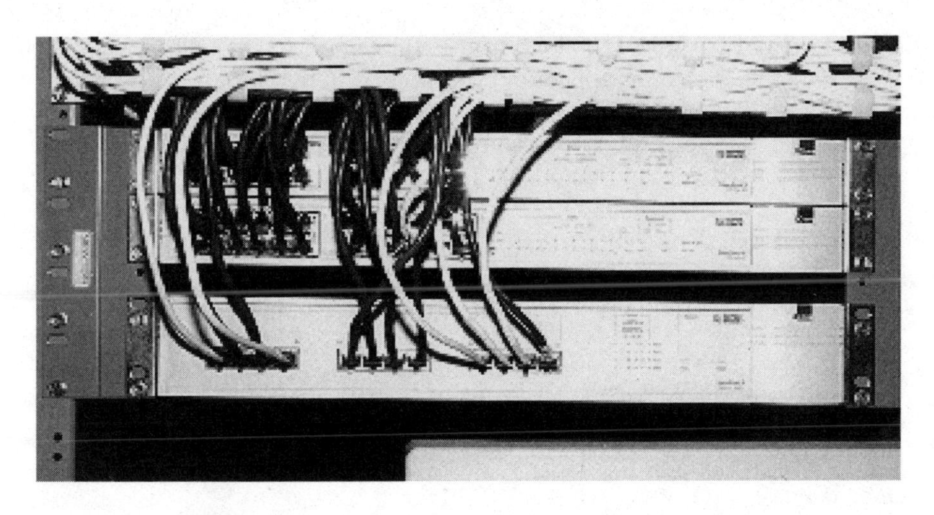

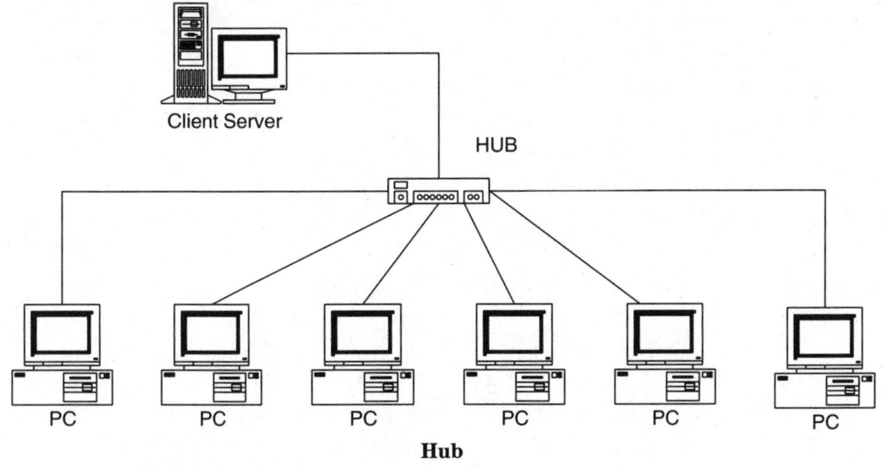

Hub

line to line until it finds a line that is not busy. If all the lines are busy, then the caller will get a busy signal.

Hunt Group A number of telephone lines that are associated together by the telephone company central office or a PBX system. When a call comes in to a hunt group, it cycles through the group of lines until it finds one that is not busy, then it rings that phone (or extension, if it's a PBX system).

Hybrid Cable A communications cable that consists of two different types of media. A cable that contains twisted-pair copper and coax, or twisted-pair and fiber optic, etc., would be a hybrid cable.

Hybrid Key System A Telephone switching system that enables the user to choose which lines appear or don't appear under specific keys of electronic telephones. This is typical of a PBX system, except in the case of a hybrid key, you still have to select a line to dial out. When using a PBX system, the line is automatically selected from a pool when you dial 9.

Hysteresis 1. A phenomenon in transformer cores or other inductive devices where an electrical current is formed in the metallic core of the device. It causes the core to heat up and increases the inductive value of the device. 2. Hysteresis is also a technique to reduce noise in digital circuitry. In this case, a comparator is designed to receive a feedback from its output. This gives it a better reference to switch from. A comparator with a hysteresis loop is better known as a *Schmitt trigger*.

Hz (Hertz) See *Hertz*.

I

I&M Abbreviation for *installation and maintenance*.

I&R Abbreviation for *installation and repair*.

I/O (Input/Output) A reference to peripheral equipment, such as a keyboard, modem, or monitor.

IBDN Northern Telecom's name for the CAT 5 horizontal wiring standard.

IC (Integrated Circuit) Another name for microchip. An *LSI (Large-scale Integration)* or *VLSI (Very Large-Scale Integration)* device that often comes in a DIP (dual in-line) package. IC circuits are designed and manufactured to do general functions or for specific applications. Some ICs are made for amplifying audio signals, some are made for storing binary memory (i.e., RAM), etc.

ICE Age (Information Communications and Entertainment Age) What the RBOCs call today's trend in telecommunications services. The striving for the ability to provide customers data connectivity, cable-TV service, and voice service on one line, with one company, with one bill at the end of the month.

IDE (Integrated Drive Electronics) An interface standard for PC disk drives. IDE is special, no additional controller card is needed to operate the disk drive—it plugs directly into the motherboard. IDE is now obsolete because of its limit to a 528MB hard disk and a medial 3MB/s transfer rate. IDE has been updated to Enhanced IDE, which has the ability to transfer up to 13Mb/s and operate a disk drive storage up to 8.4 GB. Other hard disk drive interface standards for PC's are ESDI and SCSI.

IDN (Integrated Digital Network) A term that refers to a network that has operability between many different devices through a standard digital protocol.

IEC (Inter Exchange Carrier, IXC) IEC is a long-distance company, like AT&T, Sprint, Worldcom, or MCI.

IEEE (Institute of Electrical and Electronic Engineers) A standards organization that develops industry standards for electrical and electronics applications.

IF (Intermediate Frequency) An amplifier stage in radio/TV receivers that is designed to amplify and isolate a large range of frequencies. The IF amplifier receives the specific frequency to amplify from the detector stage of the tuner. The tuner is the device or circuit with which the user of the receiver selects the desired channel.

IISP (Information Infrastructure Standards Panel) A charter of the ANSI to develop *GII (Global Information Infrastructure)* standards.

Impedance A term used to replace resistance in AC or signal-serving circuits. For example, an 8-ohm speaker is a measure of impedance (resistance to a signal being fed into it). If you measure the DC resistance of the speaker with an ohmmeter, it is about 1 ohm. This is because impedance is a measure of resistance to an AC signal, not a DC signal (ohmmeters use a DC battery to measure resistance with). If you put an impedance meter on the speaker, it will measure the resistance of the speaker with an AC signal, which is resisted by reactive components, such as coils (a speaker is a coil of wire electrically) and capacitors. The impedance meter will read an impedance of 8 ohms.

In-Band Signaling In telephone circuits (DS1 to be specific) signals can be sent in two different ways: in band and out of band. Signals are digits that you dial, dial tone, the phone being off hook, ringing, etc. An in-band telephone line is like the one in your home; the digits that you dial and the ringing are carried within the channel you talk on. Out-of-band signaling is a method that telephone companies and businesses use for larger PBX applications and data-transfer applications. An out-of-band, signaled DS1 has 24 multiplexed channels. The 24th channel carries the signaling for the other 23 channels or phone lines. The advantage of out-of-band signaling is that each channel has an increased capacity to carry data (8Kb/s more) and the 23 channels are not used to find out if a line is busy (both directions, in and out). The off-hook sensing is and busy signaling is done in the 24th channel. If you have a system that gets thousands of calls per day, this can reduce traffic.

Incoherent Light Light that consists of many frequencies and wavelengths. A light bulb emits noncoherent light. Coherent light is light that consists of only one frequency or very close to one frequency. Coherent light looks to the human eye as a very pure color. Lasers and LEDs (like the one that lights when the hard drive in your computer is running) emit light that is very close to being coherent.

Index Of Refraction A reference to how much light bends when it travels through a specific substance. When light travels through a swimming pool it refracts and makes everything appear wavy and distorted. When fiber optic is designed, it is planned to have different types of glass that have different refractive indexes. As the light travels from the core of the fiber toward the outer edge (cladding), it is bent back inward because of the increasing indexes of refraction of the glass it is passing through.

Inductance A physical characteristic of conductors, semiconductors, and other electronic components. Inductance is the formation of an electromagnetic field around a device as electricity flows through it. Inductors are coils of wire that add this effect with each winding of the coil. Inductance is measured in henries (H). An inductor, or coil of wire, is a reactive device. *Reactance* is the resistance that a component gives to an AC or fluctuating DC current. The two components that cause reactance are inductors (coils) and capacitors. The difference between resistance and reactance is that resistance is always the same, regardless of the voltage amplitude or frequency applied to the resistive device. The reactance of a component changes along with frequency changes, or the speed at which an AC current changes direction. The higher the frequency applied to an inductor, the higher the reactance or resistance to that frequency. Coils of wire cause reactance because as electricity flows through them, they force the electricity to create a magnetic field every time it changes direction. A perfect inductor has zero reactance to a DC current and has a specific reactance or resistance to every frequency of AC current. Each coil or inductor has a value in henries. The higher the number of henries, the more it will resist AC or fluctuating DC. Coils are used to filter out ("choke out") DC fluctuations in power supplies. They are also used to help tune in radio or other frequencies.

Reactance is also caused by other electronic conditions where it is not useful. All wire and electronic components possess a small amount of reactive properties (i.e., the reason that twisted-pair wire causes attenuation of signal strength is because of the inductance of the copper wire and the capacitance of the two adjacent wires.

Inductive A reference to the electromagnetic (as opposed to the electrostatic of capacitors) reactive properties of a device. See also *Inductance*.

Inductive Coupling The use of an inductor (coil of wire) to connect one amplifier or circuit stage to another. Inductive coupling is advantageous if low frequencies are the crucial part of the composite signal to be passed. High-end audio/home-stereo equipment is designed with inductive coupling.

Inductive Pick Up A microphone or sensing device that uses the changes in a magnetic field or creation of a magnetic field within a component because of the vibrations in the air or of a nearby object.

Inductive Tap A tap on a telephone line that does not connect to the pair of wires. Instead, it picks up the tiny magnetic field created around the pair and amplifies it, the same way that a radio transmitter receives electromagnetic waves from the air. An inductive tap must be within an inch or two of the pair to successfully receive an electromagnetic signal.

INE (Intelligent Network Element) A network element, such as a router, node or hub, that has the ability to be electronically reconfigured (manually or remotely) or perform additional functions, such as protocol conversions.

Information Technology The study of improving information and data processing with the use of newer and better devices/machines.

Infrared (IR) Light waves that humans cannot see. Their wavelength (or frequency) is just below that of red light. Heat radiates infrared light-wave radiation.

INIC ISDN Network Identification Code.

Inner-duct A flexible plastic conduit that is placed within larger conduits. Inner-duct is used where multiple communications companies use or lease conduit space within the same conduit.

Reels of Innerduct

Inside Dial Tone The dial tone provided by a PBX system. When you pick-up the handset of an electronic telephone that is served by a PBX, you get an inside dial tone, which allows you to dial an extension that begins with a number other than 9 (internal extensions never begin with a 9). When you dial 9, you get an outside dial tone or a dial tone that is served by a telephone company central office.

Inside Plant Electronic equipment located inside buildings, including central-office switches, PBX switches, broadband transmission equipment, distribution frame, -power supply/rectification equipment, and anything else you can find inside a central office. *Inside plant* does not include telephone poles, cable, terminals, cross boxes and cable vaults, or anything else you might find outdoors.

Inside Wiring (IW) The telephone wire that is on the customer side of the Telephone Network Interface. IW includes jacks and any wiring in or attached to the outside of the house, as long as it is electrically on the customer side of the network interface.

Installer's Tone Also called a test tone. A small box that runs on batteries and is used to put an *RF (radio frequency)* tone on a pair of wires. If

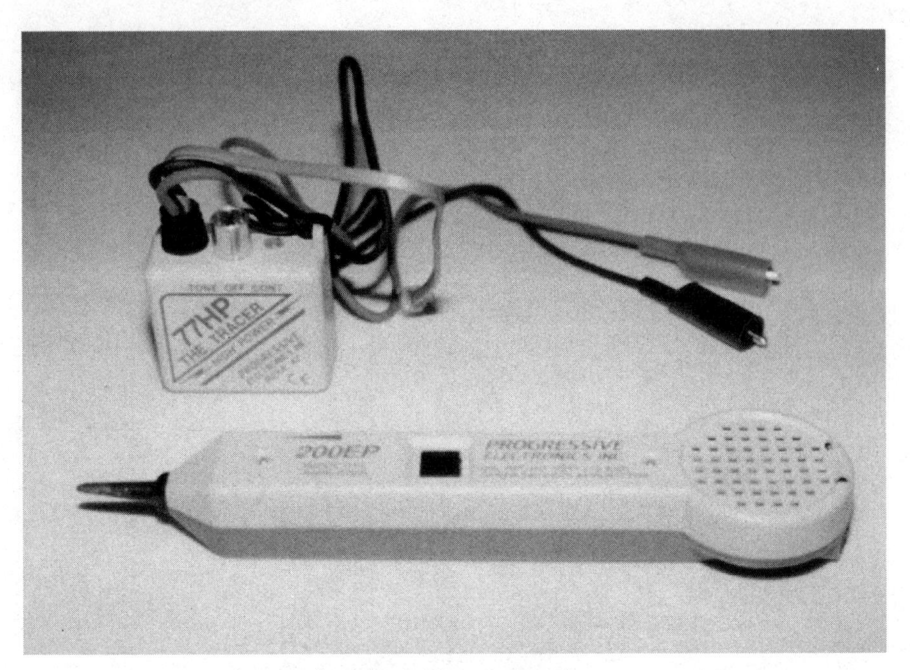

Installer's Tone and Probe

a telephone technician can't find a pair of wires by color or binding post, they attach a tone to one end, then go to the other end and use an inductive amplifier (also called a *banana* or *probe*) to find the beeping tone.

Integrated Circuit (IC) See *IC*.

Integrated Services Digital Network (ISDN) ISDN is a service that first evolved in 1979. It brings the features of PBX systems and high-speed data-transfer capability to the telephone network. The only thing that makes ISDN complicated is the many available features. The two kinds of ISDN lines are *Primary Rate Interface (PRI)* and *Basic Rate Interface (BRI)*. Two types of channels are contained within an ISDN circuit. The B (bearer) channel carries the customer's communications, and a D (data) channel provides control and signaling for the B channels. The *BRI (Basic Rate Interface)* ISDN line has two B channels and one D channel. A PRI has 23 B channels and one D channel.

The separate control of the ISDN line over the D channel is what enables the broad flexibility and features available with ISDN. When you are talking or sending a data transmission over an ISDN line, the voice and/or data is carried by the B channels. While you are talking on your ISDN line, you can still dial digits (signal the central office) to change or alter the state of your service because of the separate D channel. For ex-

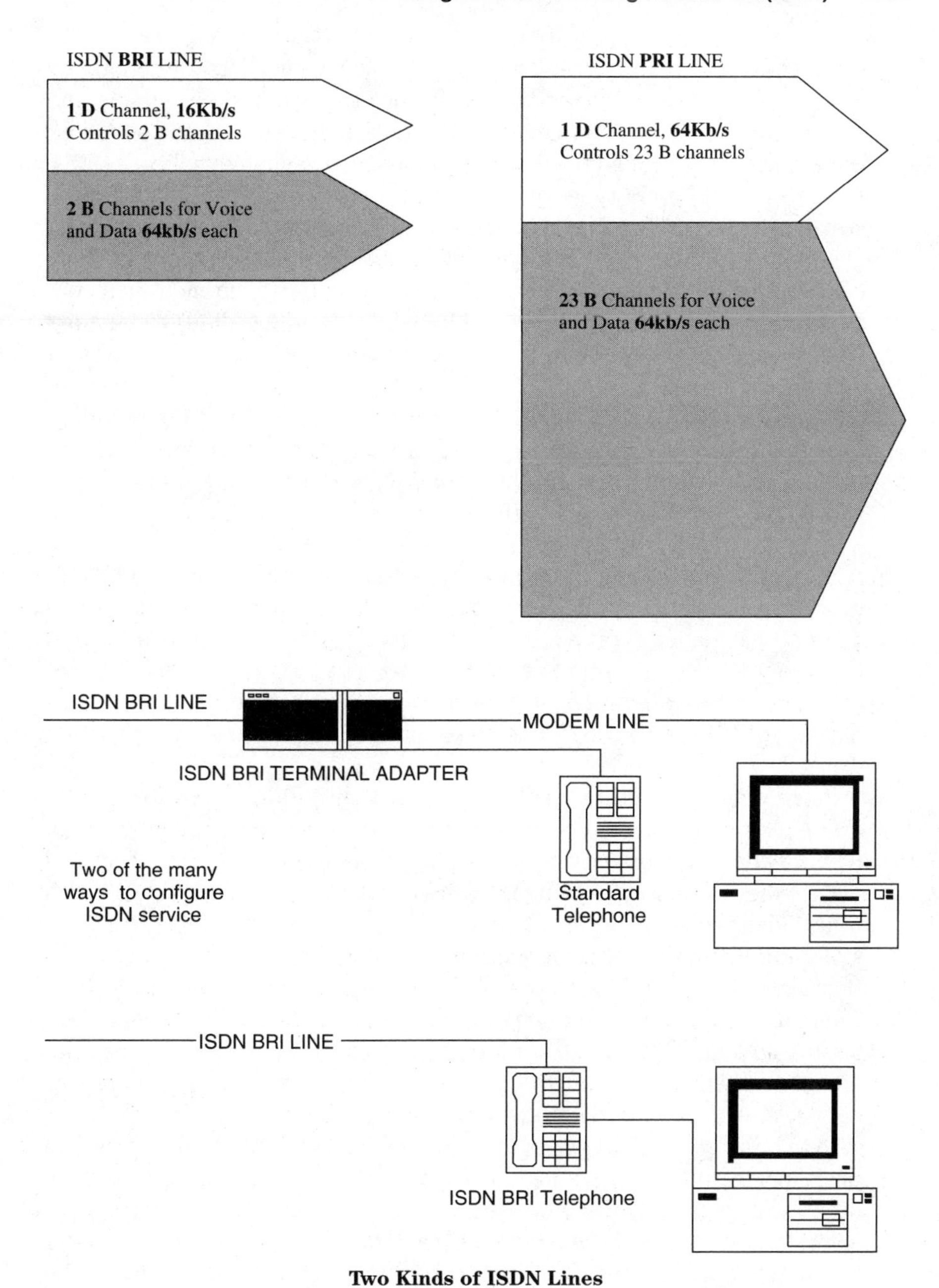

Two Kinds of ISDN Lines

ample, imagine you want to arrange a meeting with a client. You dial the client's telephone number on your ISDN telephone to reach the client. While you are speaking with the client, you can dial up an Internet access on your computer and put two baseball tickets in at the ticket counter using the same phone line. Then you can fax your client directions by downloading a map provided by the baseball ticket office, disconnect and redial your client's fax number. All of this occurs while talking to your client the entire time. Through the advanced convenience and flexibility of ISDN, you can send different types of data and messages to different places at the stroke of a few buttons, and at a much faster speed than a regular telephone line. If you are interested in ISDN, call is your local phone company. They can help you decide on what kind of terminal adapter (equipment that connects your computer and phone equipment to the ISDN line) to buy and what kind of features to subscribe to. ISDN is not yet available everywhere.

Intelligent Hub A hub that has the ability to be electronically reconfigured (manually or remotely) and perform additional functions, such as protocol conversions and bridging functions. For a diagram of where a hub fits into a network, see *Hub*.

Interactive Voice Response (IVR) A telecommunications and data-processing technology that interfaces a person to information held in a computer by using a phone line. If you have ever called your bank and entered your account number, a password, and a prompt so that a computerized voice can read back your bank account balance, then you have used IVR. IVR systems are capable of sending fax information as well.

Interconnect Agreement This is also known as a *co-location agreement*. The two types of co-location agreements are physical and virtual. A *physical co-location* is an interconnection agreement and a physical place where telephone companies hand-off calls and services to each other. This is usually done between a CLEC and an RBOC. The CLEC installs and maintains interconnection equipment usually consisting of optical carrier (SONET) equipment and a digital cross-connect system. There are other types of co-locations. Alarm companies like to have their alarm-signaling equipment located in the local central office for security and convenience of connecting alarm circuits. Long-distance companies co-locate with local telephone companies as well.

A *virtual co-location* is an interconnection agreement and a physical place where telephone companies hand off calls and services to each other. This is usually done between a CLEC and an RBOC. A virtual co-location is when telephone company A (the CLEC) requests that their phone company's network be connected to telephone company B's (the RBOC's) network. Telephone company B charges company A lots of money. Company B owns, installs, and maintains the equipment. To company A, the interconnection is virtual, because they never physically do anything to it when and after it is installed. Company B likes this, because company A does not get free access to their premises.

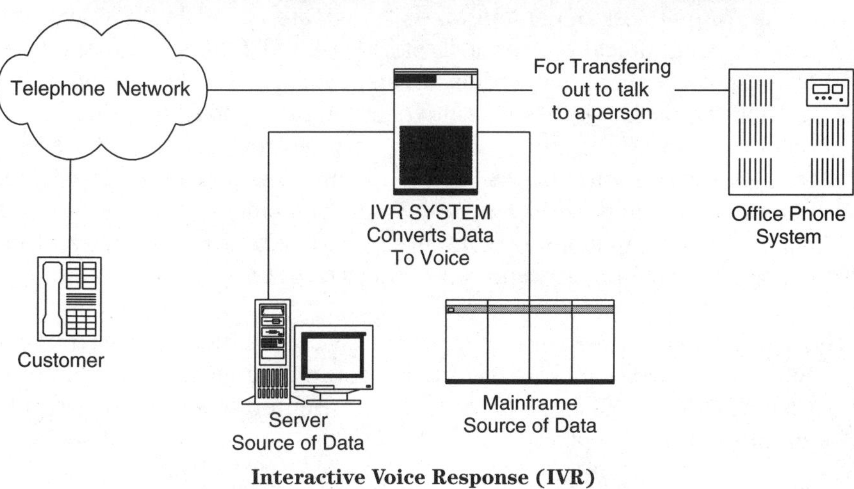

Interactive Voice Response (IVR)

Interface A device or software program that connects two separate entities. The two entities can be virtual (software), hardware/electronic devices, or distinguish a separation of responsibility between two parties (telephone network interface).

Inter LATA (Inter Local Access Transport Area) Simply stated, a LATA is an area code. Inter LATA refers to services that go from one area code to another, like long-distance telephone calls. *Intra LATA* refers to services that originate and terminate in the same area code.

International Dialing To dial international long distance from the United States, dial: 011-county code city code' number. For a listing of country codes see appendix B. To dial the United States from another country that is a part of the NANP (North American Numbering Plan), simply dial the area code the same way you would call long distance to another state. To call the United States from another country that is not a part of the NANP, consult your long-distance company. The United States has different country codes/access codes for almost every country that is not a part of the NANP.

International Gate Ways International telecommunications are done through gateway central offices. Gateway central offices (class 5 central offices) connect communications to other countries. The gateway does the translation from T1 to E1, T3 to E3, and vise-versa. The United States has five international telecommunications gateway cities. Any international call is routed through one of these cities: New Orleans, New York, Washington, D.C., Miami, or San Francisco.

Internet A network of computers that originated as ARPANET, an information communications project of the United States Department of Defense. Over time, many other organizations, private and public, have utilized the project by connecting their computers to it. Its primary protocol is TCP/IP. Today, many Internet service providers can offer access to the Internet for as little as $15 per month. The Internet is growing exponentially as more service providers and customers gain access to it. It currently links millions of computers, with which users find and exchange information, buy and sell services or products, and play games.

Internet Server A server that users access for Internet services. Popular Internet services include access to the Internet, e-mail, news updates of the subscriber's choice, Web pages, etc. An Internet server is owned by an Internet service provider.

Internet Service Provider (ISP) An Internet service provider purchases direct access to the Internet through an Internet company, such as UUnet and resells the service to smaller subscribers via dial-up modem (or to large customers via frame relay or private line T1). The ISP adds other services of their own, such as E-mail, news updates of the subscriber's choice, Web pages, etc.

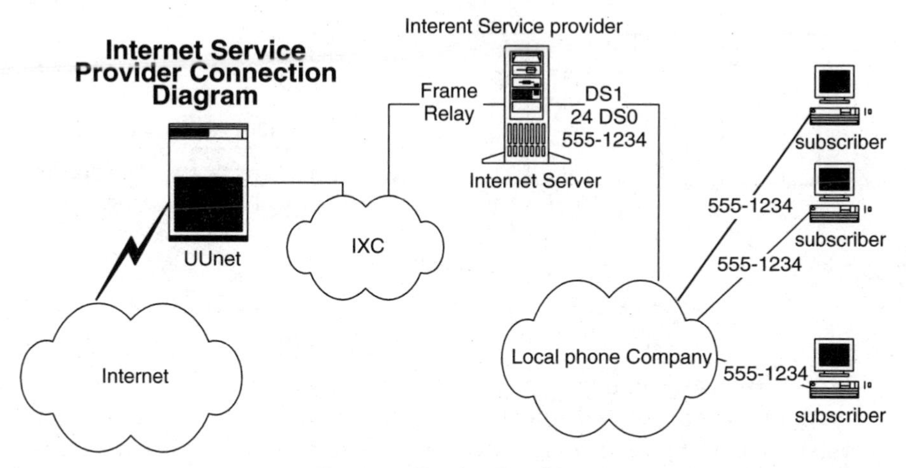

Internet Service Provider

Inter Office A reference to a telephone connection, a call, or service that originates in one central office and terminates in another central office within the same area code.

Interstate Long Distance A long-distance service or call that originates in one state and terminates in another state.

Intrastate Long Distance A long-distance service or call that originates and terminates in the same state. A call from San Francisco, CA to San Diego, CA is an intrastate call.

Intra Office A reference to a connection, a call, or service that originates and terminates in the same central office. If you call your neighbor, you are making an intra-office call.

Inverter 1. A device that converts DC to AC. Inverters are commonly used where data or computer equipment is used in a central office. The central office is equipped with –52V DC power for telephone and transport equipment. The computer equipment, or any other equipment re-

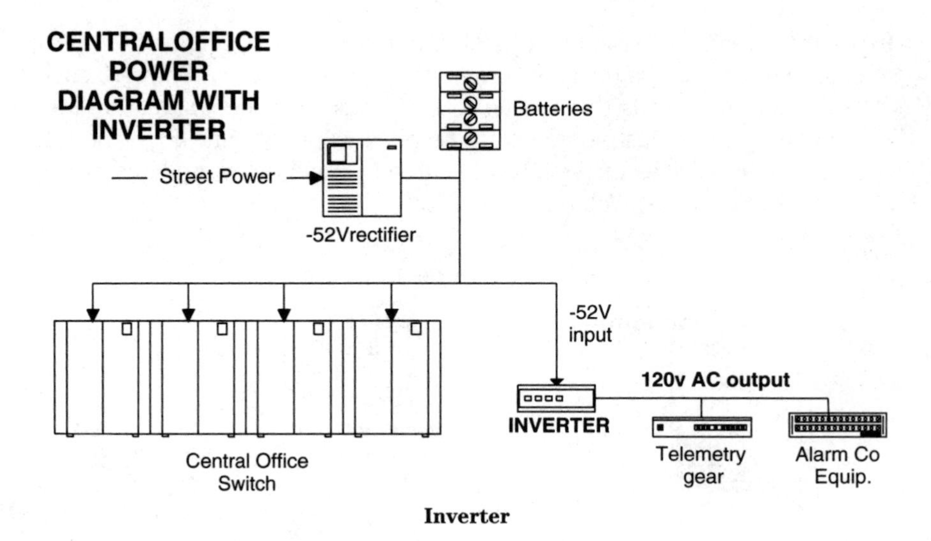

Inverter

quiring 120V AC is connected to an inverter, which converts the power. By using an inverter instead of just running a 120-V outlet, the equipment is protected from power outages on the same system as the telephone equipment. 2. A logic circuit that reverses a positive (logic 1) state to a 0 state.

Inward Trunk A shared phone line that provides a path for incoming calls to a PBX system. A true inward trunk subscribed from a local phone company cannot be dialed out.

In-WATS (Inward Wide-Area Telephone Service) An inward toll-free dialing service (800/888 lines) offered by telephone companies. In-WATS lines are priced and set up for incoming-only calls, and usually calls from a certain area. You can also subscribe to Out-Wats service as well. In-WATS can be for interstate and intrastate long distance. If you call an 800 number, you are most likely calling an in WATS service line that a company has set up for customers. The time to start checking into WATS service is when your long distance to or from a specific area exceeds $200 per month.

IOD (Identification of Outward Dialing) This is a call-accounting system feature of PBX and some key systems that captures every number dialed by a specific telephone extension and prints it out on a report for accounting and cost-tracking purposes.

Ion An atom or molecule that has lost or gained one of its valence electrons and is no longer electrically neutral.

Ionosphere A region of thin air that exists from 60 to 600 miles above the Earth's surface. Radio waves between 2 to 50 MHz are reflected (actually refracted in the same manner that light is refracted through a graded-index fiber optic) back to earth 500 to 3000 miles from where the transmission originated. The radiation from the sun causes air to be ionized. The ionosphere is different at night than it is during the day, causing some radio signals to be refracted only during the night, or vise versa. Because of ionospheric refraction, it is not unusual to pick up a Utah AM radio station in northern Mexico.

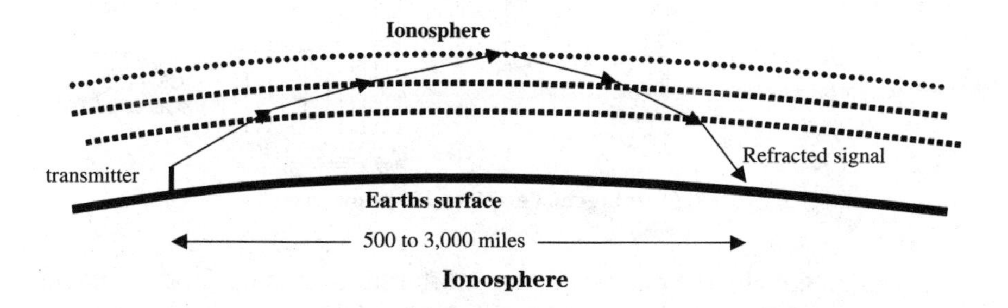

Ionosphere

IP (Internet Protocol) Internet protocol was originally developed by the US Department of Defense to enable communications of dissimilar computing platforms around the country. IP is modeled very much like the public telephone network in regards to addressing. Simply put, each computer has a "phone number" that includes an "area code," an NNX, and an extension number. An IP address (phone number) is 32 bits long, which equates to four 8-bit characters.

IPE (Intelligent Peripheral Equipment) One of Northern Telecom's terms that refers to a card that interfaces with PBX phones. Other references to equipment are *CM* (*Core Module*, which is the CPU) and *CE* (*Common Equipment*, which interfaces the different parts of the PBX switch together).

IR (Infrared) Light waves that humans cannot see. Their wavelength (or frequency) is just below that of red light. Heat produces infrared lightwave radiation.

ISDN (Integrated Services Digital Network) See *Integrated Services Digital Network*.

ISDN Terminal Adapter A device that interfaces an ISDN line to a customer's equipment. The *Terminal Adapter (TE)*, is purchased and usu-

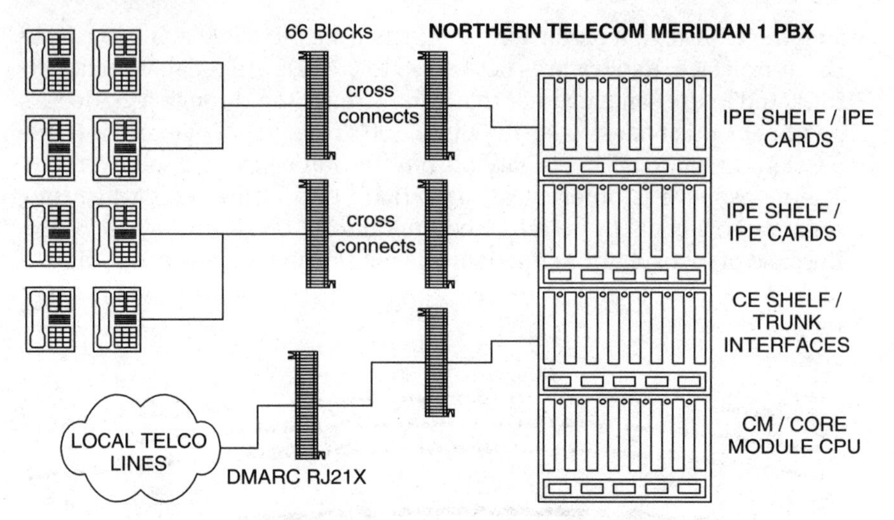

IPE (Intelligent Peripheral Equipment)

ally connected by the customer. Many different terminal adapters enable the use of different features of an ISDN line. TEs come in a wide price range as well. For a diagram showing where an ISDN terminal adapter is connected, see *Integrated Services Digital Network*.

Isochronous A transmission with no delay, such as a voice conversation between two people. Sending a letter through the mail is non-isochronous and so is frame-relay.

ISP (Internet Service Provider) See *Internet Service Provider*.

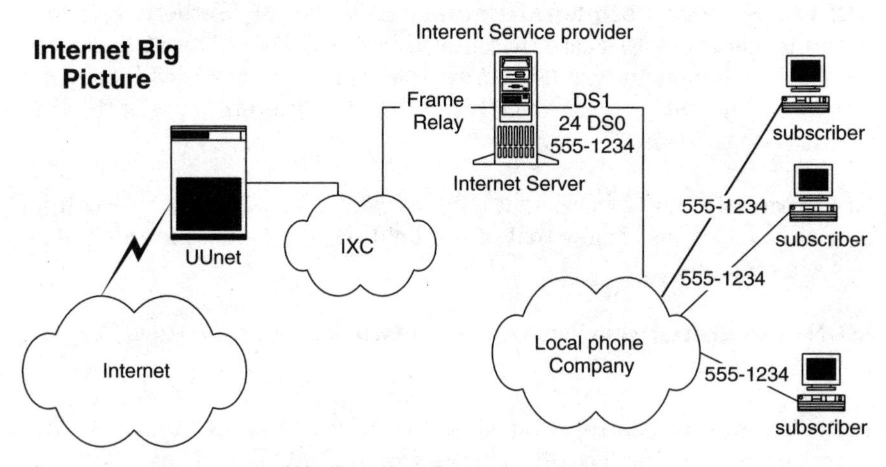

ISP (Internet Service Provider)

IT (Information Technology) The study of improving information and data processing with the use of newer and better devices/machines.

IVR (Interactive Voice Response) See *Interactive Voice Response*.

IW (Inside Wire) The telephone wire that is on the customer side of the telephone network interface. IW includes jacks and any wiring in or attached to the outside of the house, as long as it is electrically on the customer's side of the network interface.

IXC (Inter Exchange Carrier, IEC) A long-distance company, like AT&T, Sprint, Worldcom, and MCI.

J

J Box (Junction Box) A metal or plastic box used as an access for cable or wire (coax, fiber, UTP, STP). When communications companies build their networks into buildings, the building management usually requires a J-box close to each entry of the building. The J-box allows other companies that want to gain access use of the same conduit.

J Hook A spike with a hook on the end, specially designed to be pounded into wooden telephone/power poles. The installed J-hook is a means to hang telephone or other aerial service wire. To hang a service wire on a J-hook, a

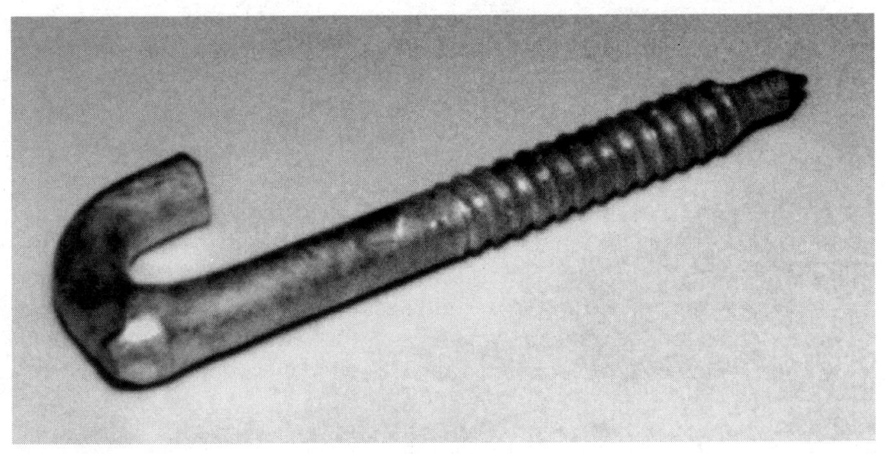

J Hook

drop-clamp or wire-vise is placed on the service wire. The drop-clamp or wire-vise is designed with a loop that fits over the J-hook.

Jack Also called a *connecting block*. A device that has a form of connectivity (a plug) used to terminate a wire/coax/fiber run. The jack provides access to the terminated wire/coax/fiber run via the connector(s)-plug(s) on its face. Some of the many types of jacks include flush jacks, which fit into a wall like an electrical outlet; baseboard (or biscuit) jacks, which are shaped like a box and mount to a baseboard near the floor; and duplex jacks and fourplex jacks, which have more than one plug on the face of them, with different types of connectors in each plug. For more information on the types of plugs available on jacks, see *RJ11*, *RJ45*, *RJ21X*, *BNC*, *RCA plug*, *ST*, and *SC*.

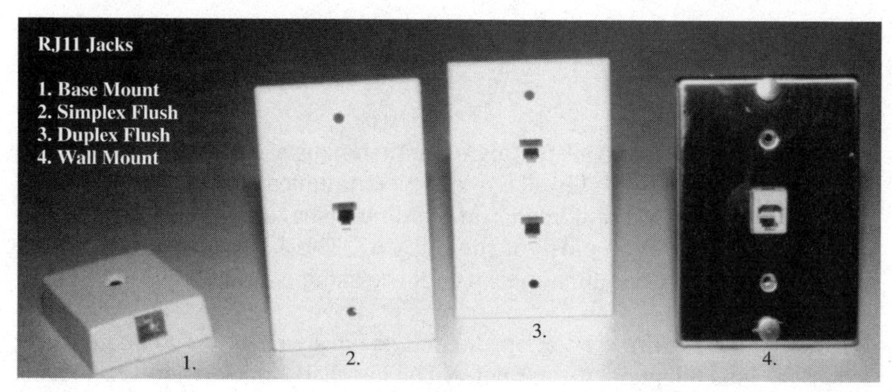

Jack

Jacket The outer covering of a multiple-wire or fiber-optic cable. The most common jacketing materials are PVC (Polyvinyl Chloride), Plenum (polyvinylidene diflouride), and ALPETH (aluminum/polyethylene). Older cable jacket types include lead and cloth. For a photo of ALPETH and lead cable, see *ALPETH*.

Jamming A reference to the intentional interference of a receiver's ability to receive, detect, or demodulate a radio signal of any frequency. A signal of identical (or nearly identical) carrier frequency to the one being jammed is transmitted with a meaningless signal modulated on it. The receiver picks up both signals, which causes a hard-to-understand blur of noise. Directional transmissions and directional antennas are less susceptible to signal jamming.

Johnny Ball A device used to link two steel-strand loops together. Also called a *Knuckle-Buster*.

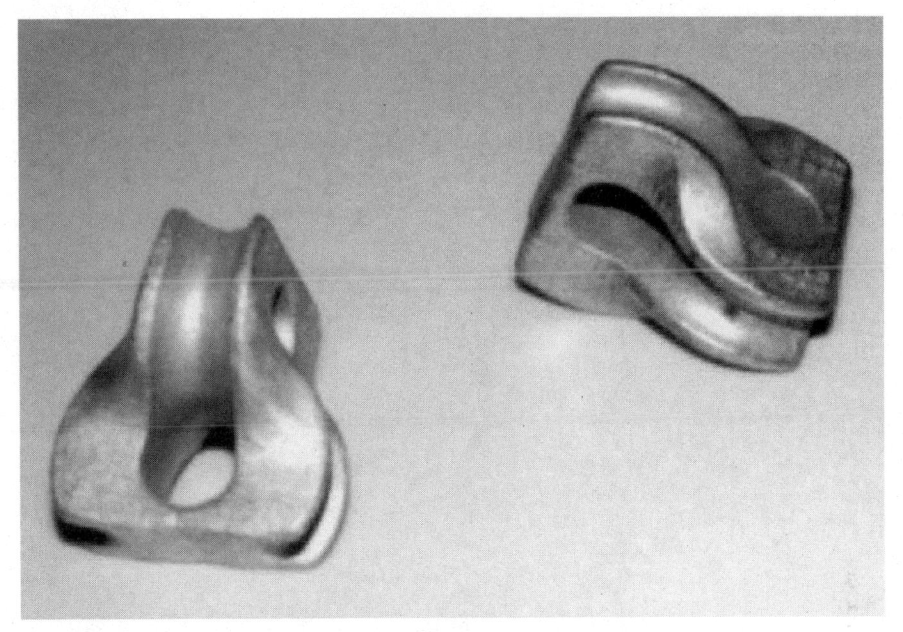

Johnny Ball

Joule (J) A unit of energy. If you combine power and time, you get energy. One Joule is equal to one watt for one second. Running a 100-W light bulb for one hour expends 360,000 Joules of energy. Running a 5-HP (five horse-power) lawn mower at full strength through tall grass for one hour expends 13.428 MJ = 13,428,000 Joules (746 watts = 1 horsepower). What is useful about Joules? Joules are a way to compare different forms of energy. After calculating the cost per Joule of gasoline energy and electrical energy, it can be determined that 1 Joule of electrical energy costs nearly twice as much as 1 Joule of gasoline energy. (Based on gasoline at $1.50/gal, the lawnmower consuming ⅕ of a gallon in one hour, and the cost of one kW/hour is 15 cents).

Jumper 1. Another name for a cross connect. A cross connect is the connection of one circuit path to another via a physical wire. Telephone cable pairs are terminated or "punched down" onto a termination block (usually a 66M150 or an AT&T 110 block) that has extra connections available for each pair so that jumper wires can be easily connected and re-arranged between them. 2. A section of coax used to connect a transmitter or transmission line.

Jumper Wire See *Jumper*.

Junction Box See *J Box*.

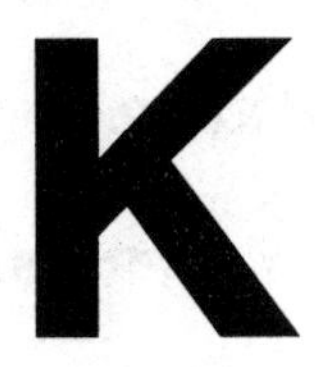

Ka Band The part of the radio-frequency spectrum that ranges from 33 GHz to 36 GHz.

KB (Kilobyte) A measure of computer memory. See also *Kilobit*.

K Band The part of the radio-frequency spectrum that ranges from 10.9 GHz to 36 GHz.

K Bit (Kilobit, Kb) 1000 bits. Not to be confused with KB, which is kilobyte. The speed of data transmission is usually measured in Kb/s and memory is measured in Kb.

Kbps (Kilobits Per Second) A reference to how fast data is being transferred on a communications path.

Kevlar A fine, stranded yellow fiber used to make bulletproof vests and built into fiber-optic cable to reinforce it.

Key A button on a telephone that executes a feature or accesses a line on a telephone.

Keyed RJ45 An RJ45 jack or plug that has a small protrusion on its side that acts as a key to prevent the wrong kind of equipment being plugged into the wrong jack. Some modems come with keyed RJ45 plugs to help prevent them from being plugged into an RJ45 that contains voltages intended for other equipment, which could cause damage.

Key Pad A dial pad.

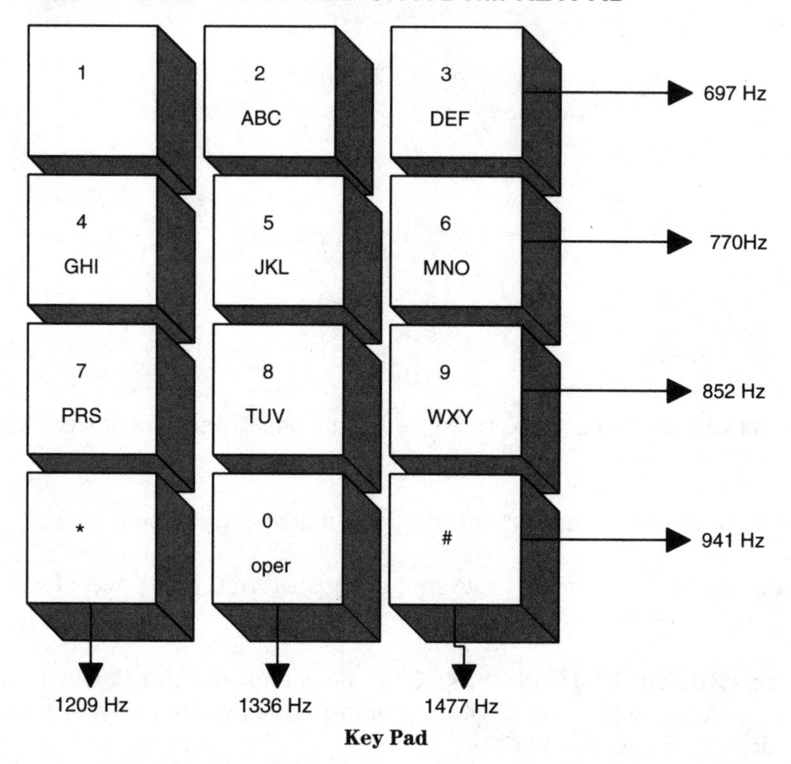

Key Pad

Key Service Unit (KSU) The main part of a key telephone system. The KSU contains the electronics that control which line is directed to which phone. The KSU is usually mounted in a closet or near the telephone companies demarcation point (where the phone lines come into the building).

Keyset An electronic telephone that works only when connected to a proprietary KSU via the correct wiring. Keysets do not work if you connect them directly into a normal telephone line. If you like the idea of having Keyset features on a normal public telephone line, see *ISDN*.

Key System See *Key Telephone System*.

Key Telephone System The less-expensive and less-flexible alternative to a PBX system. On a key system, each telephone line appears under a key (button) on the phone. To access an outside line or answer an incoming call, you press the key associated with that line. Key systems are digital and have six major parts, the *KSU (Key Service Unit)*, line interface, station interface, power supply, connectivity, and the key sets (telephone sets). Most key systems are very user friendly, and can be installed by a user that knows little about telephony. The KSU is the cabinet that contains the electronics that controls the switching between key sets and phone lines. Some key systems have detachable line and station interfaces that plug into the KSU. The system usually comes with a specific number of incoming line interfaces (6) and a specific number of station interfaces (16). This size system is commonly referred to as a *4-16* (four/sixteen). The power supply is often a "power-adapter" type.

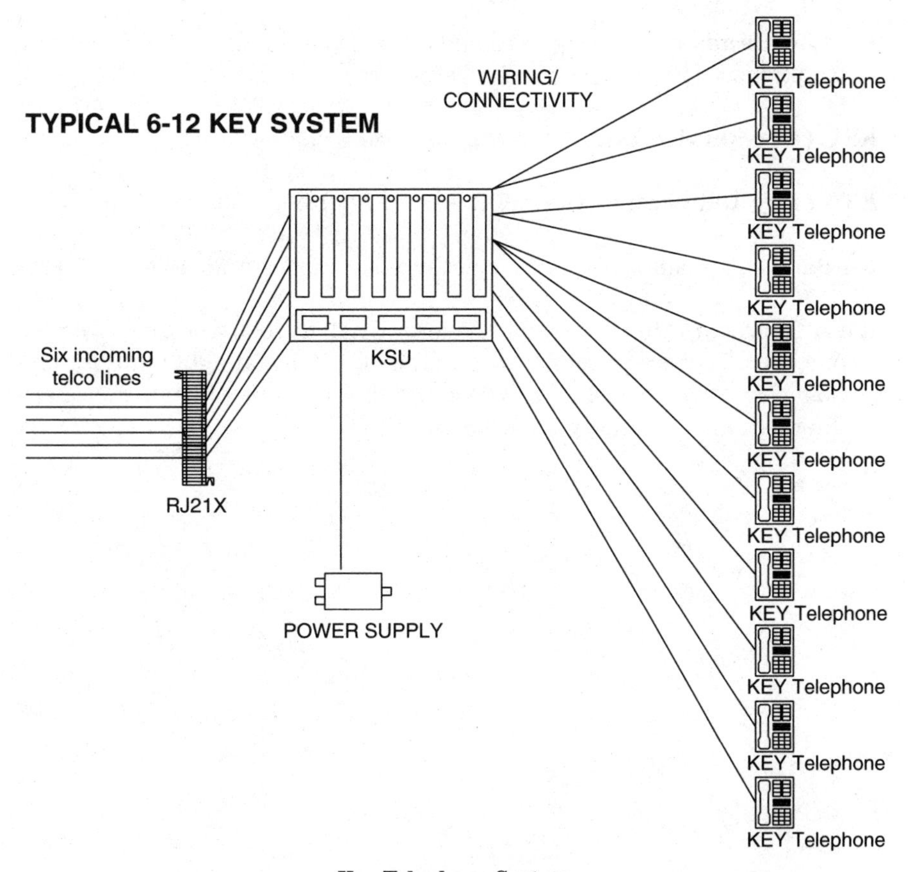

Key Telephone System

Some manufacturers offer a UPS back-up option specially designed for the key system. The key-set telephones are sold individually, in a variety of choices that include 10-key (10 button), 20-key (20 button), display, and hands-free.

KHz (Kilohertz) One kHz is equal to 1000 Hertz, or 1000 Hz. Kilo is just an easier way of saying "one thousand," and k is a shortcut to having to write three zeros.

Kilo (k) Kilo is just an easier way of saying "one thousand," and k is a shortcut to having to write three zeros. 25 kHz is equal to 25,000 Hz; 64Kb/s is equal to 64,000 bits per second.

Knuckle-Buster A device used to connect two steel strand loops together. Also called a *Johnny Ball*. For a photo, see *Johnny Ball*.

K-Style Handset The type of handset typically found on PBX and key telephones. The ear piece and receiver are square shaped.

KSU (Key Service Unit) See *Key Service Unit*.

KTS (Key Telephone System) See *Key Telephone System*.

Ku Band The radio-frequency spectrum that ranges from 12 to 14 GHz.

KWH (Kilowatt Hour) A unit of energy equal to 1000 watts of power for one hour. This is equivalent to operating ten 100-watt light bulbs for one hour. The average cost of a KWH in the USA is about 14 cents. For more information about energy, see *Joule*.

L

L Band The part of the radio-frequency spectrum that ranges from 390 MHz to 1550 MHz.

L Carrier A long-haul TDM carrier that is transmitted over two coax cables. It is still in use in sub-oceanic applications. L carrier has a capacity of 13,300 voice channels, 3 kHz in bandwidth (each).

LADT (Local-Area Data Transport) A reference to digital carrier (any carrier, T1, 56K, etc.) in the local-carrier twisted-copper plant network.

Lag Current A reference to the current flow in a capacitive electronic circuit carrying a signal (or AC power), the voltage builds up before the current actually flows. This is caused by *capacitive reactance*. In a purely capacitive electronic circuit, the current lags behind the voltage by 180 degrees.

Lambda (λ) The symbol that is famous for representing the wavelength of a frequency. The wavelength of a frequency. Wavelength is equal to: λ = (300,000,000 m/s)/*frequency* (Hz). 300,000,000 m/s is the speed of light in a vacuum.

LAN (Local-Area Network) A group of computers connected together within a building or campus. LANs are the most detailed of computer networks because they deal with the applications and operating systems of computers. The distinguishing thing among LANs is the way that the computers are connected and the protocol at which they communicate over the media that connects them. The two major LAN protocols (logical topologies) are Ethernet (star and bus) and Token Ring. Both Ethernet and Token Ring have evolved into Switched Ethernet and Switched Token Ring, which are different in operation and speed than their predecessors. Larger LANs use an operating system to control the LAN environment, such as Novell or Windows NT. Another component of LANs is the server, which can either run programs or store data for computers connected to the network. Servers are simply other computers (usually with more processing power and memory) that are configured by an operating system, such as Novell or Windows NT, to perform a specific function. For more information, see *Client Server*, *Token Ring*, *Ethernet*, *Switched Token Ring*, *Switched Ethernet*, *FDDI*, and *DQDB*.

LAN Adapter Also known as a *NIC (Network Interface Card)*. Typical LAN adapters are made by Bay Networks, SMC, US Robotics, and others. They are a circuit card that plugs into an expansion slot located on the mother board of a PC. They have an RJ45 plug and usually a BNC for connecting to the twisted-pair or coax network wiring in an office area (or wherever the network is).

Land Line A regular POTS switched telephone line over twisted copper or other land-based facility. Land lines are noncellular or radio.

LASER (Light Amplification by Stimulated Emission of Radiation) A laser is a device that emits only coherent light that is in phase. Coherent light is of one frequency (one pure color, not always visible to the human eye). Lasers create light similar to the way a fluorescent light does, by exciting a gas that emits photons (light particles) with electricity. When the particles are exited, they are trapped between two mirrors. One mirror has a small spot that allows light travelling in a very straight line to escape. This light is the laser beam.

LASS (Local-Area Signal Service) A signaling feature that incorporates the # and * keys, which allow central-office switches to provide features to residential phone-line subscribers that could only be found on PBX systems before. The services (at an additional cost) include: Automatic Call Back (also called *Last Call Return*), which is a service that by dialing * and two digits the customer can hear a message that tells them what the last number was that tried to call their phone. It also gives them the option of dialing back the number or returning the call automatically. Automatic recall or last number redial allows the caller to redial the previously dialed number by entering a code. Nuisance-call trace or last-call trace, allows the user to trace the last call made to them and automatically file a report against the caller. Each trace usually costs about $2. This is done so that people only use it when needed, not when they want to play a joke on a friend. Caller ID or incoming call ID, allows a telephone-service subscriber to connect one caller-ID unit to their phone line to view the calling part's telephone number (and often the name) before answering the phone. Many other services offered by different names vary, depending on the telephone company offering them. Different names are required for each service by different companies—even though the services are identical. The FCC implemented this rule so that the RBOCs couldn't trademark a well-known name for each feature, thus allowing the smaller phone companies to be equally as competitive by having to use a different name for each service.

Last-Number Redial A feature incorporated into PBX systems, Key systems, and also included on home use single line type telephones.

LATA (Local-Access Transport Area) Simply stated, a LATA is an area code. *Inter LATA* refers to services that go from one area code to another, like long-distance telephone calls. *Intra LATA* refers to services that originate and terminate in the same area code.

Layered Network Architecture Communications is accomplished in steps. Each layer in a network architecture is a step toward the goal of moving data or voice from one place to another. The *OSI (Open Systems Interconnect model)* is a layered architecture.

LCD (Liquid Crystal Display) LCDs are electronic display devices that operate by polarizing light so that a nonactivated segment appears invisible against a background. An activated segment does not reflect light (absorbs it) and therefore appears darker than the background. TTL is not generally used to drive LCD displays because it does not completely "deactivate the segment". A few tenths of a volt are present—even when a TTL logic device is completely in the "off" state. CMOS is the best type of device for driving LCD displays because of its ability to turn completely off and have no remaining bias voltage. LCD displays consume very little power, in contrast to other display methods (CRT, LED), but are sensitive to heat and need external light to be viewed in the dark. LCD displays consume little space compared to CRTs and are used in laptop computers.

LCR (Least-Cost Routing) A feature of PBX systems that enable them to be programmed to associate a dialed area code with a specific trunk.

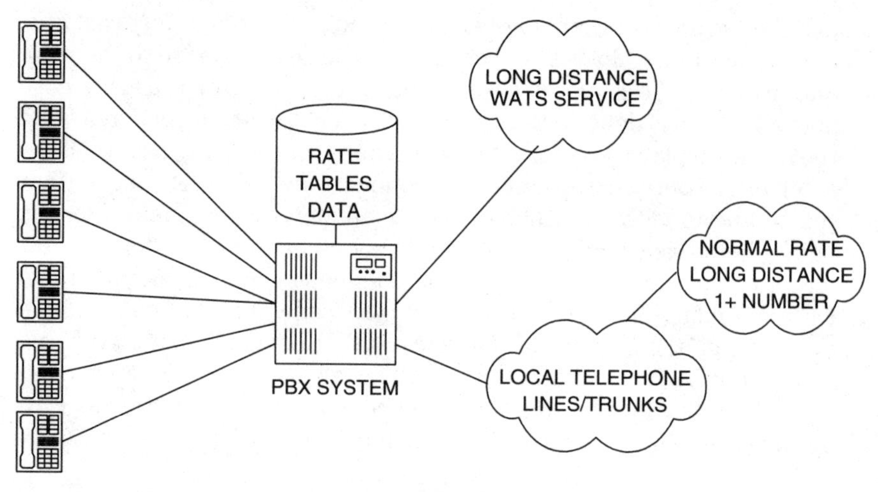

LCR (Least-Cost Routing)

That trunk will be the least-cost route for that area code. Some PBX systems are capable of having a rate-table data base with a call-accounting system. The rate-table database provides the PBX with cost per minute information for dialing certain area codes. If a user has an agreement with MCI (e.g., WATS service to Santa Clara, CA area code 408), then the PBX is programmed to connect anyone that dials "9-1-408-xxx-xxxx" to the MCI WATS trunk. The WATS TRUNK is cross connected to trunk interface/port number X. (X could be any number that the PBX uses in its trunk-numbering scheme).

LD (Long Distance) An abbreviation commonly used in PBX and key-system operator manuals.

LDM (Limited-Distance Modem) Also called a *short-haul modem* or "line driver." Short-haul modems are commonly used to extend the distance of a printer or other *DTE (Data-Termination Equipment)* device from its host. One example is to extend the printer dedicated to printing call-accounting records from a PBX to an accountant's office. For a diagram, see *Limited-Distance Modem.*

Lead Cable Before plastic (polyethylene) was invented, telephone cable was insulated with paper and jacketed in lead. The RBOCs still have some of this cable in use, and it does have one advantage over ALPETH (aluminum polyethylene) jackets. It is very heavy, so it is nonbuoyant in under-water applications, even when pressurized (see *Air-Pressure Cable*). Lead cable is currently being removed when at all possible be-

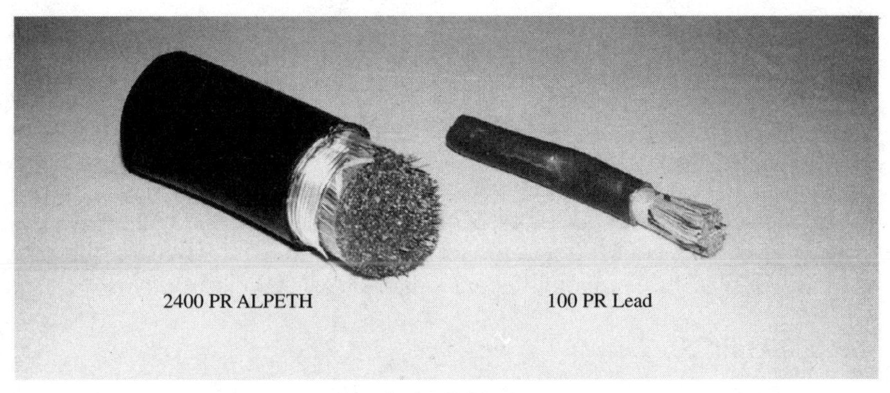

Lead Cable

cause of the poisonous effects of lead on the environment (the lead cable shown has pulp-insulated pairs).

Lead Current A reference to the current flow in an inductive circuit. Believe it or not, in an inductive circuit carrying a signal (or AC power), the current flows before the voltage is actually built up as a result of "inductive reactance". Current lead is measured in degrees or radians. In a purely inductive circuit, such as an AC motor, current leads voltage by 180 degrees.

Leased Circuit Also called a *leased line* or *private line*. A leased line is a telephone service that is permanently connected from one point to

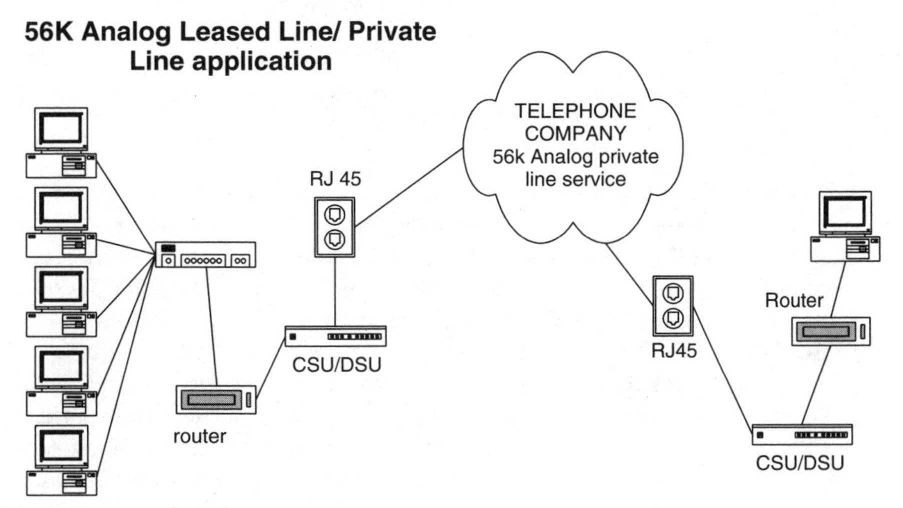

Leased Circuit

another. Leased circuits include 56K analog and DS1. A leased circuit acts like a pipeline that carries data from one point to another. If you put a bit in one side, the same bit pops out on the other side. It can carry data across town, across the country, or around the world. Leased lines are relatively expensive. Because leased lines have been offered, new services, such as frame relay have evolved. Frame relay does the same job as a private line, except that it is not isochronous (real time), and you need a private line to put your frame relay service on. Frame relay is a cost-effective solution for long-haul/long-distance data-transfer applications.

Leased Line See *Leased Circuit.*

Least-Cost Routing (LCR) See *LCR.*

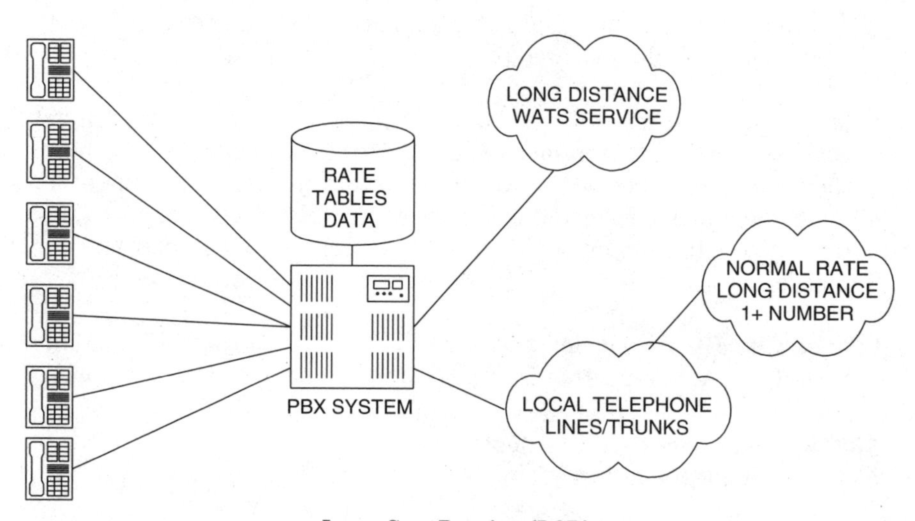

Least Cost Routing (LCR)

LEC (Local Exchange Carrier) Most commonly a reference to one of the seven RBOCs. A telephone company that provides telecommunications services to end users and bills them for it. A local exchange carrier has switching networks and outside plant or cellular service to serve its customers. Examples of local-exchange carriers are: NYNEX, USWest, GTE, PAC BELL, Southern Bell Telephone, and a myriad of cellular-telephone service providers. The newest breed of local exchange carriers are CLECs (Competitive Local Exchange Carriers). They have all the same services as the RBOCs, except that they are generally only located in large metropolitan areas. Some larger CLECs include TCG (Teleport Communications Group), ELI (Electric Lightwave Inc.), Brooks Communications, and MFS

(Metropolitan Fiber Systems). CLECs use (almost exclusively) SONET in conjunction with DCS as the foundation of their network architecture.

LED (Light-Emitting Diode) See *Light-Emitting Diode*.

Leg Iron Also called *spurs* and *climbers*. A device that network technicians wear on their legs to climb wooden telephone poles. Leg irons are a hook-shaped stirrup/shank, with an iron bar that straps to the technician's inner shin. Each shank is equipped with a spur that points out from the inner ankle.

LEOS (Low Earth-Orbit Satellite) A nonstationary satellite that orbits the earth at an altitude range of 300 to 500 miles. LEOS are used for many communications services, including paging, mobile radio communications, and data uplink. Dick Simon Trucking (based in Salt Lake City, UT) purchases satellite communications services to track the location and condition of their trucks all over the United States.

LF The ASCII control code abbreviation for line feed. The binary code is 1010000 and the hex is A0.

LIFO (Last In First Out) A method of clocking memory bits into and out of a memory bank. It means that the last bit in is the first one out. You can imagine this concept by stacking books into a box for storage. The last book you put in the box is the first one out when you remove them.

Light-Emitting Diode (LED) A diode that emits light when it is forward biased. If you have a PC, your hard-drive indicator and power indicator on the front are both LEDs. LEDs emit coherent light, which is light that is very close to one frequency (one color).

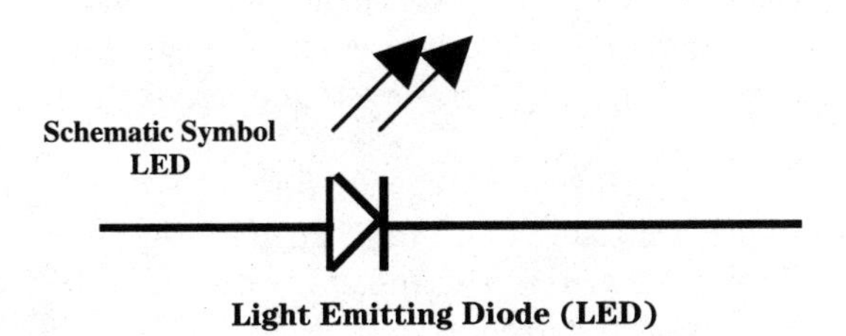

Light Emitting Diode (LED)

Lightning Protector A device used in telephone-company network interfaces that provides an easier path for lightning to travel to ground, compared to a telephone user, or inside wiring. Before lightning protectors, houses burned down because of lightning striking the telephone

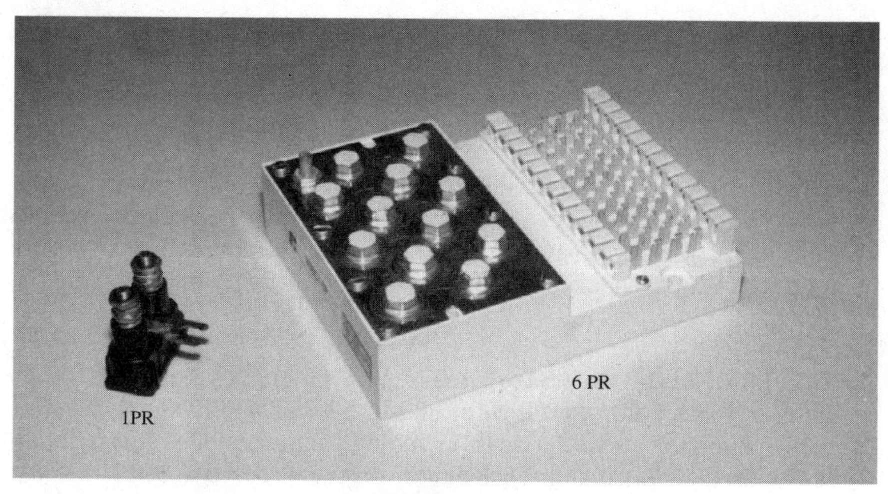

Lightning Protector

lines. The two types of lightning protectors are carbon and gas. The carbon protectors are simply a piece of carbon that connects tip and ring to ground. The gas protectors are the same, except that a gas is inserted instead of solid carbon. The good thing about gas-lightning protectors is that after they are hit by lighting, they do not need to be replaced and are re-usable.

Limited-Distance Modem (LDM) Also called a "short-haul modem" or "line driver." Short-haul modems are commonly used to extend the distance of a printer or other *DTE (Data-Termination Equipment)* device from its host. One example is to extend the printer dedicated to printing-call accounting records from a PBX an accountant's office.

Line *Line by itself* refers to a POTS (plain old telephone service) line. Any other telephone service line is preceded by the line type, such as, 56K private line, ISDN line, and 56K switched-service line.

Line Card A circuit board that is inserted into a PBX or Hybrid Key system so that additional telephone lines can be interfaced with the network. Typical line cards give an expandability of 4, 8, or 16 additional line ports on the PBX system. Most PBX and Key KSU cabinets come with a standard number of card slots, and the users can buy however many cards they need. After inserting a new card into a system, each new trunk needs to be configured. Usually the CPU in the system does not recognize the additional line ports until it is programmed or "told" to. After it is programmed it will traffic outgoing/incoming calls accordingly.

SHORT HAUL MODEM APPLICATION

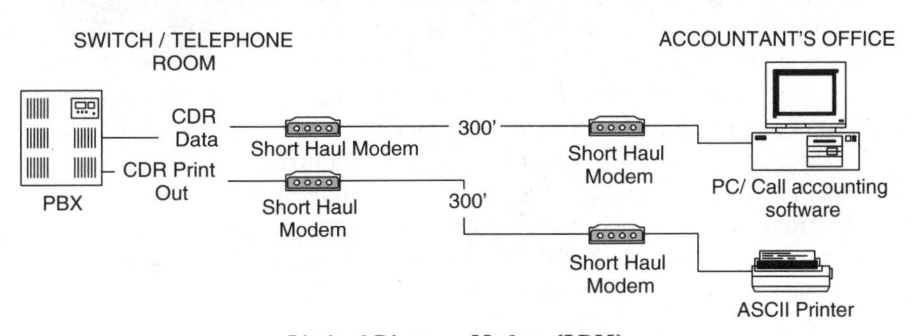

Limited Distance Modem (LDM)

Line Conditioning A term that refers to modifying a twisted copper pair in an outside plant network so it can carry a digital data signal instead of an analog voice signal. Twisted pair circuits (a circuit is a local loop, which is the pair that connects the central office to the customer) are conditioned by adding noise filtering electronic components to them.

Line Current The average current of a telephone line when the receiver is off hook is about 35 milliamps, which is 0.035 amps.

Line Driver Also called a *short-haul modem* or *limited-distance modem.* Short-haul modems are commonly used to extend the distance of a printer or other *DTE (Data-Termination Equipment)* device from its host. One example is to extend the printer dedicated to printing call-accounting records from a PBX an accountant's office. For a diagram, see *Limited-Distance Modem.*

Line Equipment Also referred to as *OE (Office Equipment)*. The line equipment is the actual interface port (from a circuit card) on the central office switch in a telephone-company central office. It is the equivalent to a station or *IPE (Intelligent Peripheral Equipment)* card in a PBX system (TN, for Nortel Specifics). Each telephone line has an associated line-equipment or office-equipment interface. That particular port is what defines the telephone service provided to the customer connected to it via the OSP network. The CPU (core) of the central-office switch associates a phone number with a line-equipment port or OE. When a customer of the phone company calls and requests that their phone number be changed, the service order eventually finds its way to a central-office technician or service translator that reprograms the line equipment with a new phone number.

Line Lock Out A reaction of a central office when a phone line is left off hook. If you leave the phone off the hook, you get a loud *Ringer Off Hook* alert signal. The signal is transmitted for about a minute, then the telephone company central office locks the line out. This doesn't put the line out of service, it just stops sending the current down the line that it normally would. When a phone is off hook, the central office sends 20 to 30 milliamps down the line. If a phone is off hook, it wastes power, so the central office locks the line out. This uses only 1 to 2 mA of current. The central-office switch keeps the line locked out until it can sense the phone has been placed back on the hook.

Line Loop Back (LLB) A troubleshooting function of CSU/DSU equipment and smart jacks, where the receive pair of a circuit is connected directly back into the transmitter (or a person manually disconnecting DTE and connecting receive to transmit). The object is to test the transmission line. If the transmission equipment transmits a signal that is "looped back" to it and it receives its own signal with no errors, then the line is ok. If there is a problem, it is inside or it is beyond the receiving equipment.

Line of Sight Another name for a terrestrial microwave link, also called an *eyeball shot*. The link is made by two radio transceivers equipped with parabolic dish antennas pointed directly at each other. Radio can carry point-to-point transmissions of many bandwidths including DS1, DS2, DS3, STS1, and OC1. Their range can vary, depending on the size of

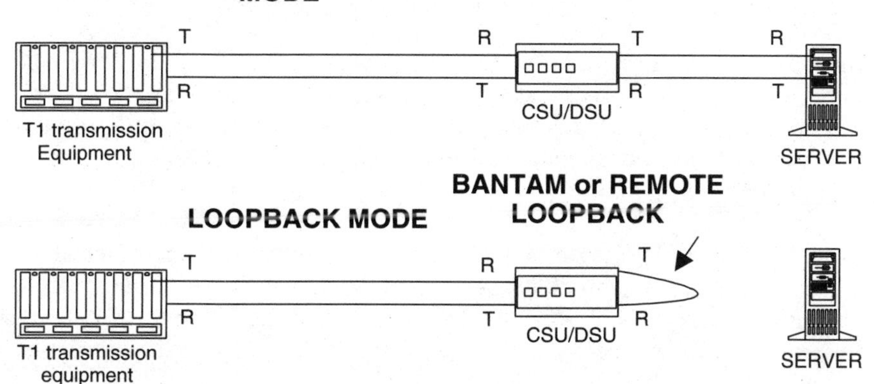

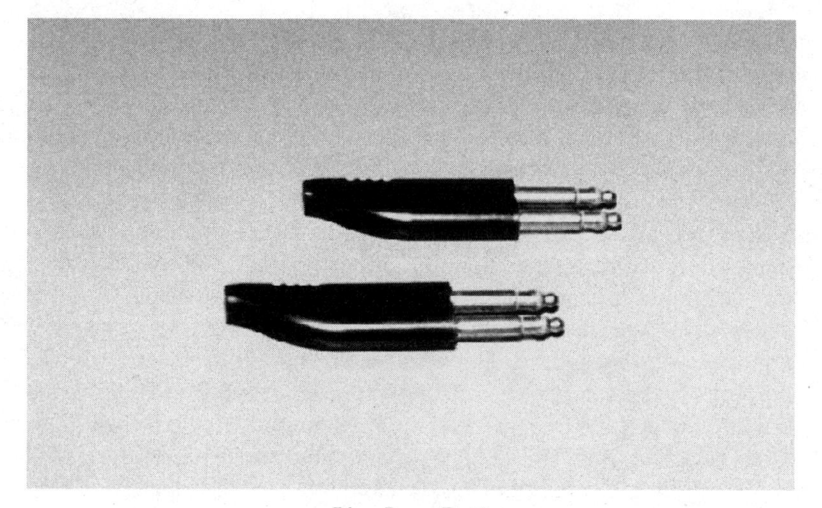

Line Loop Back

the antenna (dish), weather in the region, and the amount of power emitted. Including all of the previous factors, a link can range from 0 to 50 miles.

Line Pool A group of telephone lines or trunks configured in a PBX (in a PBX, a telephone is assigned to a pool, rather than a button) or key system for specific call purposes. Some specific purposes include grouping or "pooling" several lines under one button on a telephone. When that button is pushed, the user will access one of the trunks from the line pool that it is assigned. The line pool could be a WATS service or a group of lines dedicated for outgoing calls, so incoming calls are not blocked because of too many agents making outgoing calls at one time.

Line Powered Telephone equipment that is powered by the telephone company central office-battery and ring voltage. Standard 2500 telephones (noncordless analog phones, such as the ones in your house that have no answering machine or are hands free) are line powered. They require no battery and no power adapter.

Line Protocol The organized processes and rules that communications equipment use to transfer bits and bytes (data). There are many communications protocols, and layers of protocols that carry other protocols (called *protocol stacks*), including ISDN, ethernet, token ring, POTS signaling, DS1, ATM, frame relay, and SONET.

Line Queuing The opposite of call queuing (as in an ACD system). Some telephone systems have a feature called *line queuing*. If you try to dial out and you cannot get an outside line, you are put in queue, a waiting line for the next available trunk. Some systems can give music as if you were on hold for the line and some can ring your phone back. Back to call queuing, ACD systems place incoming calls in queue for the next available agent.

Line Ringing A feature of telephone systems that enables a user to enable a phone to ring when specific lines are called. When the user programs or configures a specific telephone extension/station/set, a prompt usually says "line ringing." The user then enters the lines that they would like to ring on that phone. The user would then enter the corresponding line ports of the telephone system. (i.e., 1, 2, and 5 for the three associated line ports. Line 555-1234 terminates on port 1, line 555-4321 terminates on port 2, line 555-1111 terminates on port 5) When any of these lines are called, the telephone extension/station/set will ring.

Line Switching See *Circuit Switching*.

Line Termination A reference to a demarcation point that gives a telecommunications customer access to their service.

Line Turn-Around Time The delay between transmit and receive in a transmission.

Line Voltage A reference to the voltage on a pair. T1 line voltage is –135 volts, loopstart line voltage (POTS) is –52 volts, and the line voltage for most PBX and key systems is –24 volts.

Line Wrench A wrench designed to fit the tool slot of a safety belt. Its primary use is to install/remove pole steps and pole attachments.

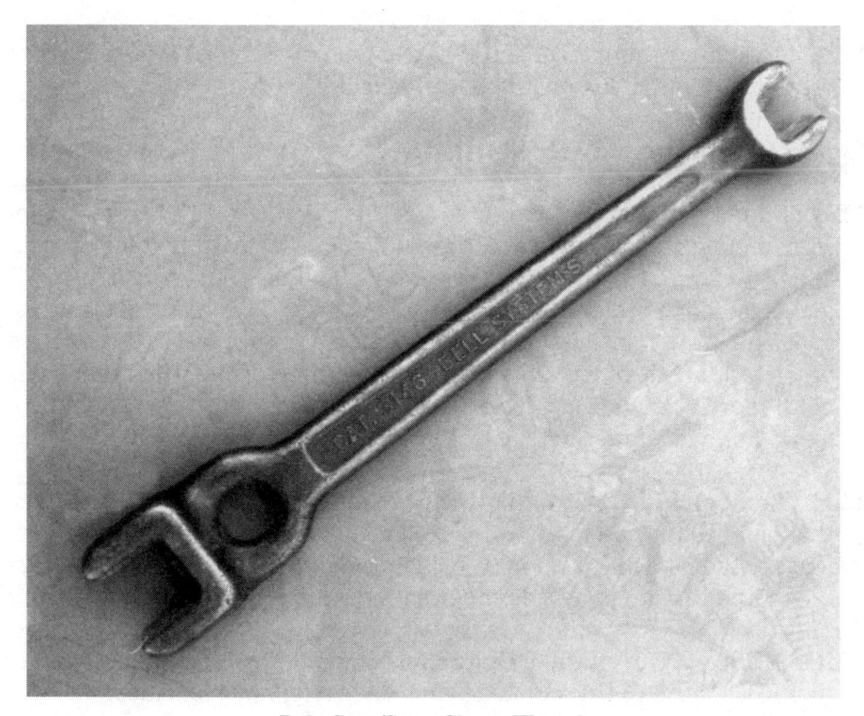

Pole Step/Span Clamp Wrench

Linear Distortion Usually referred to as Non-Linear Distortion. Nonlinear distortion is caused in amplifier circuits when the collector current in the amplifying transistor is insufficient to make the transistor work within the optimal range (the linear range) of its transconductance curve. The resultant distortion is that the top half of the signal (or positive portion) is larger than the bottom half (negative portion). In some cases, the negative portion of the signal can actually be attenuated, rather than amplified.

Linearity The consistent regeneration of a signal through an electronic circuit. If a circuit amplifies a 5-volt signal two times, 10 volts should be the resultant output. If the circuit is truly linear, raising the 5-volt input to 7 volts would give a resultant output of 14-volts.

Lineman An obsolete term that used to refer to a person that maintained and installed telephone lines and services. The new term is *network technician*.

Lines of Force Imaginary lines drawn around a magnetic or magnetized object that represent the direction and polarity (north and south) of the magnetic field around it.

Link An active communications path between two devices that is physical or virtual (a multiplexed/switched channel).

Liquid Crystal Display (LCD) See *LCD.*

LLB (Line Loop Back) A troubleshooting function of CSU/DSU equipment and Smart Jacks where the receive pair of a circuit is connected directly back into the transmit. (or a person manually disconnecting DTE and connecting receive to transmit). The object is to test the transmission line. If the transmission equipment transmits a signal that is "looped back" to it and it receives its own signal with no errors, then the line is ok. If there is a problem it is inside or it is beyond the receiving equipment. For a diagram see Line Loop Back.

LLC (Logical Link Control) The part of a communications protocol that handles the assembly and disassembly of packets. The function belongs to the data link layer in the OSI model. LLC is a very important part of ethernet, frame relay, and other packet-switching protocols.

LMOS (Line-Maintenance Operating System) A computer program that RBOCs use to track outside and inside telephone facilities.

LMSS (Land-Mobile Satellite Service) A satellite communications service that utilizes low-level satellites for communications over a widespread geographical area.

Load Balancing A reference to the designing/engineering a PBX or central-office switch so that each network group shares the traffic work load or the design/engineering of a data network to evenly share multiple communications paths. Some equipment (in both telephone and data) will automatically compensate for a lost communications path or network group.

Load Coil A load coil is a voice-amplifying device for twisted-pair wire. A load coil is usually placed on each twisted pair used for a voice line every 3000 feet past a central office. Coils are usually located in vaults, with twisted-pair splices. A typical load coil has an inductance of 30 mH. Other coils, used for other applications are usually referred to as *choke coils*.

Load Coil Detector A test device used to detect unseen load coils on a pair of wire.

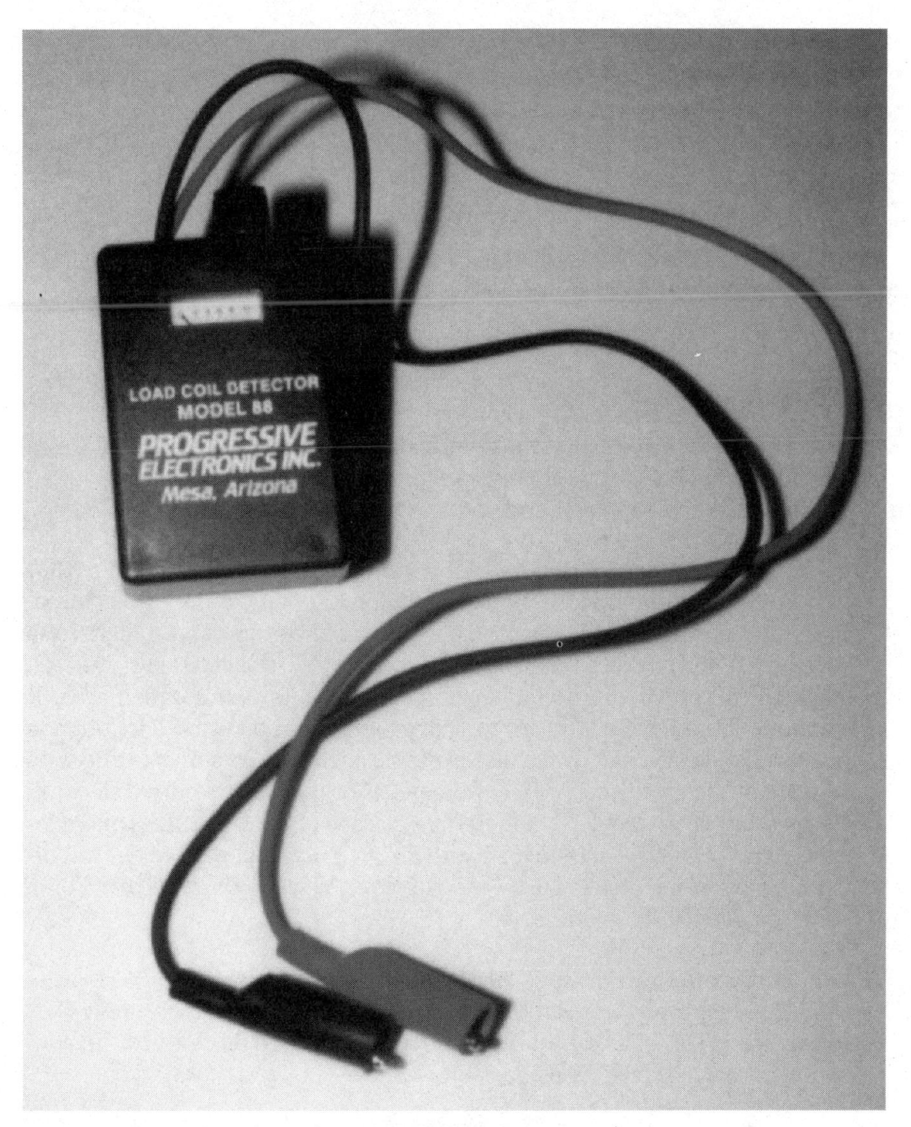

Load Coil Detector

Loaded Line A twisted copper pair that has load coils in place. Loaded lines are for voice telephone lines. To make a digital signal, such as a T1, work on a twisted pair, the telephone company must first remove the load coils.

Local-Access Transport Area (LATA) Simply stated, a LATA is an area code. *Inter LATA* refers to services that go from one area code to another, like long-distance telephone calls. *Intra LATA* refers to services that originate and terminate in the same area code.

Local Air Time Detail The list of phone calls itemized on a cellular or PCS telephone bill.

Local-Area Network (LAN) See *LAN*.

Local-Area Signal Service (LASS) See *LASS*.

Local Call A telephone call that originates and terminates within the same carrier's network or generally within the same area code. A call that does not incur additional charges for long distance. The determining factor between whether calls will be long distance or local is the tariff (laws that regulate cost of telephone service made by the FCC and local governing PUC) that the local telephone company operates.

Local Central Office (LSO) Local Serving Office. A central office that performs telecommunications switching for a specific number-plan area. The number-plan area is currently defined by the first three digits of a seven-digit telephone number. When number portability takes effect, a local central office will no longer be defined as its number-plan area. It will be defined by the laws set forth by the PUC and the area its outside plant reaches. Typical switching systems installed in central offices in North America are Lucent Technologies' 5ESS and Northern Telecom's DMS family of switches. There are five classes of central offices and a local central office is a class five. There are five major parts to a central office. As a whole, these parts are referred to as *inside plant*. For a diagram, see *Central Office*.

Local Distribution Frame Also called *Main Distribution Frame (MDF)*. The place where all the wire, fiber optic, or coax for a network is terminated. The distribution frame is usually placed as close to the central-office switch or PBX as possible.

Local Exchange A reference to the serving area of a central office. Until number portability takes effect, the first three digits of your seven-digit telephone number defines the exchange (specific central office) you are located in. After number portability takes effect, an exchange will be defined as the tariffs and laws set forth by the local governing PUC. Your telephone number will be associated to you, rather than a central-office equipment port or address. You and your phone number will be tracked in a national database that allows you to take your phone number anywhere you move to, or any phone company you switch to, within an area code. Eventually, you will even be able to transfer your number to a cellular or PCS phone.

Local Exchange Carrier (LEC) Most commonly, a reference to one of the seven RBOCs. A telephone company that provides telecommunications services to end users and bills them for it. A local exchange carrier has switching networks and outside plant or cellular service to serve its customers. Examples of local exchange carriers are: NYNEX, USWest, GTE, PAC BELL, Southern Bell Telephone, and a myriad of cellular telephone-service providers. The newest breed of Local Exchange Carriers are CLECs (Competitive Local Exchange Carriers). They have all the same services as the RBOCs except they are generally only located in large metropolitan areas. Some larger CLECs include TCG (Teleport Communications Group), ELI (Electric Lightwave, Inc.), Brooks Communications, and MFS (Metropolitan Fiber Systems). CLECs use (almost exclusively) SONET in conjunction with DCS as the foundation of their network architecture.

Local Loop The pair of wires that extends from the local telephone company's central-office main-distribution frame to the customer's premises.

Local Service Area The geographical area that a customer can make calls without being billed additionally for long distance.

Local Tandem A telephone company central-office switch that has the ability to connect or "switch" calls from two different central offices.

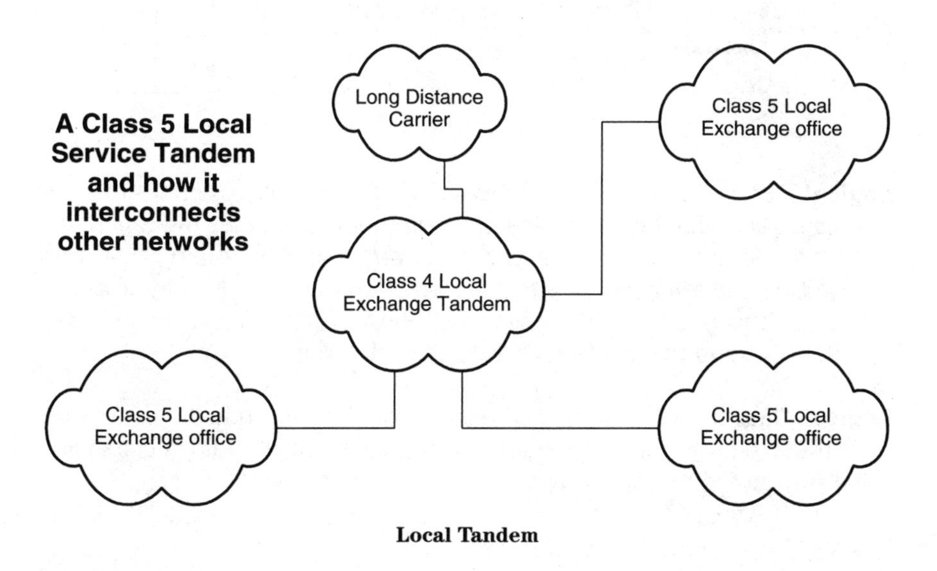

Local Tandem

Local Trunk A trunk that is fed by a local central-office switch, as opposed to a long-distance carrier WATS service.

Lock Code A code that a cellular telephone user can dial into their phone to prevent any calls from being made on it. After the code is entered again, the phone is able to make calls.

Logic A mathematical process first developed by the Irish mathematician George Boole in the 1850s. The premises of logic is to know if a certain statement is true or false. An example is "the light is on." This statement can only be true or false. Logic couples this statement with others, such as "the switch is on, the power is on; therefore, the light must be on." If the switch is off and the power is on, then the light is off. If the switch is on and the power is off, then the light is off. These statements depict the truth table for an AND electronic logic gate, which is a primary building block of microprocessors. In the table depicted, the light switch would be "A," the power would be "B," and the light bulb would be "C." Truth tables are written in ones and zeros, rather than ons and offs. The science of this math is called *Boolean Algebra*. It is a book in itself and is usually explained quite well in textbooks that cover digital electronics.

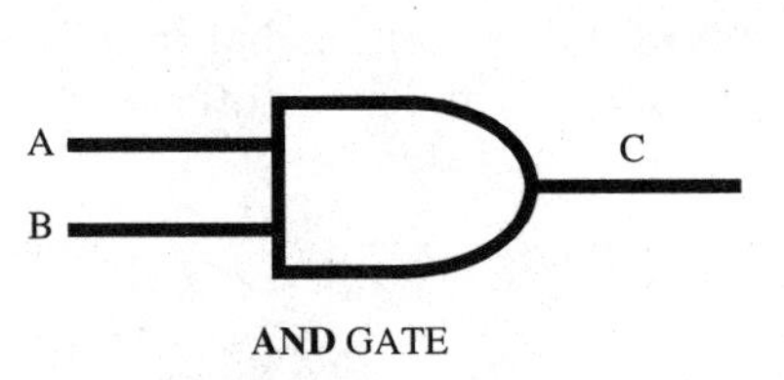

TRUTH TABLE

A	B	C
0	0	0
0	1	0
1	0	0
1	1	1

Logic

Logical Bus A LAN logical topology. The logical topology defines the way that a LAN communicates. The physical topology defines the way that a LAN is physically wired. For example, even though an ethernet network might be physically wired into the formation of a star, it really works as though it were a bus. The wire is just physically laid out and connected differently and the electronics are a little different.

Logical Channel A multiplexed channel or frame-switched media (such as frame relay), where there is no physical wire, fiber, radio or coax path, but within the protocol stack is a communications path.

Logical Link See *Logical Channel*.

Logical Link Control (LLC) The part of a communications protocol that handles the assembly and disassembly of packets. This function belongs to the data-link layer in the OSI model. LLC is a very important part of ethernet, frame relay, and other packet-switching protocols.

Logical Ring A LAN that operates as a ring-type protocol (logical topology)—even though it is physically wired as a star (physical topology).

Long Distance A telephone call or telecommunications service that originates within one local service area (usually an area code) and terminates in another.

Long Haul See *Long Distance*.

Long Reach A reference to SONET fiber-optic spans or links longer than 25 kilometers.

Longitudinal Redundancy Check (LRC) A method of checking for errors in communications transmissions by combining vertical error checking and longitudinal error checking. A transmission device sends data in bytes that are logically stacked on top of each other. The stack forms a block. The last bit of each line is used to form a check sequence. LRC is about 85% accurate in detecting and retransmitting blocks that contain errors. The newer method of error checking is *CRC (Cyclic Redundancy Checking)*.

DATA BLOCK

Even parity

byte 1	1	0	1	1	0	1	1	1	0
byte 2	1	0	0	1	1	0	1	0	0
byte 3	0	1	1	0	0	0	0	1	1
byte 4	1	1	1	1	1	1	1	1	0
byte 5	0	0	0	0	0	0	0	0	0
byte 6	1	1	1	1	0	0	0	0	0
byte 7	0	0	1	1	0	0	1	1	0
byte 8	1	0	1	0	1	0	1	1	1
byte 10	1	1	0	0	1	0	1	1	

Longitudinal Parity Sequence (byte 9) SHADED
Vertical Parity sequence (byte 10) bottom row

LRC BIT STREAM

byte 1	byte 2	byte 3	byte 4	byte 5	byte 6	byte 7	byte 8	byte 9	byte 10
10110111	10011010	01100001	11111111	00000000	11110000	00110011	10101011	100001	11001011

Longitudinal Redundancy Check (LRC)

Look-Up Table A translation table in a PBX system. Translation tables convert "dialed number" protocols into numbers that the public telephone network can recognize.

Loop Also called a *Local Loop*. The pair of wires that extends from the local telephone company's central-office main-distribution frame to the customer's premises.

Loop Antenna A directional antenna used mostly for UHF receptions. Some older TVs use small UHF loop antennas to receive broadcast channels from 13 to 83.

Loop Back Also called *Line Loop Back*. A troubleshooting function of CSU/DSU equipment and smart jacks, where the receive pair of a circuit is connected directly back into the transmit (or a person manually disconnecting DTE and connecting receive to transmit). The object is to test the transmission line. If the transmission equipment transmits a signal that is "looped back" to it and it receives its own signal with no errors, then the line is ok. If there is a problem, it is inside or it is beyond the receiving equipment. For a diagram and photo, see *Line Loop Back*.

Loop Extender An add-on device for a PBX switch or central-office switch that allows operation over an abnormally long loop or twisted pair, usually over 12,000 feet for a central-office switch and 1500 feet for a PBX. Loop extenders are also called *OPX (Off Premises Extension) adapters* for extending a station/extension to a remote location (over 1500 feet).

Loopstart Line A line that comes from a central office. The type of line determines which type of signaling the line requires to work. If a line is dedicated to one phone or group of phones (like in your house), it is a "line." If the line is going to be shared among many devices connected together by a PBX or key system, then the line is called a *trunk*. A loopstart line is a two-wire central-office trunk or dial-tone line that recognizes an "off hook" situation when a telephone switch-hook puts a 1000-ohm short across the tip and ring when the handset is lifted. This is the most common type of line. It is also called a *POTS line* and *plain-service line*. Other types of lines or trunks are: ground start and E&M trunks (ear and mouth, an old 6-wire version of a T1), ISDN PRI, and ISDN BRI.

Loop-Start Trunk A trunk is a line that comes from a central office. The type of trunk determines which type of signaling the line requires to work. A loop-start trunk is a two-wire central-office trunk or dial-tone line that recognizes an "off hook" situation when a telephone switch hook puts a 1000-ohm short across the tip and ring when the handset is lifted. This is the most common type of line. It is also called a *POTS line* and *plain-service line*. Other types of trunks are ground start, and E&M trunks (ear and mouth, an old 6-wire version of a T1), ISDN PRI, and ISDN BRI.

Loop Up/Loop Down To loop up is to put a CSU/DSU in loopback mode. To loop down is to remove the loop back and resume normal operation.

Loose Tube Buffer A PVC tube that is about as big around as a drinking straw that has up to twelve optical fibers within it. The idea behind a loose-tube buffer is that when a cable is bent, the fibers inside will have slack and freedom to move and naturally adjust to the bend. If a filler is inside the loose tube, it would crush or crack (fracture) the optical fibers when the cable is bent. This would render them useless.

Loss The reduction of a signal's voltage level as it travels down a line, measured in decibels. Attenuation is also called *Loss* because some signal is always lost through resistance and reactance. Optical lightwave signals are also attenuated when they traverse through a fiber optic because of impurities in the fiber optic and the fact that light intensity decreases with distance.

Lost Call A call that did not complete or was blocked because of a lack of switching facilities.

Loudspeaker Paging A feature of PBX and key systems that allows a user to connect the telephone system to an external paging amplifier and speakers. The interface is usually broken out onto a 66 block, then cross connected to the input of a paging amplifier. When a person wants to page someone, the telephone system will prompt the user to choose *internal* or *external* (or zone 1, 2, or 3). If the user chooses *external*, then the page is heard over the loudspeakers driven by the external amplifier instead of the telephone set speakers.

Low Frequency The range of frequencies between 30 and 300 kHz.

Low-Pass Filter An electronic device that eliminates frequencies above a specified frequency. The two categories of frequency filters are active and passive. Active filters use active devices that require power, such as transistors and 541 op amps to amplify the desired signal and attenuate the undesired signal. Passive filters are made with components that do not require external power, such as capacitors and inductors. Capacitors and inductors have reactive properties that cause them to resist or pass an AC signal.

LPT Port A logical designation for a group of I/O addresses that "tells" a computer which "plug" to send the printer communications to. LPT ports are usually designated LPT 0, LPT 1, and LPT 2.

LRC (Longitudinal Redundancy Check) See *Longitudinal Redundancy Checking.*

LSB (Least-Significant Bit) The bit in an octet that carries the least value. You can better understand this by comparing it to our base-10 numbering system. Imagine a "least-significant number." If you are 43 years old, the 4 is the most significant number and the 3 is the least-significant number. If the 4 were lost, then you would only be three (a very significant difference), it is the most significant compared to the 3. If the 3 were lost, you would still be 40. LSB is another way of saying "least-significant digit," which is used to round numbers off in elementary mathematics. In T1 in-band signaled circuits, least-significant bits are robbed from the bit stream of the 6th and 12th sample in each channel. The voice-sample bits are replaced with signaling information and maintenance information bits.

LSI (Large-Scale Integration) Microchip ICs are classified as *SSI (Small-Scale Integration), MSI (Medium-Scale Integration), and LSI (Large-Scale Integration)*. SSI ICs contain 12 or fewer devices, such as logic gates or transistors. MSI ICs contain 13 to 99 devices and LSI ICs contain 100 or more devices. The typical CPU, such as a Pentium (Intel trademark) microprocessor contains hundreds of thousands of devices. For a photo of a VLSI device, see *Microchip*.

Lug More frequently called a *binding post*. A lug or binding post is a small threaded bolt with a nut used to attach wires. The binding post usually has a number, which is a reference used to identify where twisted copper pairs are terminated in access points, cross boxes, and terminals. When a technician looks for a specific pair in a cable (called a *cable pair*), they refer to documents that list the pairs and which binding posts they are spliced to.

M

M1 A reference to the Meridian 1 *PBX (Private Branch Exchange)* switching system manufactured by Northern Telecom.

Ma Bell A reference to AT&T—the company that is said to have given birth to the Baby Bells, better known as the RBOCs.

MAC 1. *Moves, Adds, and Changes*. An abbreviation for modifications to a network. MAC-type requests include everything, except maintenance or repair of equipment. Within the operation of most phone companies and network-service companies are two types of requests made by customers: MAC orders and maintenance orders. 2. *Media Access Control* is a part of the data-link layer in the OSI model. It interfaces the bits to be transmitted with the physical transmission media.

MAC Address The address for a device that is tracked by the data-link layer in the OSI model. It is usually a 48-bit ID code that is a part of a network interface card.

Machine Language The lowest-level programming language. PROMs are the key to machine language. They contain instructions that assist the microprocessor in decoding the 8-, 16-, or 32-bit instructions into functions that the microprocessor executes. The instructions (machine-language scripts) are burned into the PROM when it is programmed. Typical machine-language instructions include: MOV A,M, which is actually entered as an OP-Code of 167, instructs a microprocessor to move the contents of memory address A to memory address M. Another machine instruction is OUT, entered as an OP-Code of 323, which instructs a microprocessor to move the contents of the previous memory address to a port that is identified in the next instruction.

Magnetic Ink Ink that is used to print information that will be read electronically. Magnetic ink is made with ferrous compounds. American Express uses magnetic ink on some of their printed materials and typical bank-account checks have the account number printed in magnetic ink so that they can be electronically processed.

Magnetic Storage A method of storing data by magnetizing a tiny section of a tape or disk for each bit. Hard disk-drives and floppy-disk drives utilize disks coated with ferromagnetic materials. Data cartridges contain thin plastic tape coated with ferromagnetic materials that are recorded and read in a similar fashion. Analog information can also be stored magnetically; cassette tapes are a common example.

Magnetic Stripe The stripe on the back of a credit card or other device. Magnetic stripes are usually used to store information, such as a name and account number, in a binary bar-code format.

Main Distribution Frame Also called a *distribution frame*. The place where all the wire, fiber optic, or coax for a network is terminated. The distribution frame is usually placed as close to the central-office switch or PBX as possible. For a photo, see *Distribution Frame*.

Main Feeder An F1 (first facility) cable from a Central Office. The feeder cable runs to cross connect points in the telephone network where F2 (second facility) cable feeds are connected/cross connected.

Main Frame A large computer capable of retrieving information from mass-storage units and calculating/processing the data in a very short time in comparison to a client-server computing process. Main-frame computers have been regaining favoritism in large data-processing environments because of their outstanding reliability and processing power.

Main PBX A primary PBX that interfaces with the public telephone network via CO (central office) trunk lines. The other type of PBX is an off-premises (remote) PBX, which is connected to the outside world or public network by switching through a main PBX.

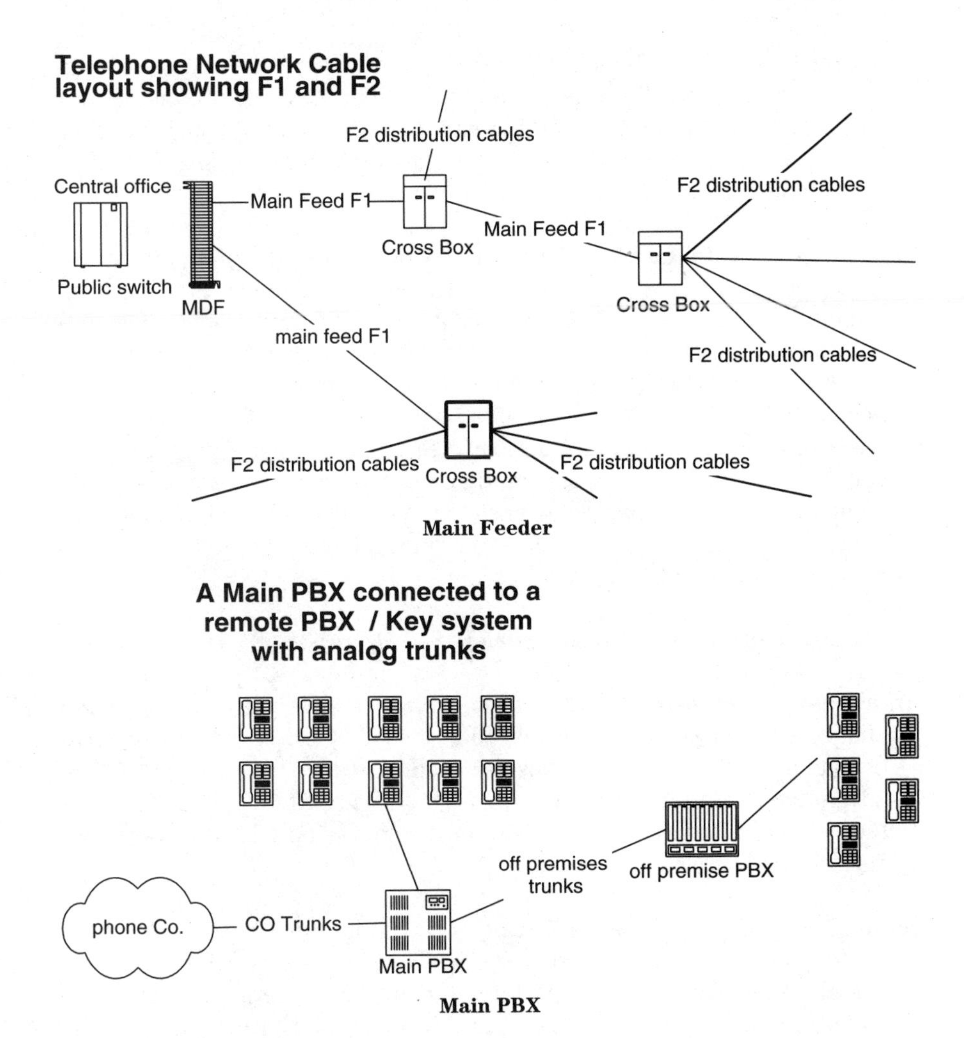

Main Feeder

Main PBX

Make Busy 1. A reference to the activation of the "busy out" feature of an ACD telephone. If the telephone is busied out, the ACD system will not transfer calls to that telephone. This is useful when an agent is on break or their shift is over. 2. A temporary fix or condition of a phone service. To "busy a line out of a hunt sequence." If a business phone line becomes defective and it is in a hunt or roll-over sequence, calls will not hunt or roll past this line. For example, four lines come into your business. The first line is the main number; if that first line is busy, then calls come in on the second line, etc. If line one goes bad, it can't be called, so it can't be busy. Because it is not busy, calls will not hunt or rotate to the next three lines. When you call the phone company repair service, they busy out the bad line, which makes it look busy to the network. Your calls then start coming in on the other three lines. When a repair technician finishes repairing the problem on the bad line, he has it unbusied. Another temporary fix is to call forward the line from the central office. The phone company can do this at the customer's request.

Malicious Call Trace A feature offered by local telephone companies. Even though all calls are kept in an archived database, it is sometimes difficult to locate a single call—even if an accurate time is given. The malicious call trace or annoyance call trace "flags" a call when a customer hangs up and dials the call-trace feature code after receiving an annoyance call. Usually, the charge is $2.00 per trace. After the trace or "flag" has been made, telephone company security officials investigate the source of the call. If there are multiple occurrences, then the telephone company will press charges against the malicious caller. The person receiving the malicious calls will never find out who the caller is unless they are summoned to a court hearing.

MAN (Metropolitan Area Network) See *Metropolitan Area Network*.

Manchester Encoding A digital transmission technique that uses voltage change instead of voltage levels to represent bits. A zero to positive change indicates a "1" and a positive to zero change indicates a "0." The advantage to Manchester encoding is that it is self timing. Its disadvantage is that it is very susceptible to noise, so it is only used in very specialized applications.

Manual Ring-Down Line Not really a phone line, but two phones connected together via a pair of wires and a talk battery (9 V to 24 V). Signaling, such as ringing, is performed manually, by flipping a switch on and off rapidly, which disconnects and connects the battery. The changing voltage imitates a weak ring voltage. Rescue teams use manual ring-down lines in cave and mine-shaft rescue operations because their radio range is very limited in underground tunnels.

Map A reference to "mapping"—a virtual tributary though an *Optical Carrier Circuit (OC-3)* over a SONET ring or end-to-end path. Mapping the tributary involves telling the SONET equipment which and how much bandwidth within the OC (Optical Carrier) will be designated a channel. The choices are DS0, DS1, DS3, and STS-1. STS-1 is an electrical version of an OC-1.

Mapping See *Map*

Marine Telephone A radio telephone that is designated specific operating frequencies by the FCC. It is not cellular (cellular is a short-distance radio application), it broadcasts with more power and a much greater distance. Tellabs manufactures marine telephone equipment.

Marker Tape A plastic label built into cables that have cable ID and specification information printed on them by the manufacturer. Marker tapes are uncommon in newer polyethylene cables, because it is much easier and less expensive to print the cable designation on the outside of the cable.

MARTians *Misaddressed or Routed Telepacket* on a LAN or WAN.

Mast Clamp A device used to attach a ram hook to a power mast. The ram hook (also called a ram horn) is used to attach an aerial service wire via a drop clamp.

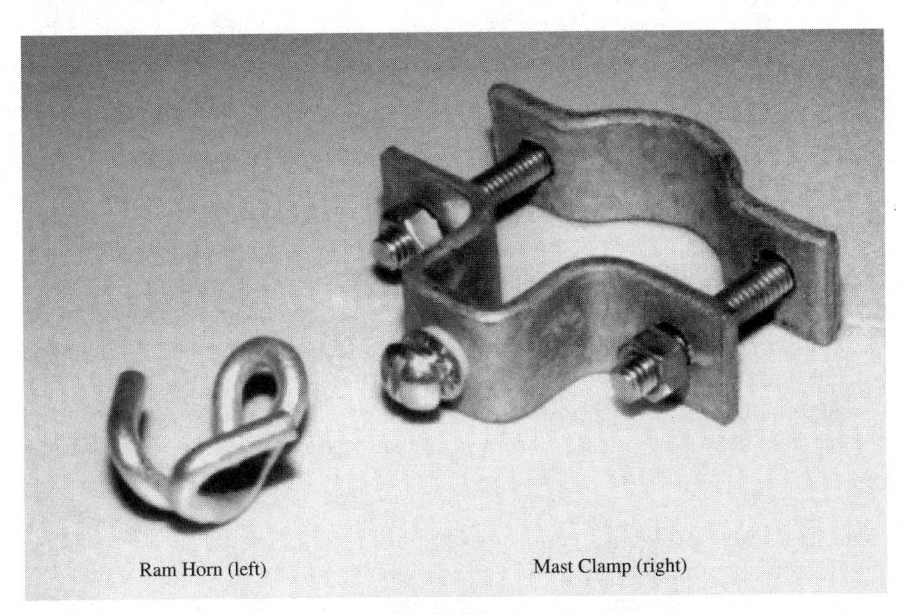

Mast Clamp

Master Clock A reference to a BITSs clock. A central timing device for synchronous networks, such as SONET networks. Bits clocks can be rack mounted, just like other telecommunications equipment.

MAT (Meridian Administration Tools) A Northern Telecom CTI application that allows a Meridian PBX system to be managed through a

GUI environment over a LAN or single PC. It enables a user to make administrative changes to the system by clicking on the picture of an item to be changed (such as a feature button on a telephone or the name display) and typing in the change. It also provides excellent traffic and core analysis tools, which graph the busy hours by network group. Call accounting is also a feature of MAT.

Matrix The part of a switch that carries and routes calls. The matrix is a virtual part of the core that commands which channels to connect with what. As multiplexed bit streams run through a digital switch, they are separated and recombined from inputs (interface cards) to specified outputs (interface cards) by the core (CPU) of the switch. The matrix is not a tangible object; it is a combination of the CPU and interface equipment.

Mbps (Megabits Per Second) Equivalent to one million bits per second. Memory or data transferred per unit of time is measured in bits. Memory storage is measured in bytes. The difference in abbreviations is that bits are lowercase (b) and bytes are uppercase (B).

MDF (Main Distribution Frame) Also called a *distribution frame*. The place where all the wire, fiber optic, or coax for a network is terminated. The distribution frame is usually placed as close to the central-office switch or PBX as possible. For a photo, see *Distribution Frame*.

Measured Rate Service Abbreviated *1MR* for residential and *1MB* for business, this type of telephone service is offered by local telephone companies. Measured-rate service means that a line is billed on a "per call basis." Telephone companies in the Southern and Western United States have tried to abolish measured service by encouraging customers to subscribe to flat-rate services, abbreviated *1FR* for residential and *1FB* for business use.

Mechanical Splice An alternative fiber-optic splice to fusion splicing. Fusion splicing equipment is very expensive ($40,000 is typical for a fusion splicer). Mechanical splices come as a kit, which connectorizes the ends of the fibers. A tool kit is required for mechanical splicing. It consists of a microscope, polishing puck, cleavers, epoxy, and polishing compound. They cost about $1,200. An oven used to "hot cure" the epoxy is also available. With a mechanical splice, you cleave or cut the end of the fiber as square and smooth as possible, then epoxy the fiber end into a connector. The epoxy takes about 12 hours to cure without an oven and about 20 minutes with an oven. After the epoxy has cured, the tip of the connector (which should be flush with the end of the fiber optic) is polished by holding it with a device called a *puck* (it is shaped like a hockey puck). The puck holds the fiber connector while it is gently rubbed against a pad coated with polishing compound. When the polishing is done, the connector is ready to be mated with another connector and the splice is complete. Mechanical splice kits cost about $15.00 per splice and are available in SC- and ST-style connectors.

Mechanized Line Testing (MLT) Also called *DATU (Direct Access Test Unit)*. MLT and DATU equipment is either added on or built into a central-office switch. DATU allows a technician or customer-service agent to dial the phone number of the DATU or MLT equipment and execute a test for shorts, opens and grounds remotely. In response to a digital voice, the technician enters a password and a choice of options. The results of the test can be read back to the technician by a digital recording or sent to them via an alpha-numeric pager. DATU units can also send a locating tone on the technician's choice of TIP, RING, or both TIP and RING. The test unit can also short lines and remove battery voltage for testing.

Media Access Control (MAC) A part of the data link layer in the OSI model. It interfaces the bits to be transmitted with the physical transmission media.

Media Interface Connector A fiber-optic connector.

Medium Frequency The range of radio frequencies from 300 to 3000 kHz.

Mega (M) The prefix for million. Sixteen megabytes is equal to 16,000,000 bytes, and would be abbreviated 16MB.

Megabyte One million bytes. Mega is abbreviated "M" and bytes are abbreviated "B." Sixteen megabytes is equal to 16,000,000 bytes and would be abbreviated 16MB.

Megahertz One million hertz. Mega is abbreviated "M" and hertz is abbreviated "H." Sixteen MHz is equal to 16,000,000 Hz (*hertz* is another word for cycles in radio frequency).

Megohm One million ohm. Mega is abbreviated "M," and the symbol for ohms is "Ω". Sixteen megabytes is equal to 16,000,000 ohms (ohms are a measure of resistance to electricity) and would be abbreviated 16MΩ.

Member Nortel's name for a trunk. For example, a T1 would contain 24 members. See also *Route*.

Memory Electronic memory comes in two families, *ROM (Read-Only Memory)* and *RAM (Random-Access Memory)*. Memory devices are made from two different technologies, bipolar (TTL) and *MOS (Metal-Oxide Semiconductor)*. Memory is stored by a technique called *writing* and retrieved by a technique called *reading*. ROM devices can only be read and are programmed during manufacture. *PROM (Programmable Read-Only Memory)* devices can be programmed at a later date by an electronics reseller or electronic assembler for a special application using special equipment. Special ROM devices called *EPROMs (Erasable Programmable Read-Only Memory)* can be electronically erased and re-used. RAM has read and write capability. The term *random* access means that any memory address can be read in any order at any time. The two types of RAM are static and dynamic. *Static RAM* can hold its memory even when power is removed. *Dynamic RAM* needs constant power to refresh its memory. The following family diagram illustrates the memory types and the technology with which they are made.

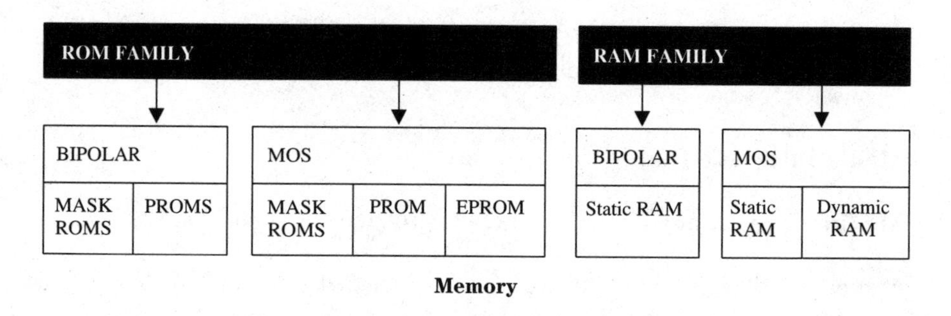

Memory

Meridian 1 A large-scale *PBX (Private Branch Exchange System)* manufactured by Northern Telecom. For a photo, see *Private Branch Exchange*.

Meridian Administration Tools (MAT) See *MAT*.

Message Transfer Part The function in SS7 networking that packetizes and de-packetizes signal data.

Message Waiting Usually a light on a telephone that indicates that the user of that phone has a voice mail or a written message left with a hotel clerk or administrative person.

Meters to Feet Conversion One meter equals 3.28 feet. One kilometer equals 3280 feet.

Metropolitan Area Network (MAN) A computer network that incorporates the local telephone company's facilities to communicate. MAN networks connect other LANs or computers in a city together. T1 private lines are popular for MAN applications.

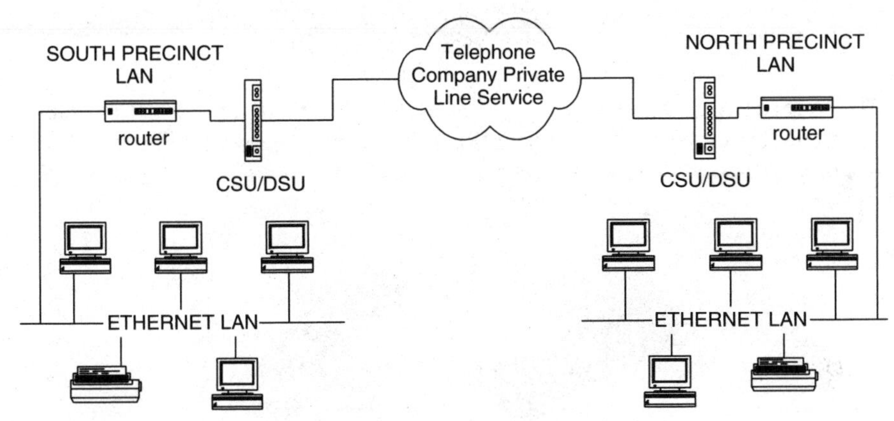

Metropolitan Area Network (MAN)

Metropolitan Statistical Area (MSA) A term that refers to the 306 metropolitan areas that the FCC manages cellular and *PCS (Personal Communications Service)* communications in. There are also *RSA (Rural Statistical Area)* markets that the FCC determined as separate from each other. 428 RSA markets are in the United States. Each statistical area, 734 in all, has at least two licensed service providers.

MFD A less-common abbreviation for microfarad (μF). See *Microfarad.*

MFJ (Modified Final Judgment) The judgment that ruled against AT&T—a telecommunications monopoly. AT&T was divided into long-distance, and local telephone service. The local telephone service part of the company (actually 22 companies) were formed into seven *RBOCs (Regional Bell Operating Companies)*. The MFJ also included rules governing the business of the telephone companies involved and set tariffs/pricing limitations on telecommunications services nationwide.

MHz (Megahertz) See *Megahertz.*

Mho Slang for Siemens, conductance. Conductance is often confused with being the opposite of resistance, which is not the case. Conductance is the reciprocal of resistance. To get the reciprocal of a resistance, simply take one divided by that number, or the resistance. If the resistance of a circuit or component is 500 ohms, then the conductance is equal to 1/500, which is 0.002, 2 millisiemens (2 mS). The higher the number in siemens, the higher the conductance and the lower the resistance.

Micro The prefix for one trillionth. Abbreviated "μ," the Greek letter mu. One microfarad is equal to one trillionth of one farad and would be written 0.000,000,000,001 Farad, and abbreviated 1μF.

Microchip A reference to a *VLSI (Very Large-Scale Integration)* electronic device. For more information on microchips see *Very Large-Scale Integration*.

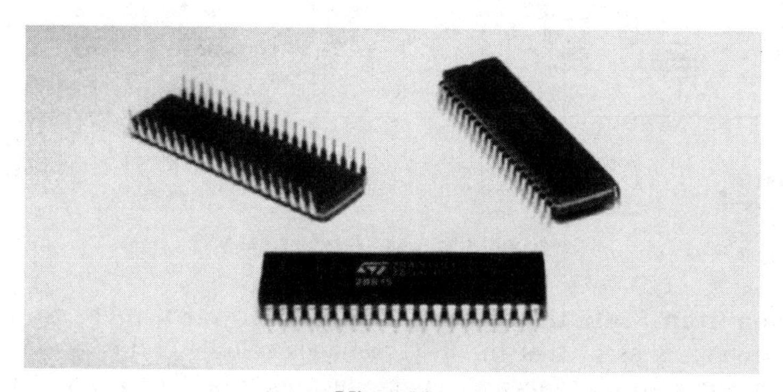

Microchip

Microfarad Usually represented as μF. Farad is the standard unit of capacitance. A capacitor is an electronic device that has two special properties. It only allows alternating current to pass through it, and it can store an electric charge. One of the many applications of capacitors is to filter alternating current (AC) out of DC power supplies and rectifiers. This is done by placing a capacitor from the DC output to ground. The capacitor appears as an easier path to voltage fluctuations and RFI, and as an impossible path to direct current (DC). Physically, a capacitor is two plates of metal, separated by an insulator (mylar is common). The physical size of a 1-F capacitor would be two sheets of tin foil the size of a football field, insulated (or separated) by a thin sheet of mylar. The farad is a huge unit of capacitance. This is why most capacitors are microfarads (μF) in value. For a schematic symbol of a capacitor, see *Capacitance*.

Micron A standard unit of measurement that is equal to 1⁄1000 of one millimeter or 1⁄25,000 of an inch. The core and cladding of fiber optic is measured in microns.

Microprocessor Also called a *CPU (Central Processing Unit)*. The device within a computer (or switch or other machine that performs complex tasks) that controls the transfer of the individual instructions from one device connected to its bus (the data or I/O bus) to another, such as ROM, RAM, subcontrollers, decoders, and I/O Ports. Some communications equipment manufacturers actually call a certain card or portion of the system the *CPU*. That is because they include all of the RAM, subprocessors, buffers, clocking circuitry, and ROM as a part of the CPU.

Microwave In telecommunications, this is usually a reference to a terrestrial microwave link. The link is made by two radio transceivers equipped

Microwave

with parabolic dish antennas pointed directly at each other. Radio can carry point-to-point transmissions of many bandwidths, including DS1, DS2, DS3, STS1, and OC1. Their range can vary, depending on the size of the antenna (dish), weather in the region, and the amount of power emitted. Including all of the previous factors, a link can range from 0 to 50 miles. For a diagram of a microwave system, see *Terrestrial Microwave.*

Mid-Span A telephone service wire that runs from a pole to a hook attached to a cable strand, then to a house or building.

Mileage of Circuit The mileage of a private-line circuit is calculated using V and H coordinates. For a table of V&H coordinates, see *Airline Mileage.* AT&T developed a grid-coordinate system that gives every telephone central office in the United States a vertical and horizontal grid number. To calculate the mileage between two cities, the Pythagorean theorem is used. For an example of calculating airline mileage, see *Airline Mileage.*

Milli Milli is the prefix for one-thousandth, abbreviated "m." Five mA is equal to five thousandths of an amp and is written as 0.005 A or 5 mA.

MIS (Management Information System) Also called *IS (Information Systems).* The part of a company that cares for data and voice communications/processing. The two have been merging together over the past decade and are becoming one entity as communications technology advances. The latest craze in MIS is *CTI (Computer Telephone Integration),* which enables users to track telecommunications events and operate telecommunications equipment on a computer, with a *GUI* (*Graphical User Interface*, such as Windows). *IVR (Integrated Voice Response)* is a form of CTI.

MLT (Mechanized Line Testing) Also called *DATU (Direct-Access Test Unit).* MLT and DATU is equipment that is either added on or built in to a central-office switch. DATU allows a technician or customer-service agent to dial the phone number of the DATU or MLT equipment and execute a test for shorts, opens, and grounds remotely. In response to a digital voice, the technician enters a password and a choice of options. The results of the test can be read back to the technician by a digital recording or sent to them via an alphanumeric pager. DATU units can also send a locating tone on the technicians choice of TIP, RING, or both TIP and RING. The test unit can also short lines and remove battery voltage for testing.

Mnemonic A computer programming command that is an abbreviation or shortened version of what the command does. PRT is a mnemonic in Northern Telecom applications software that makes a switch "print" a specified list of information. LOGI is a mnemonic for "log in."

Mobile A communications link made by portable radio.

Modal Dispersion As light travels down a fiber optic, each individual light ray/particle takes a different path. Imagine that a bunch of small rubber balls are shot down a long tube at the same time. Each ball will bounce differently as they make their way around curves. At the end of the tube, the balls will come out at different times. Light behaves the same way. A sudden pulse of light on one side of a fiber optic will disperse itself as it traverses down the fiber, causing the pulse of light at the far end to be more of a "blip."

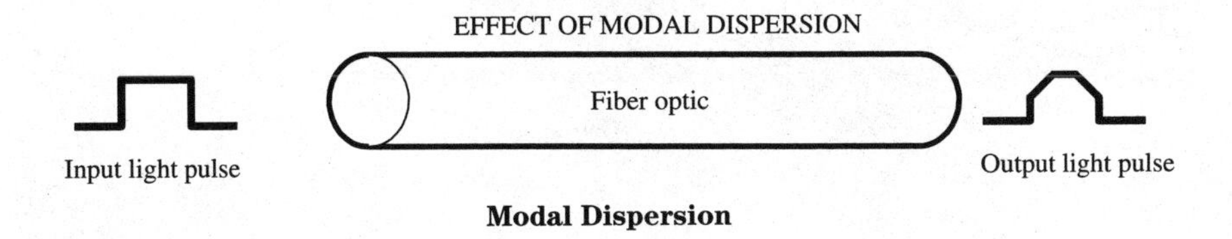

Modal Dispersion

Modal Loss The attenuation of a light signal as it travels through a fiber optic because of tight bends. See *Modal Dispersion.*

Modem (Modulate/Demodulate) A device that transmits digital information over a telephone line (standard POTS line) or a private circuit (56K line). Modems modulate the digital information before transmitting it. One standard of modulation is *FSK (Frequency-Shift Keying)*. Each positive (1) bit is sent as a frequency or "pitch" of sound and each (0) bit is transmitted as a different frequency or "pitch" of sound.

Modem Standards

Standard	Baud rate	Modulation	Duplex
V.21 / Bell 103	300	FSK	Full duplex
V.22	1200	DPSK	Full duplex
V.22 bis	2400	QAM	Full duplex
V.23 / Bell 202	1200/75	FSK	Half duplex
Bell 212A	1200	DPSK	Full duplex
V.32	9600	QAM	Full duplex EC
V.32 bis	14,400	TCM	Full duplex EC
V.32 ter	19,200	TCM	Full duplex EC
V.34	28,800	TCM	FULL DUPLEX EC
			EC=Error Correction

Modular A reference to equipment that is equipped with plug-in type interfaces, rather than being hard wired.

Modular Adapter A device used to interconnect one wire/cable type with another, without the use of termination blocks. Also called *harmonica adapters*. See also *258A Adapter*.

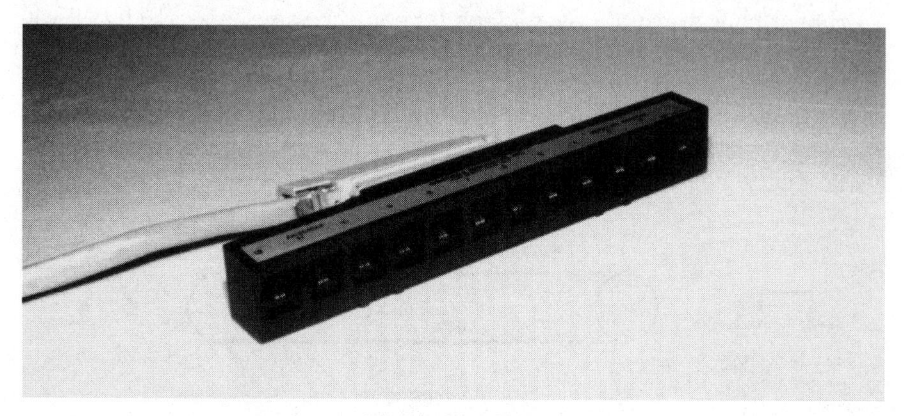

Modular Adapter

Modular Jack A jack that is equipped with a plug so that devices can be easily attached and detached. Old jacks, still found in old homes, are hard wired, which means that the telephone cord had to be permanently affixed to the terminals inside the jack with screws. The same went for nonmodular or hard-wired telephones. If you wanted to have a longer cord, you couldn't buy one at the store and just plug it in. You had to call the phone company and they would send a telephone technician out to install a longer cord for you.

Modulation A method of varying a radio carrier frequency so that a signal (the variations) can ride on it. After the carrier signal has the variations imposed on it, it is amplified and transmitted. The variations in the signal are then detected by the receiver. The variations in the carrier signal are actually voices, music, or whatever is to be transmitted. The different methods of modulating a carrier frequency are *AM (Amplitude Modulation)*, *FM (Frequency Modulation)*, and *PM (Phase Modulation)*.

MOH (Music On Hold) A feature of PBX and key systems that allows an audio signal from a tape recorder, radio, or other audio device to be fed to callers that are on hold. The PBX or key system user manual designates how to cross connect the audio signal into the system.

Monopole Antenna An antenna mast of one pole extending from the ground. These are popular with cellular and PCS wireless services.

Monopole Antenna

MOP (Method Of Procedure) When network engineers design an addition to a network (a new node in a SONET ring is a good example), they write a MOP. The MOP instructs technicians step by step which circuits to reroute, which circuit cards to swap and when, and when to activate the new node. MOPs are a crucial communication tool between engineers and technicians.

Morse Code In 1836, Samuel F.B. Morse built the first working telegraph. He also derived a code that enabled people to exchange information. The Morse Code is still in use today. It is used by amateur radio operators, ships at sea, etc.

MORSE CODE

A • —	N • —	1 • — — — —
B — • • •	O — — —	2 • • — — —
C — • — •	P • — — •	3 • • • — —
D — • •	Q — — • —	4 • • • • —
E •	R • — •	5 • • • • •
F • • — •	S • • •	6 — • • • •
G — — •	T —	7 — — • • •
H • • • •	U • • —	8 — — — • •
I • •	V • • • —	9 — — — — •
J • — — —	W • — —	0 — — — — —
K — • —	X — • • —	. • — • — • —
L • — • •	Y — • — —	? • • — — • •
M — —	Z — — • •	- — • • • • —

Morse Code

MOS Another reference to *CMOS (Complementary Metal-Oxide Semiconductor)*. The reason why many computer and other high-speed components are static sensitive. CMOS' largest advantage over TTL is their low power consumption (less than 1⁄10 of TTL); they switch on without drawing very much current in contrast to TTL technology. Because very little current is drawn, very little power is consumed and very little heat is given off. This allows the devices to be much smaller.

MPEG (Motion Picture Experts Group) Mostly known for the video compression methods MPEG 1, 2, and 4. MPEG compresses video by transmitting an initial picture, then transmitting only the changes in that picture, with CD-quality stereo sound. MPEG 1 is designed to carry a standard TV-quality picture, 30 frames per second with 352 × 240 pixels of resolution. The bandwidth required is a minimum 128Kb/s, (ISDN Line). MPEG 2 has a higher resolution that is compatible with HDTV. It transmits 30 frames per second with 720 × 480 pixels of resolution. The bandwidth required for MPEG 2 is variable. Different motion video pictures have different amounts of movement, so MPEG 2 lets the user decide how much bandwidth they want to use.

MQA (Multiple Queue Assignment) A reference to the ability of an ACD system to allow agents to log into multiple queues. A Northern Telecom Meridian PBX with MQA software is capable of allowing agents to log into and receive calls from five separate queues. When properly used, MQA makes call centers more efficient by allowing agents to share the incoming call load more effectively. See also *Skills Based Routing*.

MSA (Metropolitan Statistical Area) A term that refers to the 306 metropolitan areas where the FCC manages cellular and PCS communications. The FCC has determined 428 *RSA (Rural Statistical Area)* markets as being separate from each other. Each statistical area, 734 in all, has at least two licensed service providers.

MSB (Most Significant Bit) The bit in an octet that carries the most value. You can better understand this by comparing it to our base 10 numbering system. Imagine a "most significant number." If you are 43 years old, the 4 is the most significant number and the three is the least significant number. If the 4 were lost, then you would only be three (a very significant difference). If the 3 were lost, you would still be 40 (not so significant).

MSC (Mobile Switching Center) A place where cellular telephone call traffic is controlled. A cellular switch is used to perform the functions of the MSC. Bandwidth and cells are switched between users, and trunks that interface to landlines are also managed here.

MTSO (Mobile Telephone Switching Office) This is where all control is done for a cellular switching network within a LATA. Smaller MSCs hand off calls to the MTSO. The MTSO is where the billing, trafficking, maintenance monitoring, and hand-offs to long distance and local land-based carriers happens.

MULDEM (Multiplexer Demultiplexer) Another name for a multiplexer.

Multi Hop A reference to microwave links that require two or more links to get to a destination. Multi-hop links can extend distance and enable a more flexible path to go around buildings or mountains.

Multimeter An electronic test device used to measure voltage levels, electric current, and circuit resistance. Some multimeters are analog and some are digital. For a photo of a digital multimeter, see *Voltmeter*. For a photo of an analog cable test meter, see *145A*.

Multi Mode The alternative to *Single-Mode* fiber optic. Multi mode has a larger core (50 to 100 micron). Therefore, it accepts more light and more frequencies of light. Multi mode is used for shorter-distance applications, such as LANs. Single-mode fiber optic has a smaller core (5 to 15 micron), but is capable of longer-distance transmissions. It is used in the public network more often and is the choice for SONET applications. Multi-mode fiber optic is made with an orange-colored tube or insulation, and single mode is made with yellow.

Multi NAM A cellular phone that is programmed to have multiple phone numbers, usually two. Multi-NAM cell phones can have numbers that are subscribed to from different cellular companies.

Multiple Queue Assignment (MQA) A reference to the ability of an ACD system to allow agents to log into multiple queues. A Northern Telecom Meridian PBX with MQA software is capable of allowing agents to log into and receive calls from five separate queues. When properly used, MQA makes call centers more efficient by allowing agents to share the incoming call load more effectively. See also *Skills Based Routing*.

Multiplex Multiplexing is the process of encoding two or more digital signals or channels on to one. Channels are multiplexed together to save money. When we use all of the wires in a cable and need more, it costs less to add electronics on the ends of a cable than to install a new one (imagine the expense from LA to NY). A T1 encodes 24 channels into 1 by using frequency-division multiplexing. In a simpler explanation, a T1 makes it possible to place 24 lines that once needed 24 pairs on only 2 pairs. When a group of signals are multiplexed together, they are all sampled at a high rate of speed, faster than the combined speed of all the channels being multiplexed. For a diagram on the multiplexing process, see *Time-Division Multiplexing*.

Multiplexer

Multiplexer An electronic device that encodes several digital signals into a single digital signal for transmission on a single medium (such as a pair of wires). For a diagram on the multiplexing process, see *Time-Division Multiplexing*. Illustrated is an Alcatel DS3 to DS1 multiplexer. For another photo, see *Mux*.

Mushroom Board Also called a *white board* or *peg board*. It is placed between termination blocks (such as 66M150 blocks) to provide a means of support for routing cross-connect wire. For a photo, see *White Board*.

Music on Hold (MOH) See *MOH*.

Mute A feature of PBX and key telephones that turns off the microphone. Mute is also used in conjunction with handsfree to prevent the other party's voice from cutting in and out during a call.

Mux A shortened name for multiplexer. For a diagram of the multiplexing concept, see *Time-Division Multiplexing*. For an additional photo, see *Multiplexer*.

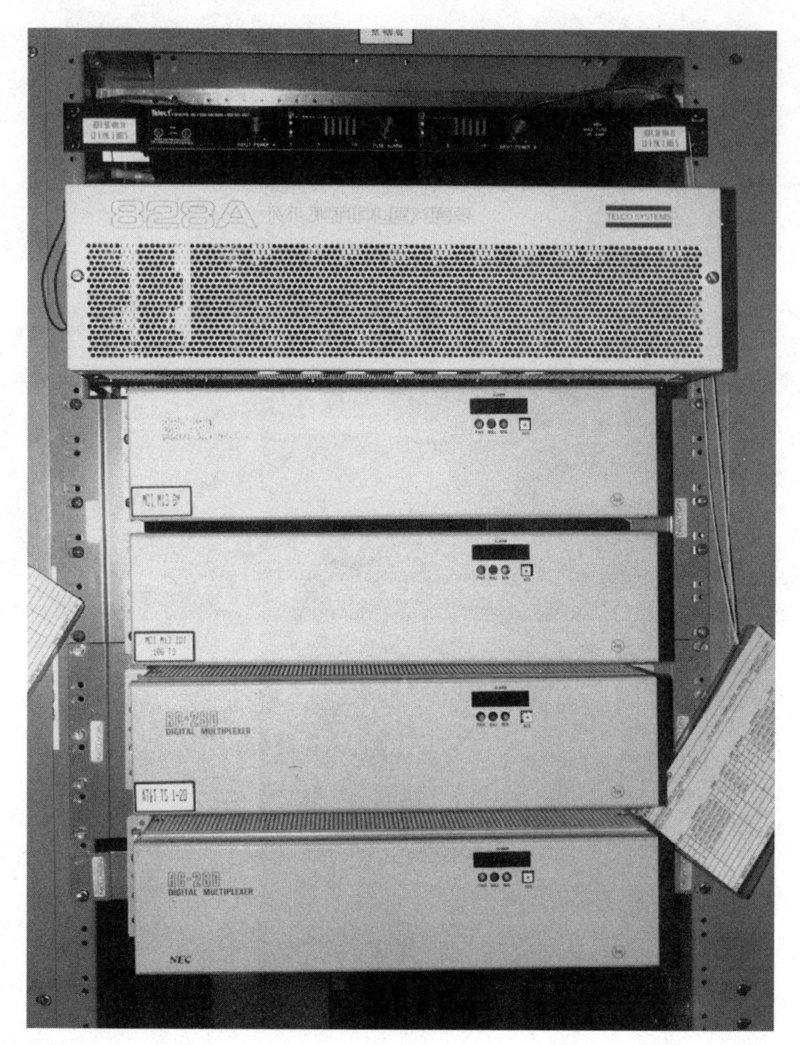
828A MULTIPLEXER
TELCO SYSTEMS
RC-280
DIGITAL MULTIPLEXER
AT&T TS 1-20
RC-280
DIGITAL MULTIPLEXER
NEC

Mux

N

NAK The ASCII control-code abbreviation for negative acknowledge. The binary code is 0101001 and the hex is 51.

Naked Call An incoming call that receives no call menu's or flexible routing before it is routed into an ACD queue.

National Access Fee A Federal tax placed on telecommunications services provided by telephone companies.

NDT (No Dial Tone) An abbreviation frequently used on telephone company repair orders and by service personnel.

NE (Network Element) A device on a network, such as a node, router, hub, server switch, DACS, modem, or PC.

Network Architecture The combination of software and hardware type of a network. Each network architecture can have one or more protocols within it.

Network Interface (NI) Also called a *Standard Network Interface (SNI)*, demarcation point, or lightning protector. The device that contains carbons to protect a phone line from being overloaded by lightning and acts as the separation point between the telephone company's wire and the customer's wire, which is also called the *IW (Inside Wire)*. For a photo, see *Standard Network Interface.*

Network Layer A layer in a communications protocol model. In general, the network layer does the job of switching and routing of the data being transmitted within the protocol. A central-office switch would be a good example of a network layer function. The latest model (guideline) for communications protocols is the *OSI (Open Systems Interconnect)*. It is the best model so far because all of the layers or functions work independently of each other. For a diagram of the OSI model and its layers, see *Open Systems Interconnection*.

Neutral Also called common or floating ground. Neutral/common is a reference point and is ungrounded. It is usually a signal return or DC reference coupling for transmission circuits.

NI (Network Interface) See *Network Interface*.

Night Service A feature of PBX and hybrid key systems that allows the lines ringing into an office to be handled differently during certain times of the day. The phone system is programmed as two different systems, usually a day system, and a night, or after-hours system. If a user would like all calls that come into the office after hours to ring to a voice-mail system, or be forwarded to security, it can be done with the night-mode feature. Some systems are equipped with software that allows the night-mode feature to activate automatically at certain times of the day.

NMC (Network Management Center) A place where large or public telephone networks are managed, monitored, and maintained from a central location.

NOC (Network Operations Center) A place where large or public telephone networks are managed, monitored, and maintained.

NOD (Network Outward Dialing) See *Network Outward Dialing*.

Node 1. A point on a communications network where two communications paths come together in a device, such as a switch or SONET node. 2. Regarding LAN networks, any device on a network that has an address, such as a PC or router (hubs don't have addresses). 3. Some CLECs call their central office's nodes to distinguish themselves from the RBOCs.

Noise Noise is any kind of distortion or unwanted signal. The two main categories of noise are electromagnetic interference and ambient noise. Electromagnetic Interference is caused by a radio signal or other magnetic field inducing itself onto a medium (twisted-/nontwisted-pair wire) or device (telephone or other electronics). The world we live in is full of radio waves that are emitted from electric appliances, such as blenders, automobile engines, transmitters, and even fluorescent lights. Even though we take preventative measures to avoid receiving these unwanted signals, they sometimes find their way into places that they are not wanted.

Electromagnetic Interference is usually caused by one of two things. The first is when a wire connected to a device acts like an antenna and receives the EMI, which is then passed on to the electronics inside the device and amplified. The second is when an electronic component inside a device acts like an antenna because of poor design, poor shielding, or because the component is defective. Ambient noise is noise caused by the random movement of electrons in an electronic circuit when the power is off or by the random movement of air.

Noise Canceling Noise canceling is accomplished by filtering a sample of the noise from a preamp stage of a circuit, then inverting the signal 180 degrees and adding the inverted noise signal to the original signal containing the noise. The noise combined with the inverted sample of the noise cancel each other out (electronically add to 0 V). When the original noise signal goes positive in its cycle the noise sample goes negative and the resultant output is 0 V. A good application of noise canceling is in the radio headsets that aircraft pilots use. The cockpit noise is sampled and fed into the radio system, inverted, re-fed into the amplification system, and the surrounding noise is canceled out.

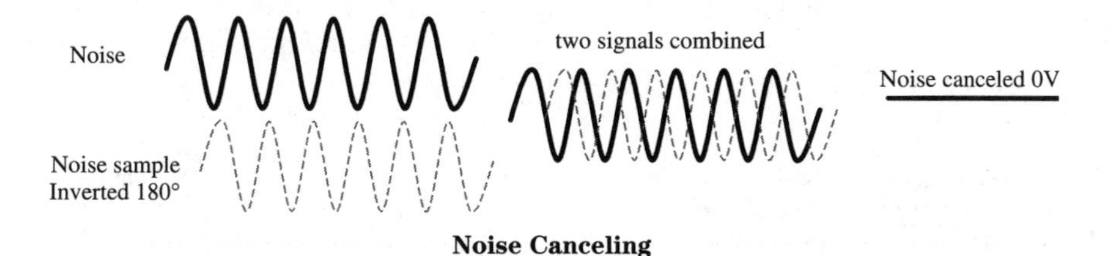

Noise Canceling

North American Area Codes See *Area Codes*.

North American Numbering Plan See *Area Codes*.

NOS (Network Operating System) See *Network Operating System*.

Notch Filter A filter that is designed to pass or block a specific band of frequencies. The three types of filters are low pass/block, high pass/block, and notch pass block. What determines if the filter is a pass or block filter is how the filter is arranged. If the filter is set in series with a circuit, then it passes the desired frequencies down the line. If it is connected to ground, it will pass the desired frequencies to ground, thus preventing them from continuing through the circuit to block them.

NPA (Number Plan Area) Also called an *Area Code*. Each area code contains central offices and each central office has a set of prefixes (first three digits of a seven-digit number) that identify that central office to all other central offices within the associated area code. Some people actually sit around and plan what numbers will belong to which central office.

NRZ (Nonreturn to Zero) A binary encoding method used to write information to hard-disk drives in computers.

NRZI (Nonreturn to Zero Inverted) A binary encoding scheme used to write information to hard-disk drives in computers.

NT1 (Network Terminal 1) Another reference for an ISDN terminal adapter.

NUL The ASCII control-code abbreviation for null. The binary code is 0000000 and the hex is 00.

Null Modem A communications cable, such as an RS-232 cable, that has the transmit and receive wires switch places in pin-out from one end to the other. These cables are used to connect *DCE (Data Communications Equipment)* with *DTE (Data Termination Equipment)*, so the transmit of one reaches the receive of the other.

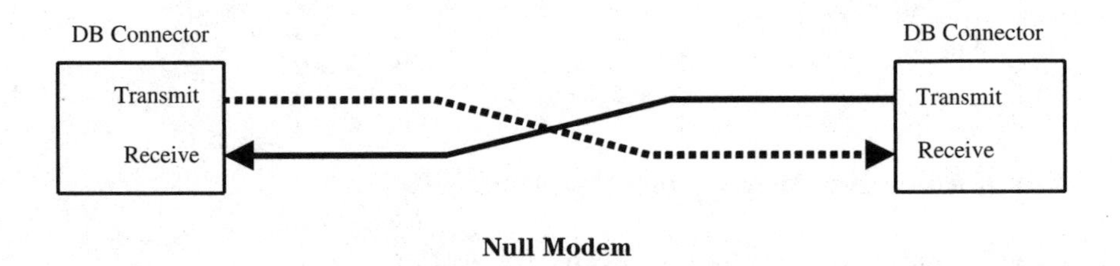

Null Modem

Null-Modem Adapter An adapter that is used to convert a straight-through cable into a null-modem cable. Null-modem adapters are available in many different pin-outs and connector types.

Null-Modem Adapter

Number Crunching A reference to data processing or data manipulation done by a PC, server, or mainframe.

Numbering Plan A plan of what numbers will be used where. In a local phone company, each central office has its own numbering plan or range of numbers. In a PBX or key system, numbering plans are implemented to ease the complexity of accounting, and sometimes they aid in remembering what a person's extension number is. For example, sales can be extensions that range from 3000 to 3999, manufacturing can be extensions that range from 4000 to 4999, off-premises extensions can range from 5000 to 5999, etc. Numbering plans can be formed any way that a user/administrator likes with the following exceptions: Usually no extension on a PBX starts with 9 because 9 as a first digit is used to access outside lines, so 9000 to 9999 is not used in a numbering plan. Zero (0) is also restricted from a numbering plan because it is often used to dial the attendant or operator. A good numbering plan will make call accounting much easier. Call reports can be sorted by department if every department has its own unique numbering plan.

Number Portability Number portability is still in the legal, financial, and architectural planning process. When it is completed (different places will be implemented at different times, the goal for starting is 1998), there will be a national data base that stores every phone number subscribed to by every user. The ultimate goal is to automate the ability of a customer to switch telephone companies and take their phone number to whichever exchange area (within an area code) they wish. If a customer decides to switch companies, their number must first be disconnected by the old company, then reconnected/reactivated by the new. If

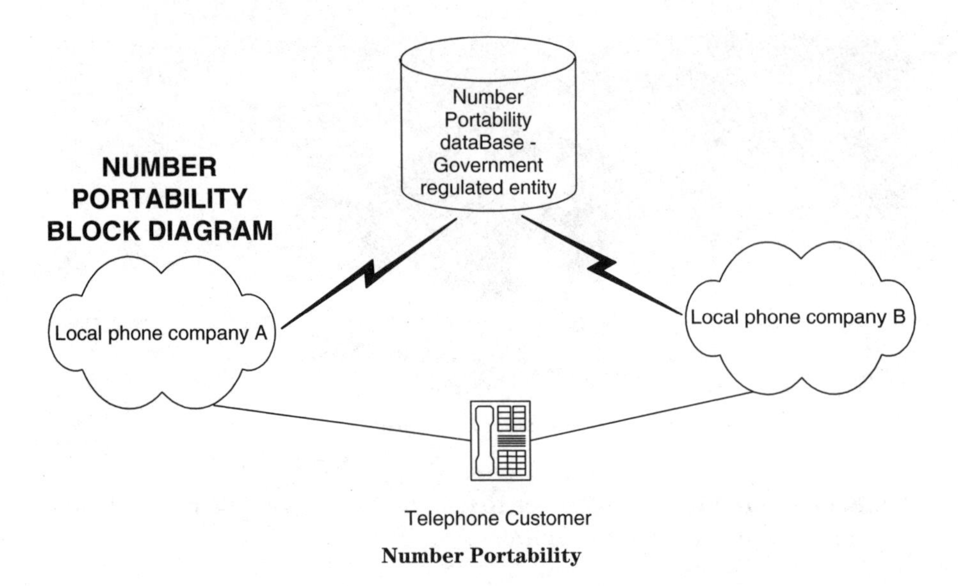

Number Portability

both numbers are active at the same time, the telephone network will become confused and most likely not complete calls. With the planning and implementation of *CTI (Computer Telephone Integration)*, it will be possible for a company to enter into a data base (operated by a third party company), the order to disconnect, and when another company enters the connect, a computer will make the actual switch with no outage of service. This will eliminate the possibility of one telephone company interfering with a customer's change-over by failing to disconnect service at the proper time.

Competition in the telecommunications industry is hampered by the fact that no customer can switch local telephone companies and take their phone number with them (1-800/888 long-distance service is a different story. Customers can transfer those numbers.). The cost to re-advertise phone numbers for business is too costly and too inconvenient for the patrons of businesses. The estimated cost to implement number portability is 100 million dollars per LATA (area code). The legal argument at the writing of this definition is that the new phone companies do not want to pay for the number portability upgrade because the cost would outweigh the profit. The RBOCs don't want to pay for it because they are regulated by the government, which means that any increase in costs of the phone network are passed on to the subscribers/rate payers. It seems unfair that telephone customers would have to pay for huge corporate investments—especially if they are not clients/patrons of the particular company that is receiving the benefit of the investment.

NYNEX (New York New England Exchange) One of the original seven regional Bell Operating Companies that was divested from AT&T.

Nyquist Theorem A theory that states that any analog signal to be converted to a digital signal (ADC conversion) must be sampled at twice the frequency of the top end of the bandwidth of the signal to be converted. If you would like to convert a high-fidelity recording to a compact disc, you would need to sample the audio at a minimum of 36 kHz (36,000 times per second) because the bandwidth of high-fidelity music is 18 kHz (18,000 cycles per second). This sample rate would give two samples per cycle at the highest frequency of human hearing, which is 18 kHz. A DSO channel in a channel bank samples a voice at 8000 times per second (8 kHz). This gives a Nyquist standard sample up to 4 kHz, which is sufficient to sample all sounds in the voice range.

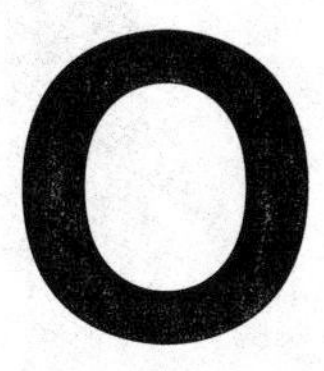

OAI (Open Application Interface) A means for a computer system and a PBX to exchange information. It is an older name for *CTI (Computer Telephony Integration)*. It allows a person in the workplace to enter information into a computer by using their telephone. Some of the information is time-reporting information, inventory information, etc.

OC (Optical Carrier) A prefix for SONET carrier hierarchies, which is followed by a number, such as OC-1, OC-3, etc.

OC-1 (Optical Carrier 1) The beginning of the SONET-level transmission speeds. An OC-1 is capable of carrying one DS-3 within its payload. It's transmission carrier speed is 51.840Mb/s. OC-1 can be converted into an electrical signal, which is called an *STS-1 (Synchronous Transport Signal-1)*. For more information on OC speeds, see *OC-N*.

OC-3 (Optical Carrier 3) A SONET level of transmission speed. It is capable of transporting three DS-3 signals, which is equal to 255.520Mb/s. For more information on OC speeds, see *OC-N*.

OC-12 (Optical Carrier 12) A SONET level of transmission speed. It is capable of transporting three DS-3 signals, which is equal to 622.080Mb/s. For more information on OC speeds, see *OC-N*.

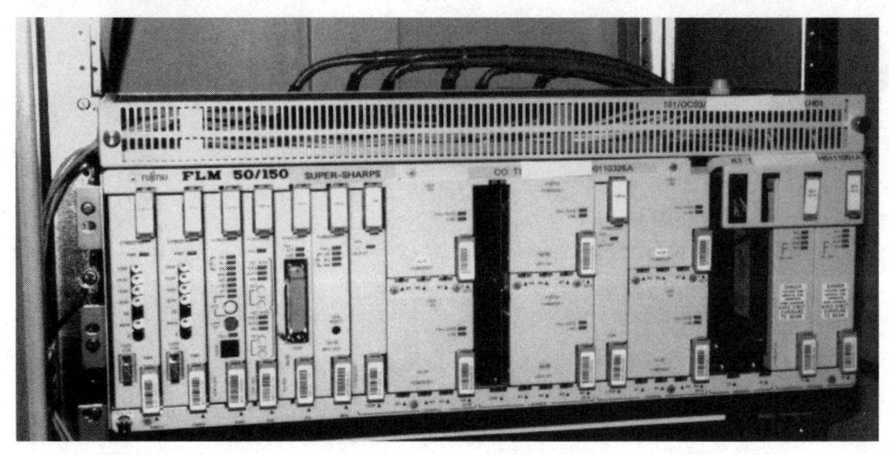

OC-3

OC-48 (Optical Carrier 48) A SONET level of transmission speed. It is capable of transporting three DS-3 signals, which is equal to 2.488Gb/s. For more information on OC speeds, see *OC-N*.

OC-192 (Optical Carrier 192) A SONET level of transmission speed. It is capable of transporting three DS-3 signals, which is equal to 9.953Gb/s. For more information on OC speeds, see *OC-N*.

OC-N (Optical Carrier N) The *N* denotes a number in the SONET optical carrier hierarchy, which now extends from OC-1 to OC-192.

Octal A numbering system. Base 8. The Base 10 system has 10 different characters to represent numbers, 0 through 9. Base eight uses only eight of those characters, 0 through 7.

Octel A company that manufactures stand-alone voice-mail equipment. For a photo of an Octel voice-mail system, see *Voice Mail*.

Octet Another term for *byte*, a string of eight bits.

Octopus Cable Also called a *Y* or *three-way cable*. An octopus cable is used to break a larger connector (usually from a bus) to two or more smaller connectors. The Nortel Meridian utilizes a cable with 50 pins on one connector that breaks out in two different RS-232 connectors. Having the single large connector plug into the back plane uses less space than three smaller connectors.

OC-12

OC-48

Odd Parity A method of bit-stream checking. Parity is used in error correction. The number of logic "ones" is counted in a bit stream. There is "odd parity" and there is "even parity." Which is used depends on if you like odd or even numbers, or if the modem you are trying to connect with

Name/Acronym	Bandwidth	Equivelent DS0	Equivelent DS1	Equivelent DS3	comments
DS0	64Kb/s	1	*	*	one phone line
DS1/T1	1.544Mb/s	24	1	*	popular service
DS1C	3.152Mb/s	48	2	*	equipment
E1/CEPT1	2.048Mb/s	32	1	*	European
DS2	6.312Mb/s	96	4	*	equipment
E2	8.448Mb/s	96	4	*	European
DS3/T3	44.736Mb/s	672	28	1	popular service
E3	34.368Mb/s	512	16	1	European
DS4	139.264Mb/s	2016	80	6	long haul radio
STS-1	51.84Mb/s	672	28	1	electrical OC1
OC-1	51.84Mb/s	672	28	1	SONET
OC-3	255.520Mb/s	2,016	84	3	SONET
OC-12	622.080Mb/s	8,064	336	12	SONET
OC-48	2.488Gb/s	32,256	1,344	48	SONET
OC-192	9.953Gb/s	129,024	5,376	192	SONET

OC-N

likes odd or even numbers. Parity is a part of error-checking protocol. It is simply the part of the protocol where the two devices are told if they are counting odd number bits or even number bits. In odd parity, if the number of ones is odd, then a parity bit is set to "one" at the end of the bit stream. This is odd parity because the parity bit is set to one when the number of "ones" is odd. In even parity, the parity bit is set to "one" when the number of "one" bits is even. See *Parity*.

OE (Office Equipment) Also referred to as *line equipment*. Line equipment is the actual interface port (from a circuit card) on the central-office switch in a telephone-company central office. It is the equivalent to a station or *IPE (Intelligent Peripheral Equipment)* card in a PBX system (*TN* for Nortel Specifics). Each telephone line has an associated line equipment or office-equipment interface. That particular port is what defines the telephone service provided to the customer connected to it via the OSP network. The CPU (core) of the central-office switch associates a phone number with a line-equipment port (OE). When a customer of the phone company calls and requests that their phone number be changed, the service order eventually finds its way to a central-office technician or service translator that reprograms the line equipment with a new phone number.

Off-Premises Extension (OPX) Off-premises extension adapter, also called a *loop extender*. An OPX adapter is an add on device for a PBX switch or central-office switch that allows operation over an abnormally long loop or twisted pair, usually more than 12,000 feet for a central-office switch and 1500 feet for a PBX. The PBX or key-system manufacturer usually offers special equipment for a long-reach application. Some OPX adapters for PBX systems can be programmed to dial digits into an outgoing trunk, which will automatically ring a telephone somewhere else, such as the CEO of a company's home office.

Off-Sight Night Answer A feature of PBX and some key systems that allows a main line to be forwarded to a telephone number programmed in by the user/administrator when the system is put into night mode.

OHD Optical hard drive.

Ohm The unit of resistance, represented by the Greek letter Omega, Ω. Resistance is just what its name depicts, resistance to electric current flow. A 100-W, 120-V household light bulb has about one ohm of resistance. The more resistance is in a circuit, the less current flows through it.

Ohm's Law A series of mathematical relationships for electronics. The relationships are based on voltage, resistance, power, and amperage. The two basic Ohm's law formulas are:

P (power in watts) = I (current in amps) × E (voltage in volts)

E (voltage in volts) = I (current in amps) × R (resistance in ohms)

Omnidirectional A reference to a microphone that receives sound from all directions.

ONA (Open Network Architecture) The architecture of the public telephone network. Under FCC rulings, the Bell operating companies must allow other companies that offer "value-added services" to connect to and offer services through the local telephone companies network. Value-added services, under open network architecture, are voice mail, operator services, and IVR telephone-shopping applications. You don't have to use the Bell companies' voice mail if another voice-mail service provider is available. If you order the other company's voice mail, all of your voice-mail connections will go through the alternative value-added service provider, on their separate equipment. The problem with ONA is that from a technology sense, access is equal, but in competition for market share, it is not.

One-Way Trunk A reference to a *DID (Direct Inward Dial)* or *DOD (Direct Outward Dial)* trunk used in PBX applications.

Ones Density A reference to the maximum number of consecutive "zero" bits can be transmitted in a row using specific transmission equipment without losing the timing of the carrier (T1). To eliminate successive zeros in T1 transmissions line/protocols, such as B8ZS, have been implemented.

ONI Optical Network Interface.

OOF (Out of Frame) A fault condition of a T1 carrier circuit. If an OOF condition exists, the circuit is down and not operational. Many T1 carrier equipment manufacturers implement recovery measures in the operating system to help systems come back on line automatically.

Open See *Open Circuit*.

Open Application Interface (OAI) See *OAI*.

Open Architecture The ability of different systems to integrate with each other, such as a PBX system and a Novell LAN. The newer term for open network architecture is *CTI (Computer Telephony Integration)*.

Open Circuit A circuit fault. Many confuse an open with a short. An open is literally an open, a "disconnection" in a circuit. A short is a "crossed circuit," an easier path to ground caused by a bad component, water, or other means for electricity to get to where it is not wanted.

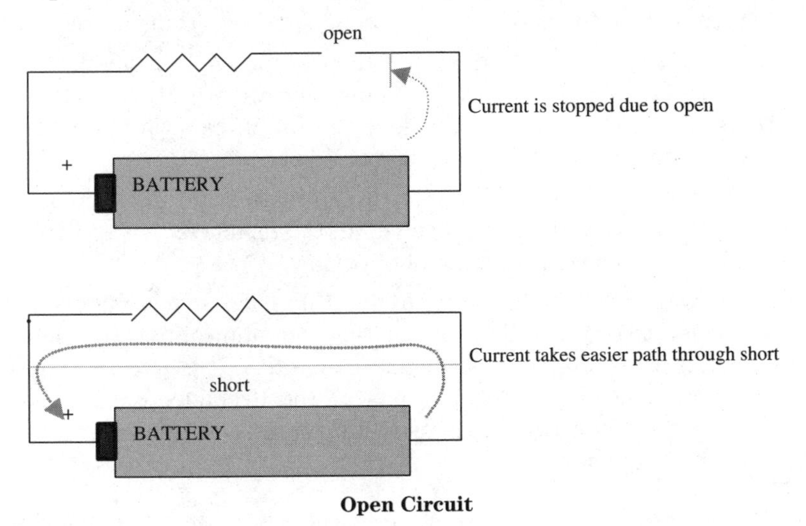

Open Circuit

Open-Ended Access A term that refers to a switched telephone line that is not restricted from any calling prefixes or area codes. Most residential customers have subscribed to service that has open-ended access, not knowing what it is called.

Open Network Architecture (ONA) See *ONA*.

Open Systems Interconnection (OSI) The latest model, or guideline for communications protocols is the *OSI (Open Systems Interconnect)*. It is the best model so far because all of the layers (functions) work independently of each other. Older proprietary communications models are shown in the figure. The OSI model is a seven-layer or "step" process for communications. The different functions are:

- *Application layer* The seventh and highest layer of the OSI communications protocol model. The applications layer is the function of connecting an application file or program to a communications protocol.
- *Presentation layer* The sixth layer in the OSI. In general, the presentation layer performs the function of encoding and decoding the data to be transmitted within the communications protocol.
- *Session layer* The fifth layer of the OSI model. In general, the session layer establishes and maintains connection to the communications process of the lower layers. It also controls the direction of the data transfer.
- *Transport layer* The fourth layer or function in a communications protocol model. In general, the transport layer performs the function of error correction and the direction of data flow (transmit/receive).
- *Network layer* The third layer in the OSI model. In general, the network layer does the job of switching and routing of the data being transmitted within the protocol. A central-office switch would be a good example of a network-layer function.
- *Data-link layer* The second layer or function in the OSI model. In general, the data-link layer receives and transmits data over the physical layer media (twisted pair, fiber optic, etc.).
- *Physical layer* The first layer in the OSI model. In general, the physical layer is the actual media of the communications transmission (twisted-pair wire, coax, air, fiber optic, etc.). It is also the types of connectors used and the pin-outs of those connectors. The 568B wiring scheme for CAT 5 wire is a physical-layer function. For a basic conceptual diagram of the OSI model, see *OSI Standards*.

Open Wire Also called "C" wire. Wire that is steel strengthened for long-span aerial-plant applications. Some C wire is uninsulated, so it is also called *open wire*. Open wire better fits the application because it is used in wide open or very rural areas. The old telegraph system was an open-wire system.

Operating System The computer software program that controls the functions of computer hardware. Examples of operating systems are Windows 95, MS-DOS, Pick, UNIX, and OS/2.

Operator An attendant that assists callers. Operators can work for telephone companies or private telecommunications service companies.

Operator Console The huge telephone used by a PBX attendant. The console is distinguishable by its large *BLF (Busy Lamp Field)* and many feature keys.

Optical Attenuator A fiber-optic attenuator works like your sunglasses, it reduces the level of light entering your eyes so that you can see more effectively. They come in various connector types. Typical fiber-optic attenuator values are 5 dB, 10 dB, and 20 dB.

Optical Attenuator

Optical Fiber Patch Panel A means of terminating fiber-optic cable. Fiber patch panels contain a fiber splice tray equipped with pigtails. The pigtails are simply fiber connectors with a piece of fiber optic connected to them so that a fiber from within a cable can be easily spliced to them. The connectors are spaced on the front of the fiber patch panel.

Optical Fiber Splice The two types of fiber-optic splices are fusion (heat) and mechanical.

Optical Time-Domain Reflectometer (OTDR) A testing device that measures the loss over a fiber optic and the distance from the tester. OTDRs look similar to oscilloscopes with a CRT display tube. OTDRs are a specialized optical version of a *TDR (Time-Domain Reflectometer)* used to test copper pairs. The way a TDR works is that it transmits a signal down a media (copper or glass), then waits for a reflection to come back. When the reflection returns to the device, the time difference is used to calculate the distance that the signal traveled. The size or power of the return signal is used to calculate loss.

Optoelectric Transducer A class of electronic components that converts light energy into electrical energy and electrical energy into light energy.

OPX Adapter (Off-Premises Extension Adapter) Also called a *loop extender*. An OPX adapter is an add-on device for a PBX switch or central-office switch that allows operation over an abnormally long loop or twisted pair, usually over 12,000 feet for a central-office switch and 1,500 feet for a PBX. The PBX or key system manufacturer usually offers special equipment for a long-reach application. Some OPX adapters for PBX systems can be programmed to dial digits into an outgoing trunk, which will automatically ring a telephone somewhere else, such as the CEO of a company's home office.

Oscillator An electronic circuit that produces an AC cycle from a DC power source. Oscillators are used as carrier references for transmitters and for the timing signal in clock circuits for digital instruments, such as PCs and telephone systems. Quartz crystal oscillators are the most reliable and inexpensive. For a photo, see *Crystal Oscillator*.

Oscilloscope A testing device that allows a user to view a waveform on a screen (CRT). The screen is graduated to show different frequencies and voltage levels. The value of each graduation (or division) is determined by the setting of the frequency/division knob. The voltage level of each division is determined by the voltage-level selector knob. Oscilloscopes range in price from about $400 to more than $8000. The features that make an oscilloscope increase in price are the ability to read/display very fast frequencies and the ability to view more than two waveforms at a time. Some oscilloscopes are capable of being connected to plotters, which gives the user the ability to print a waveform displayed on the screen.

OSI (Open Systems Interconnect) See *Open Systems Interconnect*.

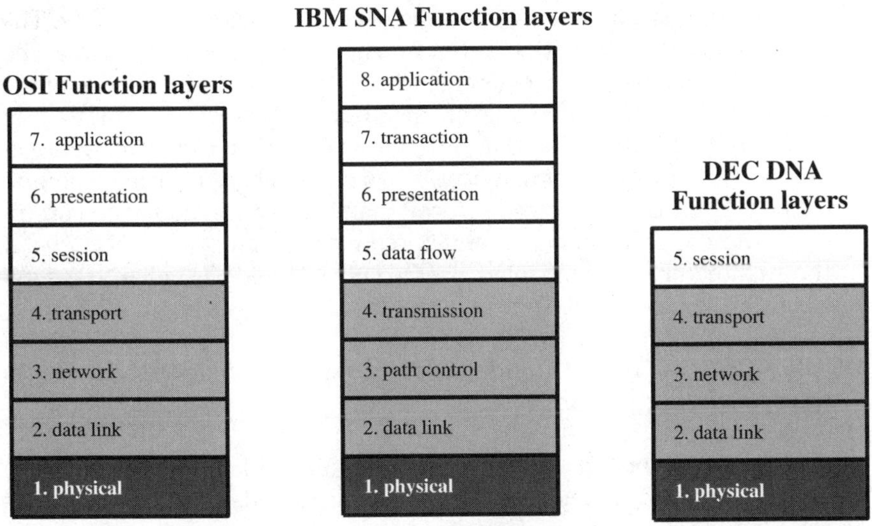

IBM SNA Function layers

8. application
7. transaction
6. presentation
5. data flow
4. transmission
3. path control
2. data link
1. physical

OSI Function layers

7. application
6. presentation
5. session
4. transport
3. network
2. data link
1. physical

DEC DNA Function layers

5. session
4. transport
3. network
2. data link
1. physical

OSI (Open Systems Interconnect)

OSI Model (Open Systems Interconnect Model) See *Open Systems Interconnect*.

OSI Standards An architecture set up by the *ISO (International Standards Organization)* that sets some broad standards for communications. The purpose of the standards is to help manufacturers make equipment that is universally compatible. The OSI is not widely embraced by the telecommunications and data communications industry,

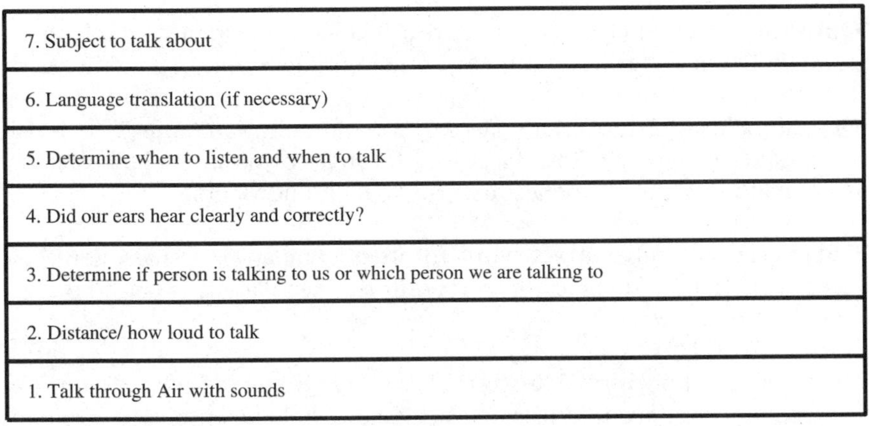

7. Subject to talk about
6. Language translation (if necessary)
5. Determine when to listen and when to talk
4. Did our ears hear clearly and correctly?
3. Determine if person is talking to us or which person we are talking to
2. Distance/ how loud to talk
1. Talk through Air with sounds

OSI Standards

but it is used as a model in the design of communications protocols. The basic idea of the OSI is that seven functions, steps, or layers are in the successful completion of a communication transmission. The goal of the OSI is to make all of these layers separate and individual "entities" in hardware and software so that different manufacturers can integrate at different levels. Data communications is modeled after voice communications. Even though humans speak many different languages (protocols), there is still a common architecture of human communications. If the architecture were modeled like the OSI, it would be similar to the following table. See *Open Systems Interconnect*.

OTDR (Optical Time-Domain Reflectometer) See *Optical Time-Domain Reflectometer*.

Out-of-Band Signaling In telephone circuits (DS1 to be specific), the two different ways to send signals are in band and out of band. Signals are digits that you dial, dial tone, the phone being off-hook, ringing, etc. An in-band telephone line is like the one in your home; the digits that you dial and the ringing are carried within the channel that you talk on. Out-of-band signaling is a method that telephone companies and businesses use for larger PBX applications and data-transfer applications. In an out-of-band signaled DS1, there are 24 multiplexed channels. The 24th channel carries the signaling for the other 23 channels or phone lines. The advantage of out-of-band signaling is that each channel has an increased capacity to carry data (8Kb/s more) and the 23 channels are not used to find out if a line is busy (both directions, in and out). The off-hook sensing and busy signaling are performed in the 24th channel. If you have a system that receives thousands of calls per day, this can reduce traffic.

Outdoor Jack Closure Closures are available that help protect telephone and other jacks from moisture and other outdoor weather conditions.

Outside Plant A term that refers to a communications utility's twisted-pair and/or coax network that winds through towns and neighborhoods. It includes terminals, pedestals, cross boxes, and vaults.

Outsource To subcontract work to other companies, usually for their construction or technical expertise in the installation of specific electronics or software.

Outward Restriction A feature of *PBX (Private Branch Exchange)* telephone systems that prevents selected telephone extensions from di-

Outdoor Jack Closure

aling outside the office/building. When a user of one of these extensions dials "9" for an outside dial tone, they will just get a fast busy signal.

Overhead The part of a transmission that contains the information/signal that controls the operation of the transmission. If you are transporting yourself across town in your car, you are the payload and your car is the overhead.

P

P Connector A 25-pair male amp connector. For a photo of the female version, called a *C connector*, see *25-Pair Connector*.

PA System See *Public Address System*.

PABX (Private Automatic Branch Exchange) The old name for *PBX*, *Private Branch Exchange*.

Pac Bell The RBOC that operates the public telephone network in the state of California, owned by Pacific Telesis, who was recently purchased by Southern Bell.

Pacific Telesis The RBOC that owns PAC Bell and Nevada Bell, which was bought out by Southern Bell.

Packet A unit of data in a transmission that contains payload (transmitted information) and overhead (addressing and error-correcting information). Packet-transmission techniques are used in different ways in frame relay, ATM, and ethernet protocols.

Packet Assembler Dissembler (PAD) See *PAD*.

Packet Buffer Memory allocated or dedicated to the temporary storage of a copy of a data packet until the original has reached its destination.

Packet Controller Another name for a packet switch. A packet switch is the central controlling device in a packet-switched network, such as switched ethernet, switched token ring, or ISDN packet switching.

Packet Interleaving To place many data packets from many data packet sources on one transmission channel.

Packet Switch A reference to a frame relay switch. Not to be confused with a switched ethernet or switched token ring network device. A packet switch provides non-isochronous services.

Packet Switching Exchange (PSE) Part of a packet-switching network that receives packets of data from a *PAD (Packet Assembler/Dissembler)* via a modem. The PSE makes and holds copies of each packet, then transmits the packets one at a time to the PSE that they are addressed to. The local PSE then discards the copies as the far-end PSE acknowledges the safe receipt of the original.

PAD (Packet Assembler/Dissembler) The device or software program in a packet-switching network that takes a large file to be transmitted (or small) and breaks it down into smaller pieces. It gives each piece an identification number in relation to the rest of the pieces (e.g., 387 of 8954) and an address, along with error-checking information (usually CRC) and other *HDLC (High-Level Data Link Control)* information. The PAD can be a part of an end users computer or a separate device. The PAD sends the packets to a *PSE (Packet Switching Exchange)* via a modem, where the packets are individually copied and transmitted. The copies are made by the PSE in case a packet needs to be retransmitted because it was lost or corrupted.

Pager A small, portable device that receives simplex messages. Pagers are small enough to be worn on a belt or wrist (as a wristwatch-type pager). Pagers come in three types: numeric, alphanumeric, and PCS. The numeric are the older type, receiving numbers that a person desiring a return call inputs to the page signal. The alphanumeric pagers have a larger LCD display and have the ability to receive text messages as well as numeric messages. PCS paging is offered as a service with PCS cellular telephones by many cellular telephone companies.

Page Zone When attendants use a page feature on a telephone system, they are prompted to input a choice of zone or "area" they want to page. This area is called the *page zone*. Most PBX systems have three separate page zone options and "all zones" is an option to the attendant as well.

Pair Two copper wires.

Pair Gain Usually a reference to a Lucent SLC96 or SLC2000 system. A system in the public network that multiplexes many conversations or phone lines into one or two copper pairs. T1 is a pair-gain system used by public telephone-service providers, such as USWest, PAC Bell, Brooks, ELI, and virtually every other local facilities-based phone company. The photograph shows a pair-gain system outdoor closure/cabinet.

Pair Gain

PAM (Pulse Amplitude Modulation) See *Pulse Amplitude Modulation.*

Parabola The curve that all projected objects travel when acted on by a force of gravity. If you watch a baseball that is hit or thrown through the air, it curves during its fall. This oblong curve is called a *parabola*, and it is very special. Ancient mathematicians discovered that this curve can be duplicated mathematically with trigonometry. Radio communications engineers later used it to focus and guide radio signals because all radio waves from a single point were reflected in exactly one direction. For a diagram, see *Parabolic Dish Antenna*.

Parabolic Dish Antenna A directional antenna. This name results from its parabolic shape, which means all radians from a single point are reflected into one direction. For a photo of a parabolic microwave antenna, see *Microwave*.

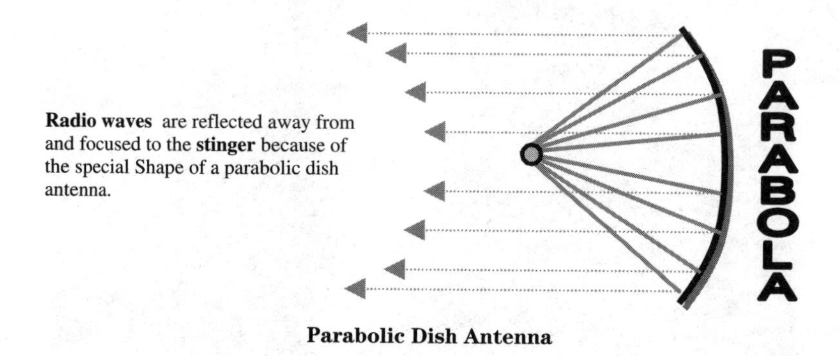

Radio waves are reflected away from and focused to the **stinger** because of the special Shape of a parabolic dish antenna.
PARABOLA

Parabolic Dish Antenna

Parallel Circuit A circuit that has more than one path for current through multiple loads or devices. The other type of circuit is a series circuit, which has only one path for current through multiple loads.

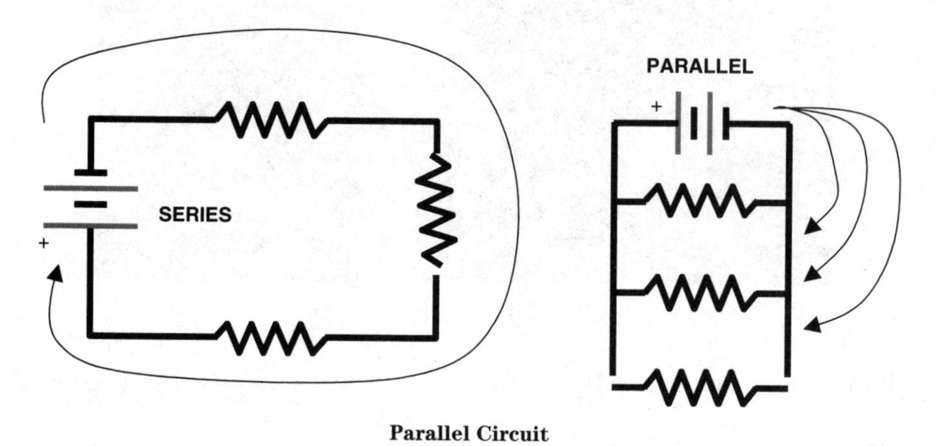

Parallel Circuit

Parallel Data The transmission of data over a media with multiple bits being transferred at one time, such as a whole byte. The other type of

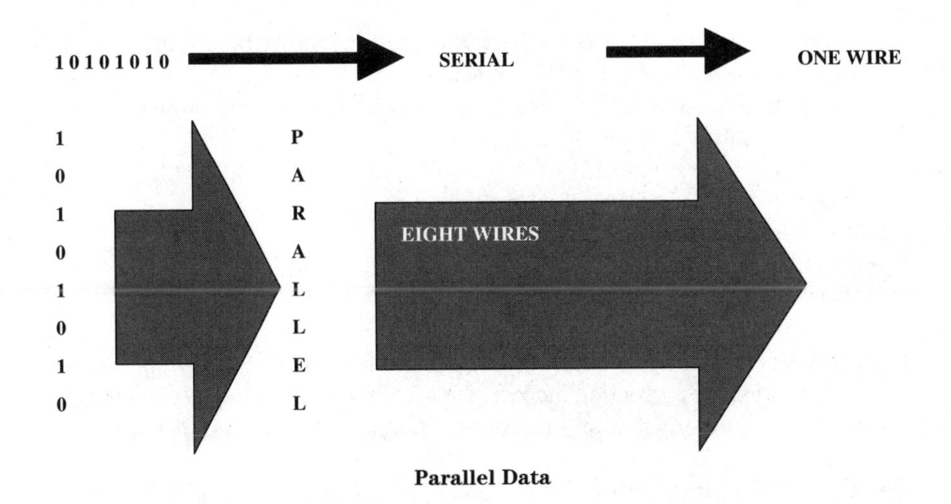

Parallel Data

transmission is serial data, which sends data one bit at a time. The illustration shows an example of 8 bits being sent in series and in parallel.

Parallel Port A connection that is designed for transmission multiple bits at one time over more than one wire. See *Parallel Data*.

Parallel Printer Cable A cable that is designed for transmitting multiple bits at one time, used on a parallel-printer port. See *Parallel Data*.

Parasite A device that gets its power to operate from the telephone line. The telephone line has a –52-V battery voltage when it is idle and –12 V when it is in use. Standard telephone sets (2500 type) are parasitic telephones. Other devices, such as RFI filters and tapping devices, are parasitic.

Parity A method of bit-stream checking. Parity is used in error correction. The number of logic "ones" is counted in a bit stream. There is "odd parity" and "even parity." Which is used depends on if you like odd or even numbers, or if the modem you are trying to connect with likes odd or even numbers. Parity is a part of error-checking protocol. It is simply the part of the protocol where the two devices are told if they are counting odd number bits or even number bits. In odd parity, if the number of ones is an odd number, then a parity bit is set to "one" at the end of the bit stream. This is *odd parity* because the parity bit is set to one when the number of "ones" is odd. In *even parity*, the parity bit is set to "one" when the number of "one" bits is even.

Odd-parity bit stream: 0 0 1 0 1 0 1 Subsequent parity bit would be 1 because the number of "ones" is odd.

Odd-parity bit stream: 1 0 1 0 1 1 0 Subsequent parity bit would be 0 because the number of "ones" is not odd.
Even-parity bit stream: 0 0 1 0 1 0 1 Subsequent parity bit would be 0 because the number of "ones" is not even.
Even-parity bit stream: 1 0 1 0 1 1 0 Subsequent parity bit would be 1 because the number of "ones" is even.

Parity Bit A parity bit is a bit added into a bit stream, usually after every seven bits. See *Parity*.

Parity Check An older and original method of error correction. Newer methods of error correction include *CRC (Cyclic Redundancy Checking)*, which are more efficient and far more accurate (99% CRC, 50% parity).

Park A PBX system feature that allows a user to transfer a call to a "ghost" extension. When the call is transferred to the "ghost" extension, it can then be retrieved from any telephone by dialing that extension. The "ghost" extension is thought of as a parking spot in the network.

Part X Usually a reference to Part 64 or Part 68 of the MFJ (Modified Final Judgment) handed to the RBOCs (Regional Bell Operating Companies) by Judge Harold Greene. It specifies the separation of customer-owned equipment (Customer Premises Equipment, CPE) and the telephone company-owned equipment, and telephone company demarcation.

Part 64 A reference to the part of the *MFJ (Modified Final Judgment)* handed to the *RBOCs (Regional Bell Operating Companies)* by Judge Harold Greene. It specifies the separation of customer-owned equipment *(Customer Premises Equipment, CPE)* and the telephone company-owned equipment, and Telephone company demarcation.

Part 68 See Part 64.

Party Line A telephone line that is shared by multiple residences. A party line is a one-pair circuit that can have as many as eight individual residences, (each with a separate phone number) share that same pair for service. If one residence is using the line, the others can't. Each residence can have its own phone number, with the use of a *SRM (Selective Ringing Module)*. The SRM is installed in the NI of each residence and contains electronics that can be configured to recognize different ringing formats using DIP switches. Some different ringing formats that an SRM would differentiate are ring voltage on the ring side, ring voltage on the tip side, ring voltage on the ring side with the tip side grounded, and ring on the tip side with the ring side grounded. The selective ringing mod-

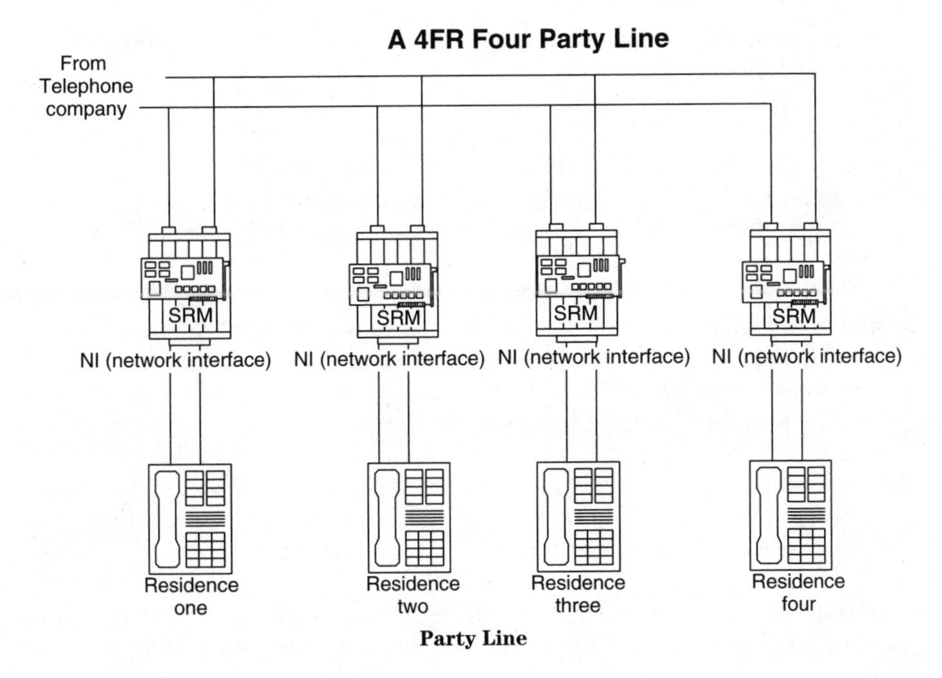

Party Line

ules are wired to recognize a certain ring. The central office sends a specific ring to reach a specific number.

Pass-Band Filter Another name for a band-pass filter. A band-pass filter is used in frequency-division multiplexing as well as the equalizer in your stereo. It is usually a capacitor/resistor/inductor network that has a resonant frequency and a rating of how well it passes one frequency (or a bandwidth of frequencies) and blocks out others (called the Q, quality) of the circuit. The resonant frequency of the circuit is the frequency that the circuit will pass.

Passive Hub As opposed to an active hub, a passive hub has no ability to amplify (extend signal transmission range) a signal on an ethernet network, so it needs to be in close proximity to the computers it is connected to. Simply stated, a hub makes a star wiring configuration look like a bus configuration to all the devices connected to it. Hubs are utilized extensively in ethernet networks. For a diagram of a hub application, see *Hub*.

Patch Panel A panel equipped with plugs, rather than terminals, for connecting wires or fiber optics. A patch panel can be used to terminate installed wire or be used as a "plug-in" test access point for communications circuits. DS0 and DS3 patch panels are very popular in central offices for testing purposes. Cat 5 patch panels are popular in computer LAN environments for the easy connection of computers to a network of pre-installed wire.

Path 1. The process of aligning a microwave radio link. Two technicians point the dishes at each other while taking AGC readings from the transmission equipment, which is often located in the dish (also called an ODU, Outdoor Unit). 2. The space between two microwave dishes that make a microwave radio link.

Pause A feature incorporated with the speed-dial feature of telephones. When speed-dial numbers are programmed, a 1.5-second pause can be inserted by pressing the # key. If a user wants to program a speed dial that rings into a PBX system where an extension needs to be input after an auto attendant answers, the user can input several pauses before the extension number in the speed-dial string that they program on their phone. When activated, the speed-call feature will then dial the number, pause while the auto attendant answers, then dial the extension.

Payload A transmission signal or packet has two components, the payload and the overhead. The payload carries the customer information, like a B Channel in an ISDN circuit. The overhead carries operational, maintenance, and synchronization information that make the protocol work. An ISDN D channel is an overhead component of an ISDN circuit.

Payphone A coin operated telephone. Many payphones are owned and operated by local telephone companies, but there are private payphone companies too. Pay or coin-operated telephones can be purchased at telecommunications distributors, such as Graybar and Anixter.

PBX (Private Branch Exchange) See *Private Branch Exchange*.

PC Board (Printed Circuit Board) The green- or brown-colored board that has copper conductive tracks etched onto its surface. Electronic components are soldered onto these boards by hand or by a method called *flow soldering*. Some PC boards are layered or sandwiched, with conductive tracks inside them and on both sides.

PCM (Pulse-Code Modulation) A concept that is similar to that of Morse code, digital signals are sent over a media (twisted copper pairs, radio, fiber optic, coax, etc.) one bit at a time, each bit being represented as a pulse or the absence of a pulse. A typical digital transmission is PCM.

PCS (Personal Communications Service) A newer form of cellular communications service that has a different transmission format. Instead of being an analog radio signal, as in cellular, PCS combines multiple customers on each radio channel. PCS operates at a higher frequency than cellular and transmits less power, thus each of the individual cells (geographical cells) are smaller and more compact. This also allows more users in the same amount of airspace. For a photo of a PCS antenna, see *Monopole Antenna*.

PDN (Primary Directory Number) An ISDN telephone number.

Peak Power A method of calculating the power consumption or power output of an electronic/electrical device. Other methods of calculating power include true power, transparent power, and RMS (Root-Mean-Square) power. Most audio applications use either peak or RMS power. A great example to demonstrate the difference between peak power and RMS power is home and car stereo amplifiers. Many people ask which is better, peak power or RMS power? The answer is both. Some stereo manufacturers put peak-power ratings on their products because it sounds better. Some put RMS power on their products because it is closer to the true power of the device. To convert from peak power to RMS power, multiply the peak-power rating by 0.707. The result is RMS power. To convert RMS power to peak power, divide the RMS power rating by 0.707. The result is the peak-power rating.

Ped (Pedestal) See *Pedestal*.

Pedestal (Ped) Usually a small green box that house telephone or cable-TV cable splices or terminals.

Pedestal

Peer-to-Peer Networking A local-area network scheme that does not use a server or host. Individual PCs are linked together via network cards and CAT-5 wire or coax. Windows 95 has its own peer-to-peer networking utility built in. You can use it, but don't forget to install your network cards.

Peg Board Also called a *white board* or *mushroom board*. It is placed between termination blocks (such as 66M150 blocks) to provide a means of support for routing cross-connect wire. For a photo, see *White Board*.

PE (Peripheral Equipment) Devices that are not a part of a system, but work with it, such as a printer.

Peripheral Equipment Devices that are not a part of a system, but work with it, such as a printer.

Permanent Virtual Circuit A dedicated (private line) channel in a multiplexed transmission or packet network used by telephone companies. Permanent virtual circuits are very common in T3 and SONET carrier networks. A virtual circuit is a switched circuit, like a plain telephone line. A permanent circuit is a dedicated twisted copper pair with a carrier, such as T1 for a private line.

Peta The prefix for 1,000,000,000,000,000. It would take five million 1GB (one gigabyte) hard drives would have the capacity of one 5PB hard drive. I don't think we will see hard drives in the PB range any time soon.

Phantom DN (Phantom Directory Number) Also called a Virtual DN. A directory number or extension on a PBX system that is used to attach a voice mailbox. The phantom DN does not really have a telephone set, but the PBX system thinks it does, so it transfers calls to that DN, which are configured to be forwarded to a voice-mail system. A user of that DN can then dial into the voice-mail system, enter their extension, and receive their messages.

Phased Array Antenna A group of small antennas placed a multiple of a wavelength in distance from each other to create one larger antenna.

Phase A reference to a sine wave and its relative cycle to another sine wave or time source. Phase is measured in degrees (0 to 360) or radians.

Phased Locked Loop (PLL) A very important electronic circuit in the world of *FM (Frequency Modulation)* and *PM (Phase Modulation)*. Phase-locked loops are used as the detector circuits in FM receivers and to create stable RF references for all types of transmitters and timing circuits.

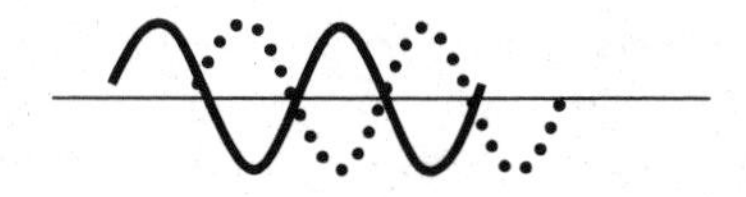

A sine wave, and a cosine wave (dashed), which is180° out of phase with the sine wave. The resultant output from combining these signals would be 0V.

Phase

Phase Modulation (PM) A method of varying a radio carrier frequency so that a signal (the variations) can ride on it. After the carrier signal has the variations imposed on it, it is amplified and transmitted. The variations in the signal are then detected by the receiver. The variations in the carrier signal are actually voices, music, or whatever is to be transmitted. The other methods of modulating a carrier frequency are *AM (Amplitude Modulation)* and *FM (Frequency Modulation)*. Phase modulation makes the phase of a carrier frequency change in conjunction with a signal that it is to carry. The color on broadcast television is sent in a PM format. A simple representation is depicted in the diagram.

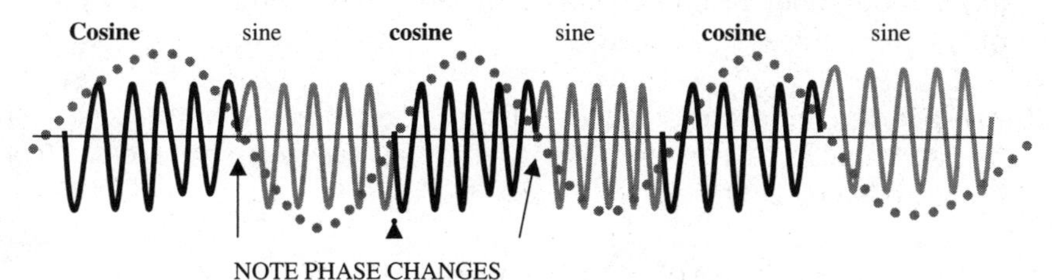

Phase Modulation

Phase-Shift Keying (PSK) A method of modulating a carrier frequency by making the carrier signal phase shift in conjunction with the digital input signal. In the diagram, a cosine phase indicates a "1" value and a sine phase indicates a "0."

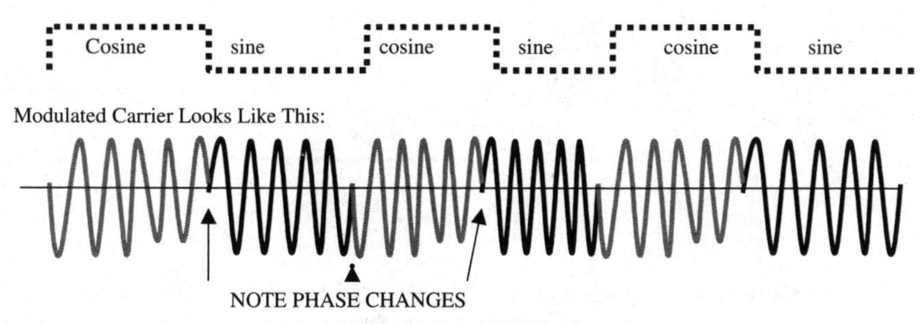

Phase Shift Keying

Photo-Conductive Cell Also called a *photo-resistor* or *photo-sensitive cell*. An electronic device that conducts electricity better when it is exposed to light. Photo conductive cells are made from Cadmium Sulfide (CdS) and Cadmium Selenide (CdSe). They are most responsive to green-colored light (5500 angstrom). They react to almost the entire spectrum of light that is visible to the human eye.

Schematic symbol for a photo conductive cell

Photo Conductive Cell

Photo-Conductor A reference to a photo-conductive cell.

Photo-Detector A photo-sensitive circuit whose main component is usually a photo-diode or photo-transistor. A photo-detector converts pulses of light into pulses of electricity.

Photo-Diode An electronic device that acts as a light-activated switch. It operates similar to a zener diode, except that the reverse current effect is controlled by light.

SCHEMATIC SYMBOLS FOR PHOTO DIODES

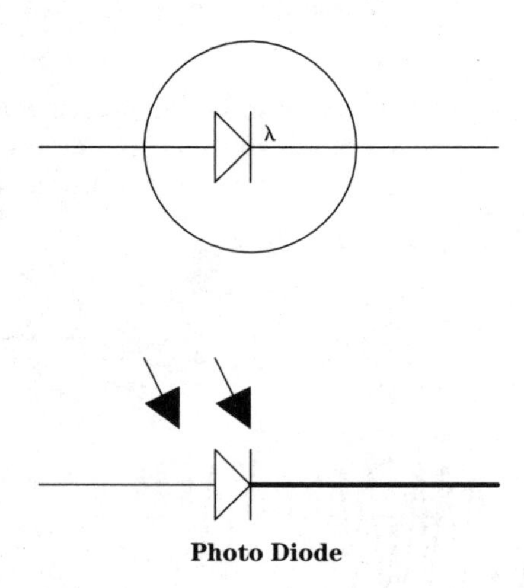

Photo Diode

Photo-Transistor A transistor that is forward biased (conducts electricity as a switch) when exposed to light. Photo-transistors are used the

Schematic Symbol for a Photo Transistor

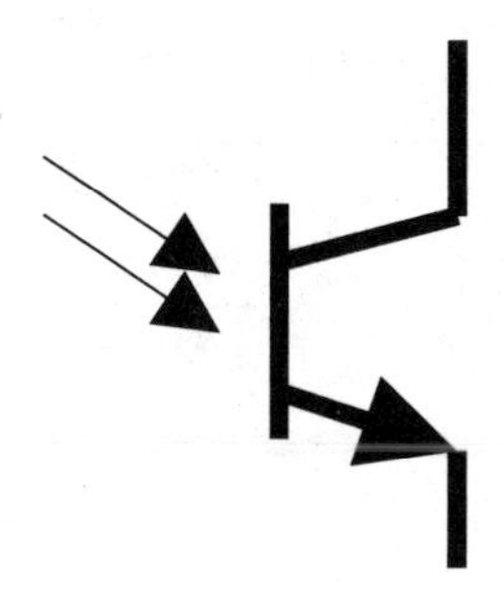

Photo Transistor

same way as switching transistors, except the base is the photo-sensitive part of the device.

Photonic Layer Fiber optic. A reference to the physical layer in the SONET protocol architecture, which is where the type of fiber optic is defined (multimode/single mode).

Physical Colocation A *colocation* is an interconnection agreement and a physical place where telephone companies hand-off calls and services to each other. This is usually performed between a CLEC and an RBOC. The CLEC installs and maintains interconnection equipment usually consisting of optical carrier (SONET) equipment and a digital cross-connect system. There are other types of colocations. Alarm companies like to have their alarm-signaling equipment located in the local central office for security and convenience of connecting alarm circuits. Long-distance companies colocate with local telephone companies as well.

Physical Layer A layer in a communications protocol model. In general, the physical layer is the actual media of the communications transmission (twisted-pair wire, coax, air, fiber optic, etc.) It is also the types of connectors used and the pin-outs of those connectors. The 568B wiring scheme for CAT 5 wire is a physical-layer function. The latest guideline for communications protocols is the *OSI (Open Systems Interconnect)*. It is the best model so far because all of the layers (functions) work independently of each other. For a diagram of the OSI, SNA, and DNA function layers, see *Open Systems Interconnection*. For a conceptual diagram of the OSI model layers, see *OSI Standards*.

Physical Topology A physical topology refers to the way a *Local-Area Network (LAN)* of computers is connected for communication. The three different types of physical topologies are: ring, star, and bus. The

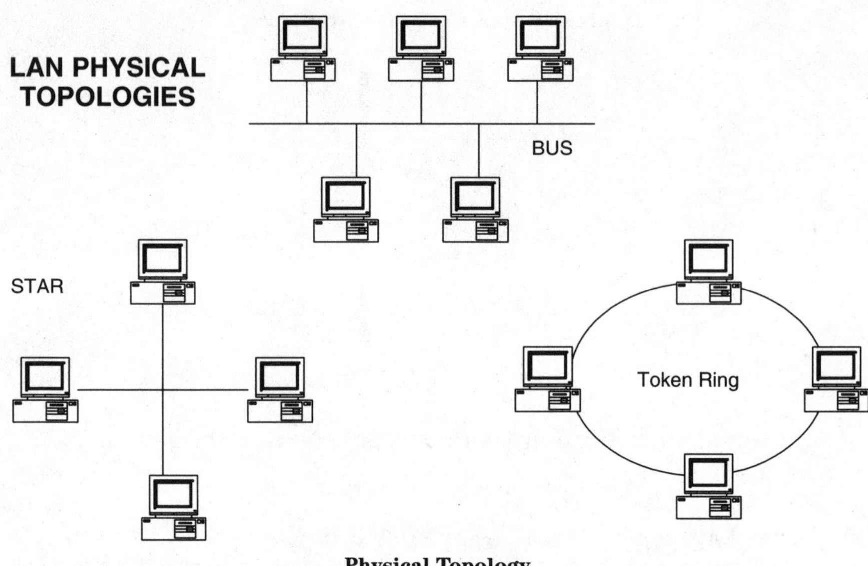

Physical Topology

star and bus topologies work very much the same. The ring topology is also called a *token passing topology*.

PIC PIC refers to color-coded cable. Icky-PIC is cable that is color coded and jelly filled to help protect the copper pairs inside from water.

Pico Prefix for 1 × 10^{-9}. One picofarad (a capacitor) is equal to 0.000,000,001 farads. It is abbreviated as 1 pF.

Picofarad A unit of measurement for capacitors. Pico is the prefix for 1 × 10^{-9}. One picofarad (a capacitor) is equal to 0.000,000,001 farads. It is abbreviated as 1 pF.

Plain B Wire Connector Most commonly referred to as *beans*. A splicing connector used to splice twisted-pair telephone wire. The connectors are crimped onto the wires to be spliced. Inside the connector are teeth, which pierce the vinyl insulation of the wire to make a good connection.

Plant A reference to telephone company equipment, poles, cable vaults, cable, central offices, and transmission equipment.

Plant Test Numbers Telephone numbers that when dialed provide a test tone or access to other testing resources, such as a quiet-line or automatic-number identification. Plant test numbers are used by telephone company personnel and are not given to the public.

Plain B Wire Connector

Plenum A reference to telephone, communications, or electrical wire that is insulated with polyvinylidene diflouride. It is made with this substance because it does not emit poison gasses when it burns, like PVC (polyvinyl chloride) does (PVC produces chlorine gas when burned). It gets its nickname because it is permitted to be placed in air ducts or plenum spaces in buildings. Plenum wiring or cable is typically three times the cost of PVC jacketed/insulated types.

Plenum Cable See *Plenum*.

Plesiochronous "Almost in time." Plesiochronous networks are those that the telephone companies use to synchronize T1 and T3 carrier signals. The electronic transport equipment at each end of the transmission does not get timing from the same source (thus being synchronous), but the timing of each individual device is very close. Stratum One clocks are used in this type of communications equipment, which provide a steady timing for each end to transmit and receive signals by. Transmissions between SONET networks over a long-haul circuit could be considered plesiochronous.

POH (Path Overhead) The overhead that is added to a signal to allow a transport network to carry it.

Point of Interface See *Point of Presence*.

Point of Presence Mostly known as *point of presence. Point of presence* is another term for demarcation or network interface. It is where the telephone company formally hands-off their services to a customer. Wire and equipment on the phone company side of the demarcation belongs to the phone company. Wire and equipment on the customer side of the demarcation belongs to the customer. Every building and home has a network interface. It is against the law (federal) to not have one.

Point to Point Usually a reference to a private-line circuit that is leased from the telephone company. It can also be a reference to a switched service, like a plain telephone line where communications links are switched from one point to another, depending on the number dialed.

Poisson Distribution A mathematical formula that used to be used in traffic engineering for calculating the probability of blocked calls in a telephone network. Now we have computer programs (CTI) that provide graphs of trunk groups and their usage. The graphs are much easier to use. When customers don't have expensive software to manage their telephone networks, they simply add more lines when they get complaints of busy signals.

Polarity A reference to positive or negative voltage potential, or to the polarization of an antenna or dish-type antenna.

Polarization The pointing of a microwave dish antenna so that the transmission dispersion is in a vertical or horizontal pattern. The headlights on cars are polarized in a horizontal manner so that the light dispersion is spread across the horizontal surface of the road. The other kind of polarization is vertical, where the transmission dispersion is in an up-and-down pattern. The two antennas or dishes used in a point-to-point application need to be polarized the same way.

Pole Attachment A lease from a utility company (usually power) that permits a telecommunications company to attach their cable facilities to power poles. For more photos of different aerial-attachment hardware, see *B Washer*, *Strand Clamp*, *Guy Thimble*, and *Johnny Ball*.

POP (Point of Presence) See *Point of Presence*.

Portability A reference to the ability to change telephone companies and take your phone numbers with you.

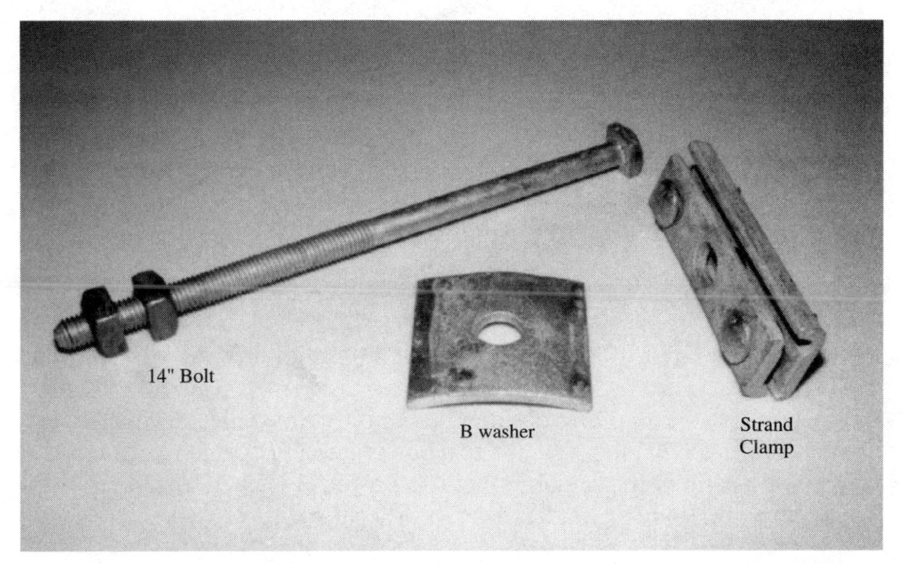

Pole Attachment

POT 1. Abbreviation for potentiometer. Also known as a *variable resistor*. Many electronic control knobs are connected to variable resistors. Variable resistors are usually made from carbon film. For a photo and schematic symbol of a potentiometer, see *Variable Resistor*. 2. *Plain-old telephone*, a reference to standard switched residential and business telephone lines.

Potato Another name for an aerial service-wire splice. Also called a *football*. For a photo, see *Aerial Service-Wire Splice*.

Potential A voltage difference. Potential is a voltage from one point to another. The voltage potential of a POTs telephone line is –52 volts from ring to tip.

Potentiometer (Pot) Also known as a *variable resistor*. Many control knobs are connected to variable resistors. Most volume controls are variable resistors. For a photo and schematic symbol of a potentiometer, see *Variable Resistor*.

POTS (Plain Old Telephone Service) A telephone line, with a telephone number, like the standard ones subscribed to by residences and many small businesses.

Pound Key The button on a telephone dial pad with the # on it.

Power Current multiplied by voltage. Power is measured in watts. If you use a certain amount of wattage over a certain period of time, then you have used energy. Energy is equal to watts multiplied by time, and the unit is joules.

Power Supply A device that converts 120-V or 220-V standard AC power to a voltage that can be useful for an electronic system.

PPSN (Public Packet-Switched Network) A reference to *Frame Relay*.

PPSS (Public Packet-Switched Service) A reference to *Frame Relay*.

Preamplifier An amplifier designed to amplify the voltage level of a very small signal so that it can be fed to a power amplifier, which amplifies the current aspect of the signal so that it is powerful enough to drive the signal current through a loudspeaker or other device.

Predictive Dialing Another term for auto dialing or progressive dialing. Instead of telemarketers dialing digits through a list or phone book all day long, the numbers are entered into a predictive dialer system. The system then dials the numbers; when a call is answered the predictive dialer transfers the call and the associated information to the computer screen of the appropriate telemarketer.

Premises Equipment Also called *CPE (Customer Premises Equipment)*. Telephones, wiring, answering machines, CSU/DSUs, and anything else you might find on the customer side of the network interface.

Premises Wire The wiring on the customer side of the communications company's demarcation point (NI, Network Interface). The premises wire is owned by the customer and is the customer's responsibility to maintain. Many communications companies sell maintenance contracts, which enable them to troubleshoot and repair the telephone wire within your home or business, at no extra charge. Typical maintenance contracts are about $2.00 per month.

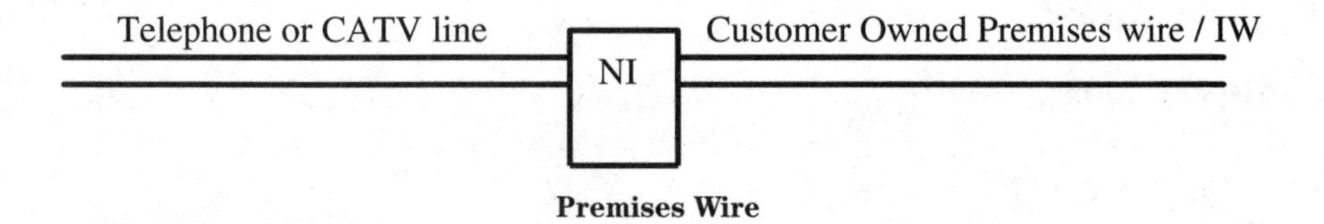

Premises Wire

Prepaid Phone Card A card that comes with an 800/888 number that the card owner dials to reach a network that allows them to dial anywhere

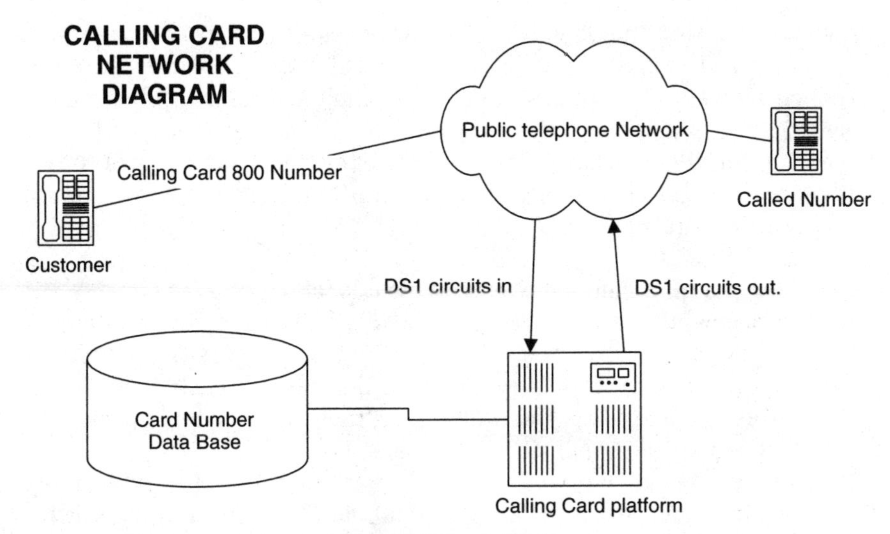

Prepaid Phone Card

they like. The service is good for the amount of time that the prepaid phone card says on its face. Prepaid calling cards are becoming bigger and bigger, especially because they don't cost much more than third-party billing to your home number when you are out of town. Typical prepaid calling cards have a rate of 30 to 35 cents per minute, flat rate, no matter when or where you call within the continental U.S. The way the system works is that a calling card company sets up a data base with card numbers in it and connects it to a calling-card platform. A calling-card platform is a computer that receives a phone call and prompts a caller to enter their calling-card number and the telephone number that they wish to dial. The calling-card platform then checks the card number to see how many minutes it has left on it (it sometimes tells the customer with a recorded message). If time left on the card, the system then dials the number on an outgoing trunk to connect the call. In reality, two long-distance calls are made, one to the calling card platform and one to the number being dialed by the customer.

Prepay A reference to a coin-operated telephone that requires a coin to be inserted before a number is dialed.

Presentation Layer A layer in a communications protocol model. In general, the presentation layer performs the function of encoding and decoding the data to be transmitted within the communications protocol. The latest guideline for communications protocols is the *OSI (Open Systems Interconnect)*. It is the best model so far because all of the functions work independently of each other. For diagrams relating to the OSI, see *Open Systems Interconnection* and *OSI Standards*.

Pressure Cable Telephone cable that is equipped with air-pressure equipment. In many cables, nitrogen is used instead of air because it is noncorrosive (air contains humidity, which corrodes copper pairs). Nitrogen is pumped into the cable and the pressure is monitored. If the cable is cut, the pressure drop notifies the telephone company of the cable problem and the nitrogen rushing out of the cable helps prevent any water from entering.

Presubscription When a customer calls a local telephone company and orders a new phone line, they are asked which long-distance company they would like to subscribe to. When the customer tells them, then the telephone company sets the customers line up in translations so that when the customer dials 1 as a first digit, they are connected directly to the long-distance company that they selected.

Prewire To install standard wiring into a building or space while it is being constructed. Standard building wiring is that all wiring from each jack terminates to a common location, usually called the *telephone closet*. Prewiring of buildings is common for telephone and CAT 5 computer LAN wiring.

PRI (Primary Rate Interface) One of two *ISDN (Integrated Services Digital Network) circuit sizes*. ISDN first evolved in 1979. It brings the features of PBX systems and high-speed data-transfer capability to the telephone network. The only thing that complicates ISDN is the many available features. The two kinds of ISDN lines are Primary Rate Interface (PRI) and Basic Rate Interface (BRI). ISDN has two types of channels within an ISDN circuit. The B (bearer) channel carries the customer's communications and the D (data) channel provides control and signaling for the B channels. The BRI ISDN line has two B channels and one D channel. A PRI has 23 B channels and one D channel.

The separate control of the ISDN line over the D channel is what enables the broad flexibility and features available with ISDN. When you are talking or sending a data transmission over an ISDN line, the voice and/or data is carried by the B channels. While you are talking on your ISDN line, you can still dial digits (signal the central office) to change or alter the state of your service because of the separate D channel. For example, imagine you want to arrange a meeting with a client. You dial the client's telephone number on your ISDN telephone to reach the client. While you are speaking with the client, you can dial up an Internet access on your computer and put two baseball tickets in at the ticket counter while on the same BRI line. Then you can fax your client directions by downloading a map provided by the baseball ticket office, disconnect and redial your client's fax number. All of this occurs while talking to your client the entire time. Through the advanced convenience and flexibility of ISDN, you can send different types of data and messages to different places at the stroke of a few buttons, and at a much faster speed than a regular telephone line. If you are interested in ISDN, call is your local phone company. They can help you decide on what kind of terminal adapter (equipment that connects your computer and phone equipment to the ISDN line) to buy and what kind of features to subscribe to. ISDN is not yet available everywhere. For a diagram that compares an ISDN BRI and ISDN PRI circuit, see Integrated Services Digital Network.

Prime Line A key telephone system and hybrid key telephone system feature. The feature enables a user to select the line that a key system connects to a telephone set to when the receiver is lifted. If you don't want people in the office using the main telephone line in the office to make outgoing calls, then don't select that line as a prime line for any of the telephone extensions.

Printed Circuit Board (PC Board) The green- or brown-colored board that has copper-conductive tracks etched onto its surface. Electronic components are soldered onto these boards by hand or by a method called *flow soldering*. Some PC boards are layered or sandwiched, with conductive tracks inside them and on both sides.

Private Branch Exchange (PBX) A telephone system used to maximize use of telecommunications services purchased from a telecommunications company. A PBX simply takes telephone lines from the outside world and makes them accessible to extensions within a certain building, home, or office. PBX systems are available in many sizes, with many software and feature options. PBX features include call forwarding, speed dial, internal/external paging, and call-detail recording (call accounting). The larger PBX manufacturers are AT&T, Northern Telecom, Siemons, Toshiba, Iwatsu, NEC, and Rolm. PBX systems have six main parts: the cabinet-backplane (also called a *KSU, Key Service Unit*), the station/telephone connectivity, the trunk/telco connectivity, the power supply, the telephones/extensions, and the administrative access.

- *Cabinet/KSU* The cabinet of the system contains the electronics that make the PBX system work. The backplane (for a photo of a backplane see, *Backplane*) that interface cards plug into is located here. The CPU or core processor (for a photo, see *CPU*) is located in here as well. Many PBX cabinets are designed to allow for additional circuit cards (trunk interfaces/trunk cards and telephone interfaces/station cards) to be added or plugged in later on as the system grows. These spaces are called *expansion slots*.

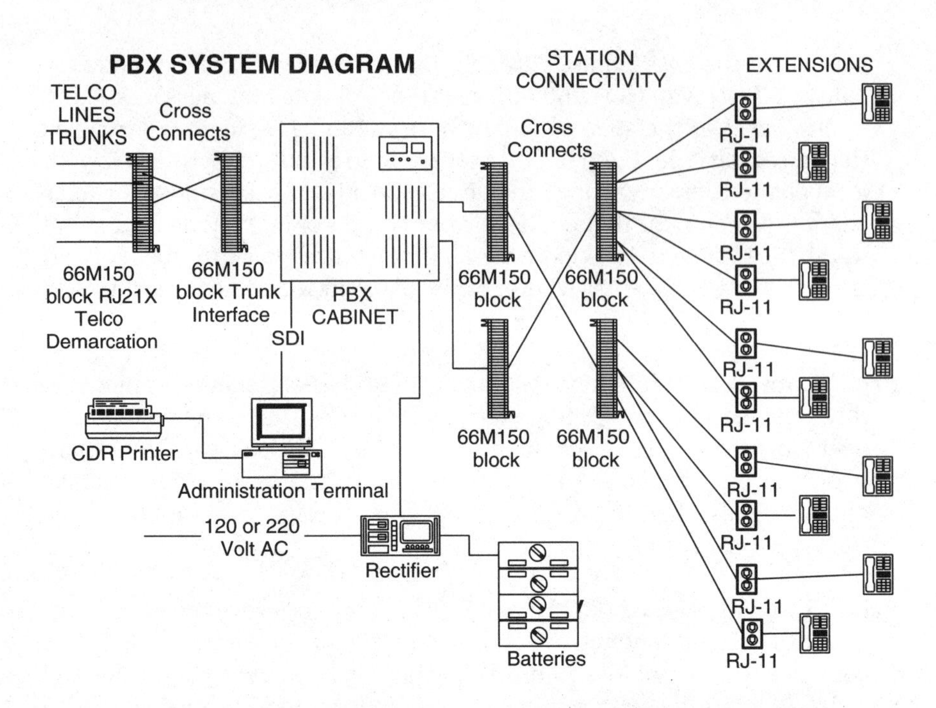

Private Branch Exchange (PBX)

- *Station-telephone connectivity* This wiring runs from each office or telephone location to the location of the PBX cabinet. Four-pair wiring is most popular because it is inexpensive and contains enough wire to add additional lines or telephones in the future (or additional wire if one or two should go bad). This wiring is installed in a "home run" method, which means that every wire installed runs directly from a jack (usually an RJ-11) directly to the location of the PBX cabinet. Next or near to the PBX cabinet, the individual pairs are neatly terminated and labeled on 66M150 or AT&T 110 (one-ten) blocks.
- *Trunk-telco connectivity* This is similar to the station connectivity, but it needs to be separately labeled from the station connectivity. This is the point where cross connects will be run from the telephone-company demarcation (or *NI, Network Interface*) to your PBX system.
- *Power Supply* The power source for the phone system is a very important consideration. If the power is interrupted, the PBX system will cease to function unless its power supply is incorporated with a UPS system or rectifier/battery system. The best way to go for power is the rectifier with battery back-up (a heavy-duty UPS system especially designed for telephone equipment). Different PBX systems can be ordered to run on 120V AC or –48V DC. The –48-V DC system is designed to be powered by a rectifier. The 120V AC system is designed to run on standard outlet power or a UPS system.
- *Telephones* The telephones for each individual PBX system will work only with that system. They will not work if they are plugged into a regular telephone line. Each phone will determine what features can be implemented. The features are enabled or disabled by the programming or administration done on the PBX system. Some systems have an interface (*SDI, Serial Data Interface*) for a computer or terminal and some are simply programmed by using the telephone stations.
- *Administrative Access* The administrative function of a PBX system can be performed by the user or a telephone-equipment service company. The administrative responsibilities of a PBX system include changing extension numbers, moving phones, changing name displays, and other programming of the system. It also includes maintaining the *Call-Detail Reports (CDR)* of the system. The call-detail reports are reports output by a call-accounting system, which is offered as an extra by virtually every PBX manufacturer. Call-detail reports summarize numbers dialed, length of calls, and incoming calls, caller ID, and their duration.

Private Carrier A telecommunications company not regulated by the rulings of the PUC; however, they are regulated by the Telecommunications Act of 1996.

Private Line Also called a leased line or leased circuit. A leased line is a telephone service that is permanently connected from one point to another. Leased circuits include 56K analog and DS1. A leased circuit acts like a pipeline that carries data from one point to another. If you put a bit in one side, the same bit pops out on the other side. It can carry data across town, across the country, or around the world. Leased lines are relatively expensive. Because leased lines have been offered, new services, such as frame relay and switched 56K services have evolved. Frame relay does the same job as a private line, except that it is not isochronous (real time), and you need a private line to put your frame-relay service on. Frame relay is a cost-effective solution for long-haul/long-distance data-transfer applications.

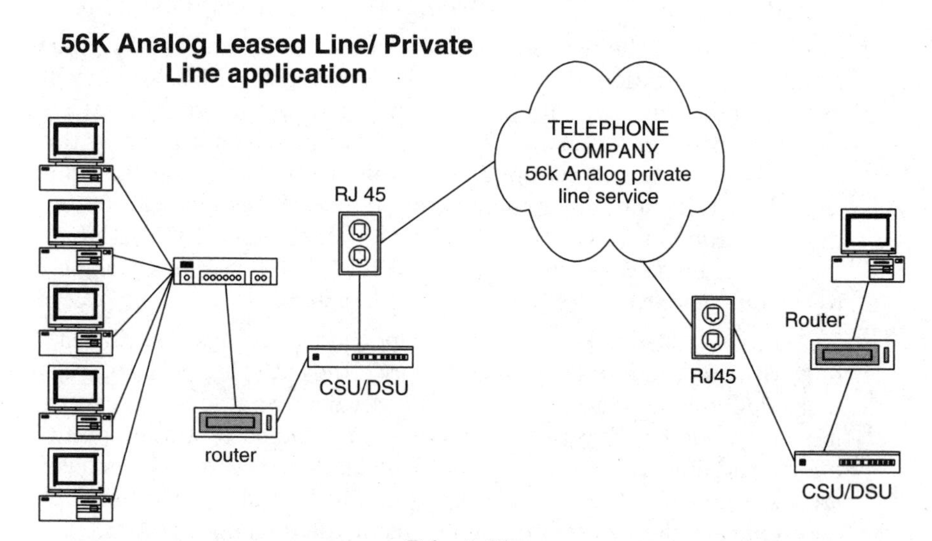

Private Line

Programmable Read-Only Memory (PROM) Electronic memory comes in two families, *ROM (Read-Only Memory)* and *RAM (Random-Access Memory)*. Memory devices are made from two different technologies: *Bipolar (TTL)* and *MOS (Metal-Oxide Semiconductor)*. Memory is stored by a technique called "writing" and is retrieved by a technique called "reading." ROM devices can only be read and are programmed during manufacture. PROM devices can be programmed at a later date by an electronics reseller or electronic assembler for a special application using special equipment. Special ROM devices called *EPROMs (Erasable Programmable Read Only Memory)* can be electronically erased and re-used. RAM has read and write capability.

The term *random access* means that any memory address can be read in any order at any time. The two types of RAM are static and dynamic. *Static RAM* can hold its memory even when power is removed. *Dynamic RAM* needs constant power to refresh its memory. For a diagram of the different types of dynamic memory, see *Memory*.

Programming PBX Many PBX systems have software that is a "prompt response" style of programming. For example, when the user inputs extension 255, the system responds with such prompts as "Hands Free?" The user then responds with "YES" or "NO." "Speed Dial?" The user responds "YES" or "NO." "Paging ability?" The user responds "YES" or "NO." "Do Not Disturb?" The user then responds with "YES" or "NO," etc. for all of the extension features.

Progressive Conference A conference-call service that enables callers to call the same telephone number, be greeted by an attendant, and then be introduced into the conference call.

Progressive Dialing Another term for auto dialing or predictive dialing. Instead of telemarketers dialing digits through a list or phone book all day long, the numbers are entered into a predictive dialer system. The system then dials the numbers and when a call is answered, the predictive dialer transfers the call to an agent, along with the associated information to their computer screen.

PROM (Programmable Read-Only Memory) See *Programmable Read-Only Memory.*

Prompt A message from a computer or interactive device that indicates that it is time for a user to input a decision, choice, or other response. Many PBX systems have software that is a "prompt response" style of programming. For example, when the user inputs extension 255, the system responds with "Hands Free?" The user then responds with "YES" or "NO."

Prompt, Response IO See *Programming PBX.*

Propagation Time The time for an electrical, optical, or radio signal to travel from one point to another.

Propagation Velocity The speed that a communications signal travels from one point to another. Electromagnetic waves (radio), electricity, and light approach 300,000,000 meters per second, which is about 160,000 miles per second.

Proprietary Specially made. All PBX equipment and other premises telephone equipment is proprietary. Northern Telecom telephones will only work with Northern Telecom PBX systems. The same goes for Lucent, Mitel, and other specialized telephone equipment manufacturers.

Protector Block A block that has many lightning protectors, used to terminate telephone cables. A protector is a device used in telephone company network interfaces that provides an easier path for lightning to travel to ground, compared to a telephone user or inside wiring. Before lightning protectors, houses sometimes burned down because of lightning striking the telephone lines. The two types of lightning protectors are carbon and gas. The carbon protectors are simply a piece of carbon that connects tip and ring to ground. The gas protectors are the same, only they are a gas instead of solid carbon. The good thing about gas lightning protectors is that after they are hit by lighting, they do not need to be replaced.

Protector Block

Protocol The organized processes and rules that communications equipment use to transfer bits and bytes (data). The many communications protocols and layers of protocols that carry other protocols (called protocol stacks), include ISDN, ethernet, token ring, POTS signaling, DS1, ATM, frame relay, and SONET.

Protocol Analyzer A test device that can plug into a hub or communications port on a LAN and monitor any address on that LAN at any protocol level. Protocol analyzers are useful for verifying that an address is

Protocol Analyzer

good through a network. Most networks are not so complex as to need a protocol analyzer to troubleshoot them.

Provisioning A term that refers to the process of allocating copper pairs, central-office ports/equipment, and programming of central-office equipment. This is what happens before a telephone company network technician installs a telephone service, such as a POTS line or a high-capacity digital service line.

Proxy Server A network server that is loaded with software and equipped with hardware to interface a LAN, MAN, or WAN to the Internet. Proxy servers make up the hardware part of a firewall, which is software that protects the LAN's interworkings from being accessed by strangers/unwanteds/hackers on the outside. Although firewalls are expensive and abound everywhere, hackers still manage to get through them.

PSC (Public Service Commission) See *Public Service Commission.*

PSE (Packet-Switching Exchange) See *Packet-Switching Exchange.*

PSI 1. *Pounds Per Square Inch*, a unit of air pressure. Telephone cables (pulp-insulated cables) that are pressurized with nitrogen are kept at a pressure of 10 to 15 PSI near the central office. 2. *Packet-Switching Interface* gives a customer a means to connect with a packet switching network, such as frame relay.

PSK (Phase-Shift Keying) See *Phase-Shift Keying.*

PTN (Public Telephone Network) Also called *PSTN (Public Switched Telephone Network)* and *PSN (Public Switched Network)*. The telephone network that we know today provides us with an open-ended dial tone, the ability to dial a telephone anywhere we wish.

Public Address System (PA System) There are different types of PA systems. High-fidelity PA systems are used in studio recording and concert productions and simple systems are used for paging/intercom and loudspeaker systems. The two main components of a PA system are the amplifier and the speakers. Different components can be attached to the input of a PA system. The PA amplifier input is a high-impedance circuit (this means that it does not draw a lot of electrical current from the source, thus transferring maximum voltage). Common source (signal input) devices include microphones, musical instruments (electric), and the paging output of telephone systems. If an amplifier is used to drive external speakers (rather than the ones inside telephones), then it is called an *external paging amplifier* or *PA amplifier*.

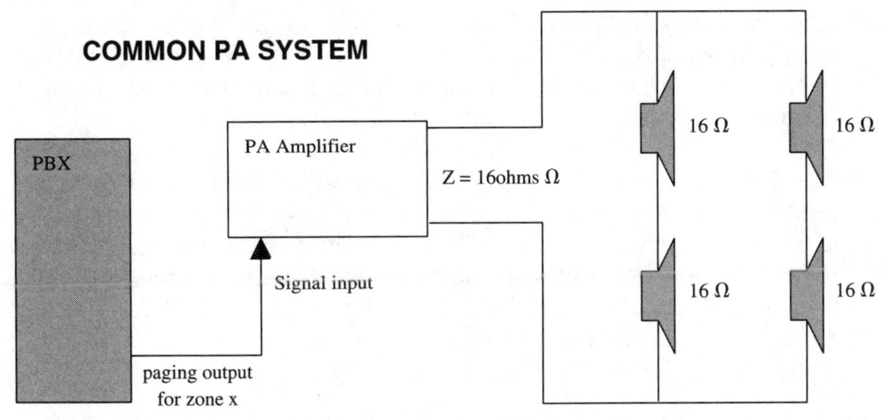

Public Address System (PA System)

The question that most people have about PA amplifiers is which one to buy. The answer is that the majority of the cost in a paging system is usually the wiring and the speakers. Most paging amplifiers are equipped with multiple inputs so that different areas or "zones" can be paged individually. The factor that affects the price of a PA amplifier the most is the power output rating. The more power that an amplifier is capable of pushing through the speaker network, the more expensive it is. A good general rule is to allot 5 watts of RMS power per speaker in an office environment and 10 to 25 watts of RMS power per speaker in an industrial environment. The crucial factor in designing the speaker network is that the impedance (Z) of all the combined speakers matches (or is equal to) the output impedance of the PA amplifier, which is usually 8 or 16 Ω. If the impedance is not matched, there is a possibility of over working the amplifier and causing it to fail or having a poor performance and sound quality. Quality PA amplifiers/paging amplifiers have instructions on how to wire and arrange the connections of speakers. See the drawing for an example of how a 16-Ω output amplifier is matched with four 16-Ω speakers.

Public Service Commission (PSC) The watchdog for the Public Utilities Commission. The Public Utilities Commission regulates the telecommunications companies under federal judgments (which change from time to time), and other utility companies. For a telecommunications company to be regulated, it must have a minimum number of customers. All the RBOCs are regulated by the PUCs of their area.

Public-Switched Digital Service A general name for switched 56K service from a local or long-distance telephone company.

Public Switched Network (PSN) Also called *PTN (Public Telephone Network)* and *PSTN (Public Switched Telephone Network)*. The telephone network that provides open-ended dial tone, the ability to dial a telephone anywhere we wish.

Public Utilities Commission (PUC) The governing body of regulated public utility service companies and the *Public Service Commission (PSC)* that watches over them. The Public Utilities Commission regulates the telecommunications companies under federal judgments (which change from time to time) and other utility companies. For a telecommunications company to be regulated, it must have a minimum number of customers. All the RBOCs are regulated by the PUCs of their area.

PUC (Public Utilities Commission) See *Public Utilities Commission*.

Pulling Strength A cable specification. The maximum pulling force that can applied to a strength member of a cable without voiding the warranty.

Pulp Cable Telephone cable used in outside plant applications that uses paper insulation on the twisted copper pairs. The other kind of widely used cable is *pick cable*, which has color-coded plastic-insulated pairs. For a photo of pulp-insulated cable, see *Lead Jacket*.

Pulse Amplitude Modulation (PAM) See *PAM*.

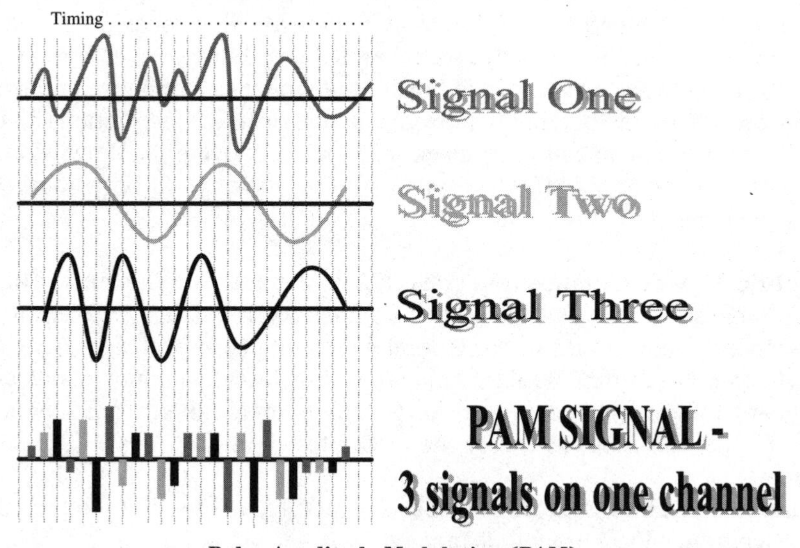

Pulse Amplitude Modulation (PAM)

Pulse-Code Modulation (PCM) See *PCM*.

Punch-Down Block A 66M150 block, AT&T 110 (one ten) block, crone block, or other wire-terminating device. A punch-down block provides connections to neatly connect and label wires.

Punch-Down Tool A tool that is used to terminate telephone wires on to punch-down blocks.

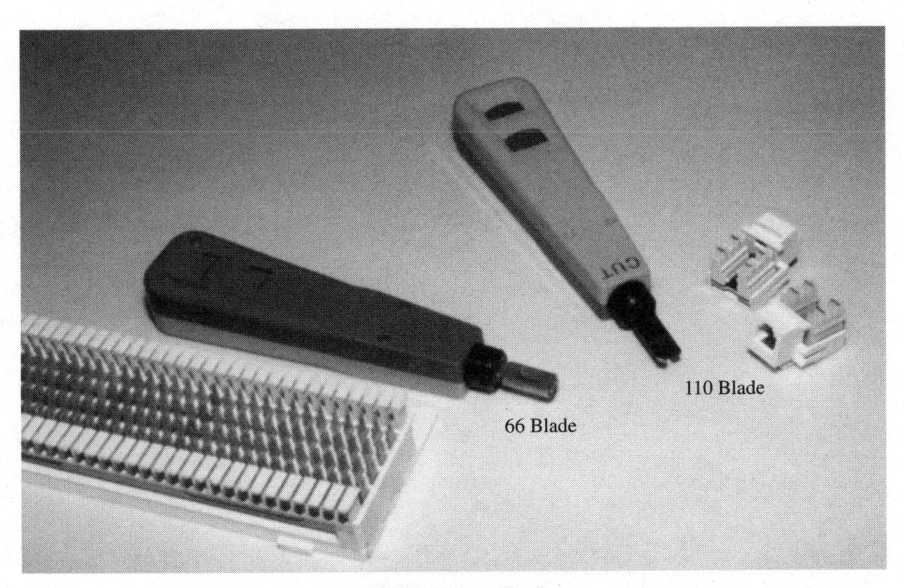

Punch Down Tool

PVC (Polyvinyl Chloride) The substance with which common telephone wire is insulated. PVC wire is available in many colors. The other more expensive option for telephone wiring is Plenum. Plenum wiring is required in many newer buildings because when it burns, it does not emit poison gasses (PVC produces chlorine gas when burned). Plenum wiring is made from polyvinylidene diflouride, and costs about three times as much as PVC does.

PVDF (Polyvinyl Diflouride) Better known as plenum wire.

Quad IW (Quad Inside Wire) Older standard telephone wire used by telephone companies. Quad has four wires, the colors are red, green, black, and yellow. Some quad wiring is not twisted, so it is susceptible to RFI. The colors for line one are green and red, and the colors for line two are yellow and black.

Quad Lock Conduit Conduit that is designed to be direct buried. The four individual conduits allow communications companies to lease con-

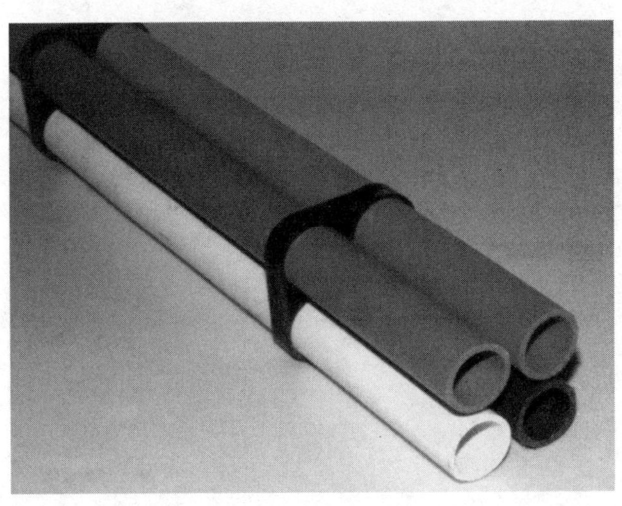

Quad Lock Conduit

duit space to each other in a way that is easy to track for fiber-optic cable installers/splicers, etc.

Query A name given to a programming string that asks a question about data in a relational data base. Queries are common in call-accounting applications.

Query Language A programming language designed for manipulating data in data bases.

Queue Queue is a way of saying *waiting line* in telephony. The two types of queues are line queues (very uncommon) and *ACD (Automatic Call Distribution System)* call queues. Some telephone systems have a feature called *line queuing*. If you try to dial out and you cannot get an outside line, you are put in queue, or in a waiting line for the next available trunk. Some systems can provide music, as if you were on hold for the line and some can ring your phone back. The other queuing is ACD specific. ACD systems place incoming calls in queue for the next available agent and evenly distribute calls among the agents so that the workload is not unbalanced and sales opportunities are fair.

Quick Connect A name given to 66M150 block and AT&T 110 (one-ten) block connectivity.

RA (Return Authorization) Also known as *RMA (Return Material Authorization)* or *RAN (Return Authorization Number)*. A reference number in the advance-replacement process. If you receive a shipment from a distributor or manufacturer and a part is defective, you call the distributor/manufacturer and they give you an RA or RMA number to place on the package when you send it back to them. They, in turn, send you a replacement immediately.

Raceway A trough designated for wiring. Raceways can be in ceilings, attached to walls, or built into floors.

Rack Also called *relay rack*. The two standard dimensions of racks used in telephony and rack-mountable computer equipment are 19" and 22" wide. The height ranges from one to seven feet. Some racks can be attached to walls (wall mount). Most racks are rated as zone 4, which means that they are designed to withstand earthquakes to a certain degree.

Radar (Radio Detection and Ranging) Radar is a means of detecting objects within the vicinity of a radio signal. Different types of objects can be detected, depending on the frequency of radio used. Radar works by sending a pulse from a transmitter: the pulse travels outward, bounces off objects, and is returned to a receiver. The time difference between the pulses departure and arrival determines the distance of the object. Any

Doppler effect on the pulse determines the speed of the object toward (or away from) the transmitter.

Radar Detector Radar detectors are famous for their use in speed-limit enforcement in the United States. Radar detectors use radar technology to measure the Doppler effect of radio signals that are sent from a transmitter, bounced off of a moving object, then returned at a different frequency. The movement of the object compresses the radio signal as the two come into contact, thus increasing the frequency.

Radio The emission of electromagnetic radiation into the air, then picking it up with a receiver. Electromagnetic radiation occurs when a magnetic field changes at the rate of a carrier frequency. The magnetic field then traverses through the farthest reach of the magnetic field, which could be many miles. One determining factor in the distance that the signal will reach is the transmitting power. Broadcast radio has a typical output power of 15 to 100 kW (15,000 to 100,000 watts) and CB radios have an output power of 4 Watts. What makes the radio signal carry a voice or music is called *modulation*. Modulation is the act of varying a carrier signal in a way that can be sensed or "detected" by a radio receiver. Those variations are then amplified and run through a speaker so that people can hear them.

Radio Common Carrier (RCC) A cellular/PCS service provider or paging company. A company that provides one-way (paging) or two-way (mobile phone) radio services to individuals, rather than communities. Broadcast TV or radio stations are not radio common carriers.

Radio Frequency (RF) Any electromagnetic frequency that is above the range of human hearing. Most licensed radio transmissions range from 500 kHz (500,000 Hz) to 300 GHz (300,000,000,000 Hz).

Radio Frequency Interference (RFI) Also called *EMI (Electromagnetic Interference)*. Interference caused by a radio signal or other magnetic field inducing itself onto a medium (twisted/nontwisted pair wire) or device (telephone or other electronics). The world we live in is full of radio waves that are emitted from electric appliances, such as blenders, automobile engines, transmitters, and even fluorescent lights. Even though we take preventative measures to avoid receiving these unwanted signals, they sometimes get into places they are not wanted. Electromagnetic interference is usually caused by one of two things. The first is when a wire connected to a device acts like an antenna and picks up the EMI, which is then passed on to the electronics inside the device and amplified. The second is when an electronic component inside a device acts like an antenna because of poor design, poor shielding, or the component is defective.

Radome A cover for a radio antenna, typically used in public broadcast applications.

Rain Attenuation The degradation of a radio signal (particularly in the microwave region) because of rain. The rainfall average and density of the rainfall is the determining factor (along with fog, which attenuates radio much more severely) in the distance that a radio (microwave link) can send a signal. Typical ranges for the dry climate regions of the United States are as much as 6 miles, and as little as one mile for the wetter regions (for a 7-watt transmitter).

Raised Floor Many computer and telecommunications rooms have a raised floor. The raised floor is a very sturdy framework of iron, with heavy 1" tiles placed into the framework. The tiles are easily removed and replaced with a suction cup. The raised floor is used as a giant "duct" to move and run connecting cables through, and it is also used as an airway to pump cool air through the equipment. Instead of cooling a room, cool air is blown under the floor, where it finds its way into the equipment through holes in the floor. The holes are cut into the floor when the equipment is installed.

RAM (Random-Access Memory) see *Random-Access Memory*.

Ram Hook/Ram Horn A hardware attachment used to hold *ASW (Aerial Service Wire)* drop clamps in aerial-span applications.

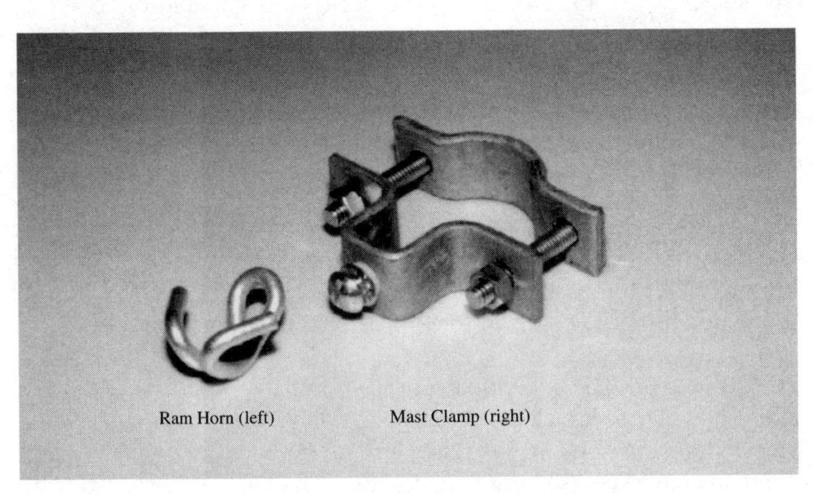

Ram Hook/Ram Horn

RAN 1. *Recorded Announcement*, a term used in IVR and ACD call-flow analysis. If you like, you can call the recorded greeting on your answering machine a RAN. For a photo of a digital announcer that stores RAN messages, see *Digital Announcer*. 2. *Return Authorization Number*, also called *RA (Return Authorization)*, or *RMA (Return Material Authorization)*. A reference number in the advance-replacement process. If you receive a shipment from a distributor or manufacturer and a part is defective, you call the distributor/manufacturer and they give you an RA or RMA number to place on the package when you send it back to them. They, in turn, send you a replacement immediately.

Random-Access Memory (RAM) Electronic memory is available in two families, *ROM (Read-Only Memory)* and *RAM (Random-Access Memory)*. Memory devices are made from two different technologies, *Bipolar (TTL)* and *MOS (Metal-Oxide Semiconductor)*. Memory is stored by a technique called *writing* and retrieved by a technique called *reading*. ROM devices can only be read and are programmed during manufacture. *PROM (Programmable Read-Only Memory)* devices can be programmed at a later date by an electronics reseller or electronic assembler for a special application using special equipment. Special ROM devices called *EPROMs (Erasable Programmable Read-Only Memory)* can be electronically erased and re-used. RAM has read and write capability. The term *random access* means that any memory address can be read in any order at any time. The two types of RAM are static and dynamic. Static RAM can hold its memory even when power is removed. Dynamic RAM needs constant power to refresh its memory. For a diagram of the different types of dynamic memory, see *Memory*.

Rate Adaptive A type of data protocol that is capable of "testing" the telephone circuit for the fastest possible transmission rate, then transmitting at that rate. This test is done using a "ping" packet similar to that used in. DSL (Digital Subscriber Loop) is a rate adaptive protocol.

Rate Design A term that refers to the way utility companies figure a way to charge money for their services. Rates are designed to be affordable for everyone (PUC/PSC requirement). A good example of rate design is the way that the telephone companies charge extra money for business lines to offset the costs of residential lines. How far the offset is and how much one rate is subsidized for the other is the rate design.

Rate Elements The individual charges and fees for a service. For instance, all of the rate elements are listed on your phone bill: the local service charge, dial tone, 911 service, etc. All of these parts of the telephone company have been separated by the FCC and billed for separately by law.

Rayleigh Fading Rayleigh Fading is a form of signal reduction or loss because of a receiver picking up the same signal from multiple directions. The signal commonly arrives from multiple directions because of reflections from buildings when there is no line-of-site path (receiving reflections of the same signal is also referred to as *multipath reception*). When the signals meet, they add or subtract each other, causing an irregular signal strength. Rayleigh fading is a common reason for geographical "dead spots" in cellular service networks.

Rayleigh Scattering The scattering of light in a fiber-optic cable because of impurities in the glass of the fiber. It has a similar effect to what a lamp shade has on a light bulb, just not as drastic.

RBOC (Regional Bell Operating Company) At the time of divestiture, there were 22 BOCs, grouped into seven *Regional Bell Operating Companies (RBOCs)*. For a listing of the BOCs and RBOCs, see *Regional Bell Operating Company*.

RCA (Regional Calling Area) The geographical area that a telephone company serves.

RCA Connector A plug first developed and used by *RCA (Radio Corporation of America)*. These plugs are very common in audio- and video-patch applications. If you have a CD player and a separate tuner/amplifier, the cord that connects the two most likely has RCA plugs.

RCA Connector

RCC (Radio Common Carrier) A cellular/PCS service provider or paging company. A company that is in the business of providing one-way (paging) or two-way (mobile phone) radio services to individuals, rather than communities. Broadcast TV or radio stations are not radio common carriers.

Reactance Reactance is the resistance that a component gives to an ac or fluctuating DC current. The two components that cause reactance are inductors (coils) and capacitors. (Reactance is also caused by other electronic conditions where it is not useful. All wire and electronic components possess a small amount of reactive properties, e.g., twisted-pair wire causes signal attenuation because of the inductance of the copper wire and the capacitance of the two adjacent wires). The difference between resistance and reactance is that resistance is always the same, regardless of the voltage amplitude or frequency applied to the resistive device. The reactance of a component changes along with frequency changes, the speed at which an AC current changes direction. The higher the frequency applied to an inductor, the higher the reactance or resistance to that frequency. The reason that coils of wire cause reactance is that as electricity flows through them, they force the electricity to create a magnetic field every time it changes direction. A perfect inductor has zero reactance to a DC current, and has a specific reactance or resistance to every frequency of AC current. Each coil (inductor) has a value in henries. The higher the number of henries, the more it will resist AC or fluctuating DC. Coils are used to filter out ("choke" out) DC fluctuations in power supplies. They are also used to help tune in radio or other frequencies.

Read-Only Memory (ROM) See *Random-Access Memory*.

Real Time A reference to the relationship of events in a communications channel, machine, or PC. *Real time* means that the inside of the machine is synchronized with real-world time that you and I live in. Another word for this is *isochronous*, which means "in time." A normal voice conversation is real time, but a frame-relay transmission is not. The individual packets are separated and re-assembled, which causes a delay. This is not real time.

Rebiller A telephone company that buys a telephone service from a facilities based telephone company and resells it. A rebiller attempts to add value to the original long-distance company's service by providing better customer service and customized technical expertise. The rebiller gets a discount from the original long-distance company, typically about 10%. Rebilling is also known as *Type-III service*, where all circuits (telephone lines) are type III.

Rebooting To restart a computer by turning it off and turning it on again. The two ways to re-boot a computer are a hard boot and a soft boot. Hard booting is manually turning off the computer to force the microprocessor to reset. Soft booting is done by pressing Ctrl-Alt-Del at the same time. This direct code sends a positive pulse to the reset of the computer. However, it will sometimes not work if the computer's keyboard is locked up with the rest of the components.

Receiver 1. The part of a telephone handset that you talk into. The receiver has a microphone inside it. 2. A radio device that is connected to an antenna and filters and detects carrier frequencies and signals modulated on them. For more information, see *Modulation* and *AGC*.

Receiver Off Hook (ROH) The condition of a telephone set being left off the hook, with no numbers dialed, or left off the hook after a conversation has been completed. This causes the central office to disconnect the voltage from the telephone line, which saves electricity. When the receiver is placed back on the hook, the telephone line does not become activated instantaneously. The dial tone can take one to two minutes to return. ROH is a common test result in telephone company mechanized loop testing.

Recorded Announcement (RAN) 1. *Recorded Announcement*, a term used in IVR and ACD call-flow analysis. If you like, you can call the recorded greeting on your answering machine a *RAN*. 2. *Return Authorization Number*, also called *RMA (Return Material Authorization)* or *RA (Return Authorization)*. A reference number in the advance-replacement process. If you receive a shipment from a distributor or manufacturer and a part is defective, call the distributor/manufacturer and they will give you an RA or RMA number to place on the package when you send it back. They, in turn, send you a replacement immediately.

Rectifier A device to convert AC power to DC power, also called a *diode*. An electronic semiconductor device that, simply put, only conducts electricity in one direction. Whether or not the device conducts depends on which direction the device is "biased." Diodes are used to change *Alternating Current (AC)* to *Direct Current (DC)*. If a more positive voltage is applied to the anode lead of the diode, then the diode simply acts like a wire. If the more positive voltage is applied to the cathode lead, then it acts like there is no connection. The following illustration shows the schematic symbols of the first diode, which was a vacuum tube, and a solid-state silicon diode. Below is an illustration of a pair of diodes converting AC to DC.

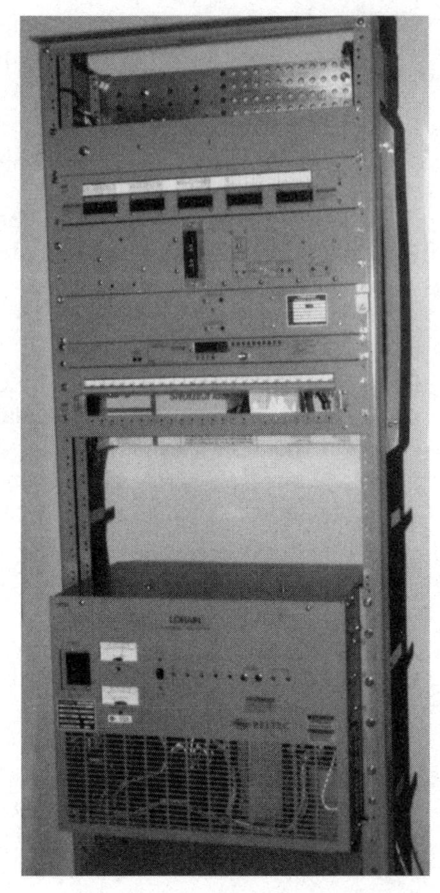

A Vacuum Tube diode

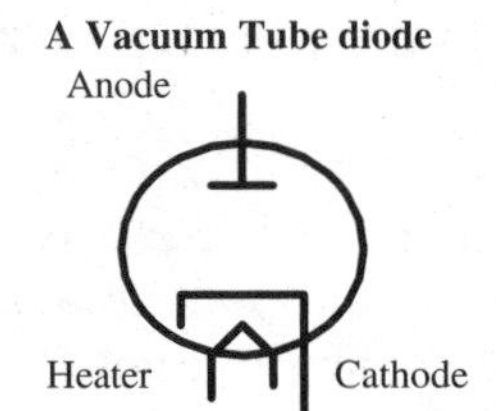

A Solid State diode

A simple one diode rectifier circuit, with a filter capacitor to eliminate DC fluctuations.

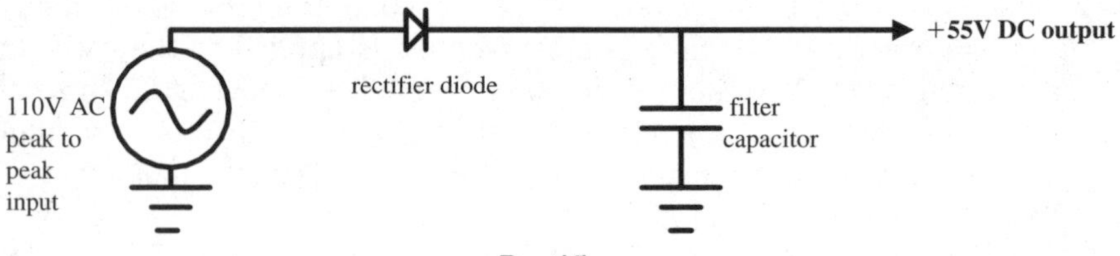

Rectifier

Redundancy To have one main and one back-up. SONET equipment is capable of being configured in a redundant manner, with two fiber-optic routes (in a ring) and duplicates of the electronic cards that control the communications transmission. Many PBX systems are capable of being configured with redundant CPU and memory cards. The idea behind redundancy is that if one device fails, the other will take over, without a loss of service.

Reference Clock Also called a *bits clock*. A device that provides a timing pulse in the form of a 1-0-1-0-1-0-1-0 bit stream. Bits clocks are used extensively in SONET networks. The bits clock provides the timing pulse that everything in the network synchronizes itself to.

Refraction *Refraction* refers to the wavelike nature of light. When light travels from one media, such as air, into another media, such as water, it bends. This is why when you look into a swimming pool, the bottom looks very distorted. Fiber-optic technology is based on the fact that light refracts (or bends) as it travels from one media to the next. A single fiber-optic strand consists of many different kinds of glass. The core is one kind and the cladding consists of many layers of glass that have different "levels" of refraction and cause light to gradually refract or bend back to the center as it travels down the fiber. A ray of light refracting as it passes through materials of different refractive indexes is illustrated.

ILLUSTRATION OF A LIGHT BEAM REFRACTED BY MATERIALS OF DIFFERENT REFRACTIVE INDEXES

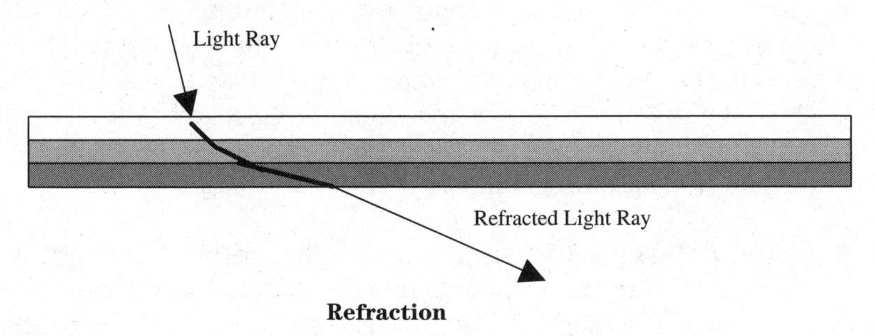

Refraction

Refractive Index The refractive index of a material refers to how much light refracts (bends) when it travels from a vacuum into the material at an angle. For a diagram and more details, see *Refraction*.

Refurbished Electronic equipment that has been repaired and cleaned, or remanufactured.

Regenerator Also known as a *repeater*. A device that is used to take a signal that has traveled a long distance and make it new again. Repeaters can be coils of wire, which are used in the public telephone network for voice (POTS) lines, or they can be electronic, taking an electronic signal that has been attenuated over a long distance, reproducing it, then retransmitting it.

Regional Bell Operating Company (RBOC) At the time of divestiture, the 22 BOCs were grouped into seven *Regional Bell Operating Companies (RBOCs)*:

BOCs: Bell Telephone Company of Nevada Illinois Bell Telephone Company Indiana Bell Telephone Company Michigan Bell Telephone Company New England Telephone and Telegraph Company US West Communications Company South Central Bell Telephone Company Southern Bell Telephone and Telegraph Company Cincinnati Bell Company Mountain Bell Telephone Company Mountain States Telephone and Telegraph Company Southwestern Bell Telephone Company The Chesapeake and Potomac Telephone Company of Maryland The Bell Telephone Company of Pennsylvania The Chesapeake and Potomac Telephone Company of Virginia The Chesapeake and Potomac Telephone Company of West Virginia The Diamond State Telephone Company The Ohio Bell Telephone Company The Pacific Telephone and Telegraph Company New Jersey Bell Telephone Company Wisconsin Telephone Company

RBOCs: Ameritech Bell Atlantic Bell South NYNEX Pacific Telesis Southwestern Bell US West

Register Also called a *shift register*. An electronic circuit used for temporarily storing memory in a serial format. Shift registers are commonly used in the serial-to-parallel conversion for data transmission. Bits are clocked into the register one at a time, then clocked out to their destination when they are needed. Each memory segment of a register is typically an RS (Reset-Set) flip-flop .

Registered Terminal Equipment Any line or telephone service that is installed by a telephone company is terminated to registered terminal equipment. The different levels of registered equipment are as simple as an RJ45 (Registered Jack 45) or as complicated as a DSX hand-off point in a colocation.

Registration Jack (RJ) The prefix to many telephone company connection and interface standards.

Regulation, Power A device that takes an unstable power source, such as public utility power, and reproduces the voltage/amperage with elec-

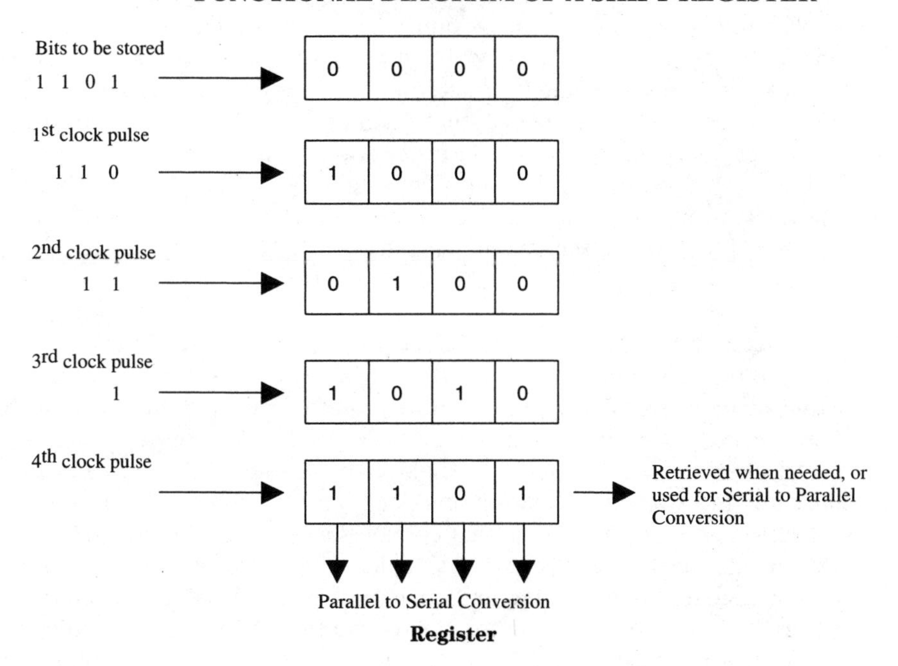

Register

tronics. The electronics in a power regulator are a controlled environment that produces the desired power signal. The street power provides the energy for the device (power regulator) to do this.

Regulation, Telco The RBOCs, CLECs, long-distance companies, and competitive-access providers are all regulated by the *Public Utilities Commissions (PUC)* of their respective areas in some fashion. The Bell companies are heavily regulated to a disadvantage to enable new telephone companies to become established. One of those disadvantages is that they are forced to charge higher rates for their services in competitive areas. This allows the new companies to attract customers with a price advantage. Many new companies do not take advantage of the price regulation, because their networking equipment is of latest technology (SONET 100%, in many cases), which therefore carries a higher value to customers. *Competitive Local Exchange Carriers (CLECs)* become regulated for pricing when they reach a certain number of customers or percentage of market share. All communications companies, new and old, must demonstrate to the PUC that they are capable and willing to provide acceptable service to the public.

Relational Data Base A data-base application (software program) that tracks data, based on relationships. It works very similar to a manual paper-filing system, with different categories and cross references. A good example is: There are many houses. Each house has many (one or more) residents. If a data base were created for this, each house would be listed and people would be related to the house. People make many telephone calls. The calls people make can be related to the people that make them. The way that a computer program would think, if queried, to find every person that dialed the phone number 1-602-555-1212, is as follows. It would scan through the data base and, as it found the phone number, it would list the relational items to it. More simply stated, it would list every person and house that dialed that phone number.

Relay An electromechanical switch. Relays are used in electronic and electrical circuits as switches. A relay consists of a coil of wire wrapped around a thin cylindrical-shaped piece of iron, called a *core*. When electricity flows through the coil, it magnetizes the core, which is close to a pair of electrical contacts. The magnetic field attracts one of the contacts to move and make a connection. A popular application for relays is between a small voltage to run a switch that connects a very large voltage. This is why you can start the motor in a large truck with a small key. The key applies voltage to a relay, which, in turn, connects a large contact from the battery to the starter of the motor. Relays were used in old central offices (some still in service), called *stepper switches*. By dialing a rotary phone, you manipulated a vast network of relays and logic circuits to connect your call. Today, relays have been replaced with transistors; in large or heavier applications, they have been replaced with semiconductor devices called *SCRs (Silicon-Controlled Rectifiers)*. However, electromechanical relays still have the special ability to physically isolate one voltage or device from another with a physical connection. They are still used in higher-end home-audio equipment, where if you turn down the volume, then turn on your stereo, after a short pause, you will hear a “click.” That “click” is a relay connecting the output signal voltage to your speakers.

Relay Rack Large racks that got their names from a time when they were used as a mounting platform for electromechanical relay circuits in telephone central offices. Their standard size is 7 feet tall by 22 inches wide. Relay racks are available in a 19" wide size as well.

Remote Call Forward A feature of PBX systems that enables users/subscribers to make calls dialed to their telephone ring to a different telephone of their choice and activate the feature from a different phone. When a user wants to activate the feature, they can dial a feature code, dial their extension, then dial the extension that they would like their calls to ring on. The great thing about this feature is that if you are in a meeting, you can pick-up a phone and forward the phone in your office to another co-worker or anywhere else within the PBX system. Some systems allow you to forward your extension to an off-premises telephone number, such as your home. In this case, you can work out of your home, and not miss any calls coming to your extension because they will ring directly to you.

Remote Order Wire A telephone line that is used to monitor an electronic system. A dial-up maintenance line for a server or mainframe is an order wire.

REN (Ringer Equivalency Number) A number that references a device's load on a telephone line when the line rings. Telephones, modems, answering machines, and other devices connected to telephone lines are required to have this number printed or stamped somewhere on the device, or placed in the device's literature. A telephone line in North America is capable of driving 5 Bells, or 5 devices with a ringer equivalency number of 1. If a telephone line has more than 5 such devices plugged into it, it may fail to ring when called.

REPACCS (Remote Cable-Pair Cross-Connect System) A remote-controlled/automated cross box. A less-sophisticated *DCS (Digital Cross-Connect System)*. In certain areas of cities that are hazardous for telco workers, REPACCS systems are implemented so that F1 and F2 cable pairs can be cross connected remotely.

Repeater Also known as a regenerator. A device that is used to take a signal that has traveled a long distance and make it new again. Repeaters can be

Splice Pedestal and Repeater Closure

coils of wire, which are used in the public telephone network for voice (POTS) lines or they can be electronic, taking an electronic signal that has been attenuated over a long distance, reproducing it, then re-transmitting it. The repeater closure illustrated is an electronic (active) repeater (right).

Repeater Coil A radio-type transformer that is used to amplify voice signals on copper twisted-pair telephone wires. Repeater coils have a typical inductance value of 33 mH and are placed every 3000 to 5000 feet.

Request to Send (RTS) A control signal that has a dedicated wire in the RS-232 protocol. When the far device places a logic "one" or 5-V voltage on this wire, it enables the near modem to initiate a transmission.

Rerouting To change the physical path or medium of a communications signal. Rerouting is a part of *SHARP (Self-Healing Alternate-Route Protection)* service from telephone companies over their SONET networks. If a cable is cut, the electronic equipment reroutes the transmission with very little or no interruption in service. If you are talking on a voice line while a fiber is cut on a SONET ring network that is very busy and fully utilized, you might hear a very light click sound.

Resale Carrier A long-distance company that leases long-distance facilities and sells service on them. Sprint and MCI are resellers in some areas; in some areas, they have their own switches, fiber-optic lines, and microwave equipment. In those areas, they are facilities-based carriers.

Reseller Also called an *aggregator*. A long-distance or cellular/PCS reseller. They sign up with a long-distance company as a reseller and all their customers are "aggregated" together for a bulk discount. The long-distance or cellular company provides the service and does the billing. The advantage to the long-distance company is that they have more people pushing their long distance. The advantage to the customer is the value-added service and personal consulting of the aggregator.

Reset See *Rebooting*.

Resistance The unit of resistance is the ohm, abbreviated/represented by the Greek letter omega (Ω). Resistance is just what its name depicts, resistance to electric current flow. A 100-W 120-V household light bulb has about one ohm of resistance. The more resistance in a circuit, the less current is allowed to pass through it.

Resistor An electronic component/semiconductor usually made from carbon. Resistors are usually used to limit current flow through a circuit or create RC/RL (resistor-capacitor/resistor-inductor) frequency filters.

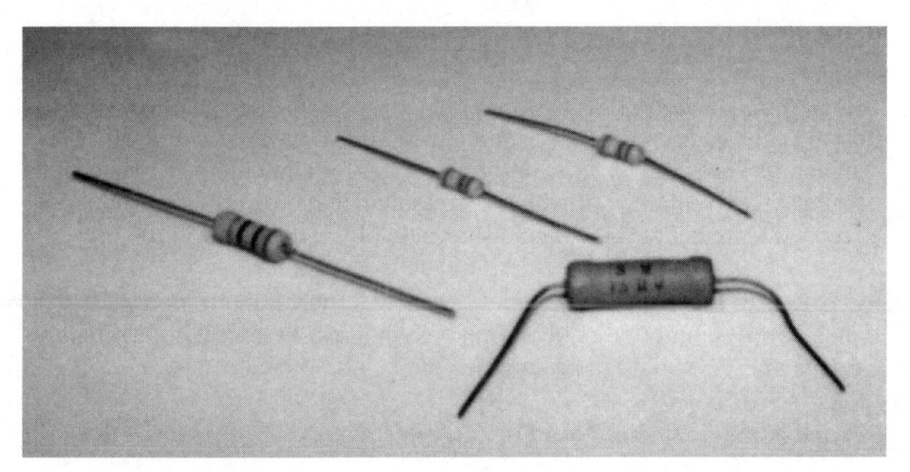

Carbon Resistors

Resistor Color Code See *Color Code*. Resistors have four color bands. They are regarded to as the first, second, third, and fourth bands. The first band is the closest to one side of the resistor and the following bands count to the inside. The first band indicates the first integer of the value of resistance. The second band indicates the second integer of the value of resistance. The third band indicates a multiplier or number of zeros to

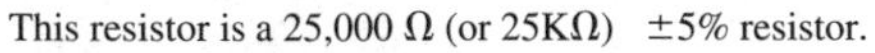

This resistor is a 25,000 Ω (or 25KΩ) ±5% resistor.

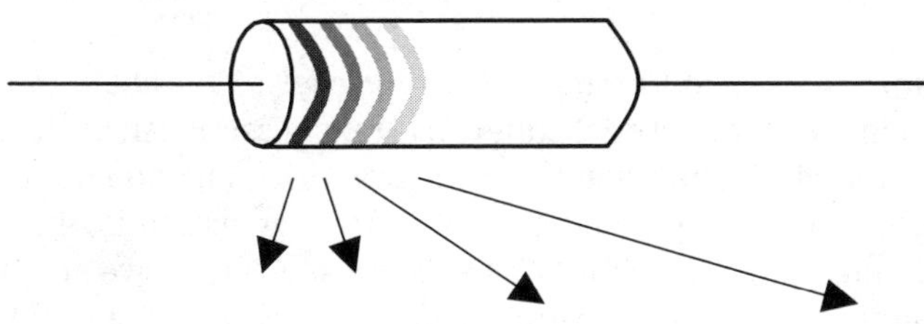

Color	band 1	band 2	x band 3
Black	0	0	X 1
Brown	1	1	X 10
Red	2	2	X 100
Orange	3	3	X 1,000
Yellow	4	4	X 10,000
Green	5	5	X 100,000
Blue	6	6	X 1,000,000
Violet	7	7	X 10,000,000
Gray	8	8	X 100,000,000
White	9	9	none

Band four is the accuracy or "tolerance" of the resistor.
Gold = ±5%
Silver = ±10%
no fourth band = ±20%

Resistor Color Code

be placed after the first two band numbers. A diagram is the easiest way to demonstrate this code.

Resonance A circuit is resonant if the inductive reactance and capacitive reactance are equal. This condition occurs for all inductor/capacitor circuits. The frequency at which the resonance happens is determined by the value in mH of the coil and the value of the capacitor in μF.

Retrofit To make older equipment work with newer equipment. *Retrofit* is a term commonly used among telephone company network technicians in reference to upgrading telephone equipment.

Return Authorization Number (RAN) See *RAN*.

Return to Zero (RZ) A transmission format where each positive bit returns or drops to a zero value during its timing period. The drop-to-zero format assists in timing/synchronizing of the transmission signal.

Reverse Battery Supervision A form of answer supervision. An in-band signaling method, if a telephone call goes from one central office to another (or PBX), the originating central office needs to know when the call has been answered so that a billing cycle can begin. The terminating central office briefly reverses the voltage on the connecting trunk line as a signal.

Reverse Channel Also called a *backward channel*. The channel that flows upstream in an asymmetrical (uneven) transmission. An asymmetrical communications transmission that is characterized by one direction being very fast, compared to the other. Cable TV is an example of asymmetrical communication. The cable TV head end sends massive amounts of video and audio information down a coax one way and the cable TV set-top decoder boxes send small amounts of ID and status information the other way back to the head end over the same coaxial connection. Sometimes asymmetrical channels are referred to as "upstream" for the slow channel and "downstream" for the fast channel, or "forward" for the fast channel and "backward" for the slow channel.

RF (Radio Frequency) Any electromagnetic frequency that is above the range of human hearing. Most licensed radio frequency transmissions range from 500 kHz (500,000 Hz) to 300 GHz (300,000,000,000 Hz).

RF Choke A coil of wire that filters out high frequencies.

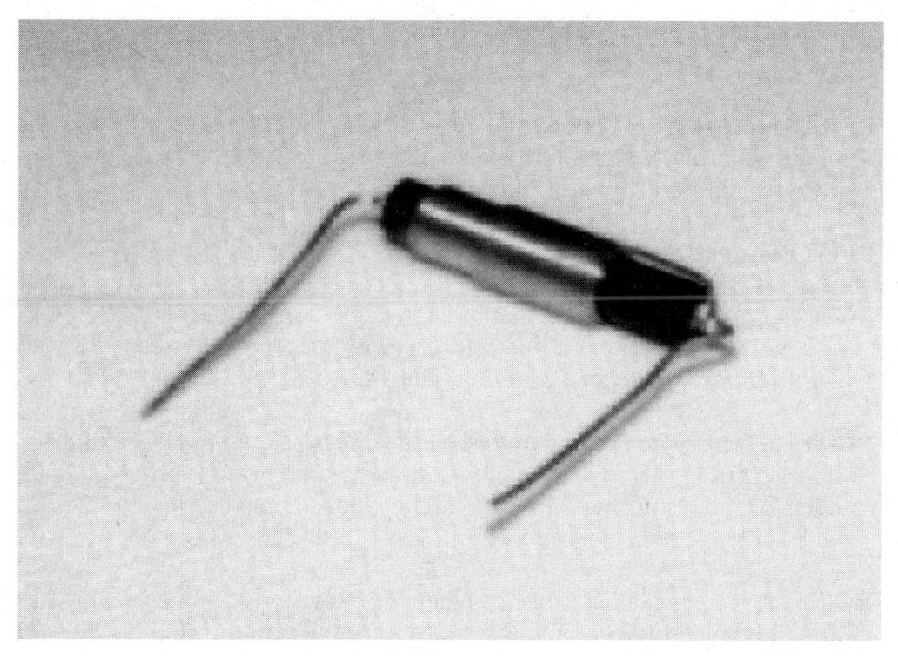

RF Choke Coil

RF Splitter Used to make a junction point or split a signal so that it will travel down multiple paths over coax. Also called a *splitter* and *UHF/VHF splitter*. Shown is a four-way splitter.

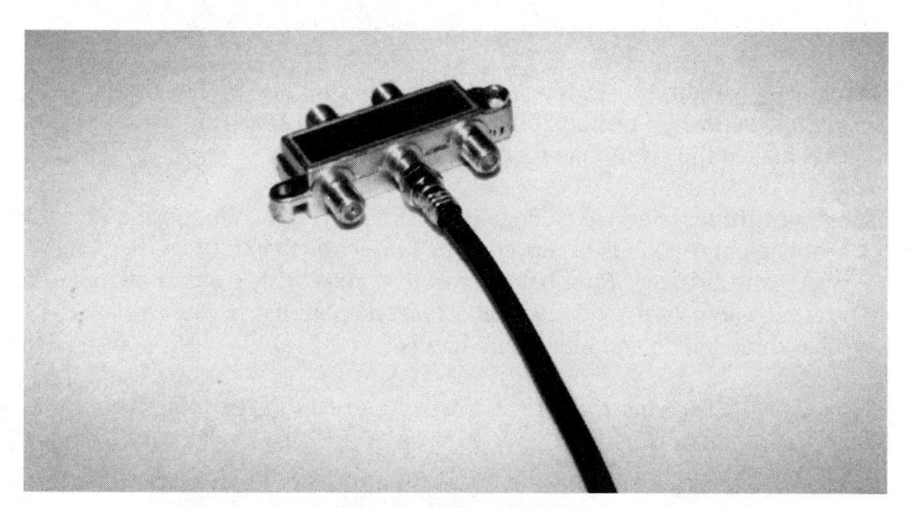

UHF/VHF 4-Way Splitter

RFI (Radio Frequency Interference) See *Radio Frequency Inter-ference.*

RFP (Request For Proposal) Also called *RFQ (Request For Quotation).* A formal invitation that a company or individual gives to other individuals or companies, to bid or price a service.

RFQ (Request For Quotation) See *RFP.*

RG-8 A type of coaxial cable that has a transmission impedance of 50 ohms. For more information on different types of coaxial cable, see *Coax* and *Characteristic Impedance.* For a photo of RG-8, see *DIN Connector.*

RG-58 A type of coaxial cable that has a transmission impedance of 50 Ω. It is used primarily in LAN applications and wired in a bus physical topology. For more information on different types of coaxial cable, see *Coax* and *Characteristic Impedance.*

RG-59 A type of coaxial cable designed for television antenna use that has an impedance of 75 ohms. For more information on different types of coaxial cable, see *Coax* and *Characteristic Impedance.*

RG-62 A type of coaxial cable with a transmission impedance of 93 ohms. It is primarily used in LAN applications and wired in a bus physical topology. ARCnet utilizes RG-62 as its transmission media. For more information on different types of coaxial cable, see *Coax* and *Characteristic Impedance.*

RG-U The military designation for general-use coaxial cable.

Ring One wire in a POTS telephone line. The ring side of the line is usually marked red when terminated and carries the 90V AC ring-voltage signal that makes the telephone ring.

Ring Banding Some pic cable comes with no ring banding, which means that the color code is determined by two wires twisted together (e.g., a white and a blue). Ring-banded cable comes with color rings painted around each wire, and the same twisted pair listed before would be white/blue bands and blue/white bands.

Ring Cycle The ring cycle for a North American POTS telephone line is two seconds of ringing, then four of quiet. Ringing cycles vary throughout the world.

Ring-Down Box Used in building ring-down circuits. A device that you put on each end of a copper twisted pair that provides battery and ring voltages that a central office would. However, no dialing is involved. When you pick up one phone, the one on the other end automatically rings. When the ringing phone is picked up, the lines are connected together with a talk battery and people can talk on both ends, just like a normal telephone call. Tellabs manufactures a wide variety of ring-down devices.

Ring-Down Circuit A simple telephone line that is made using ring down boxes. See *Ring-Down Box*.

Ring Generator A ring generator is the part of a PBX or central-office switch that provides the source of the ring voltage that rings telephones. Ring generators are an individual circuit card in many PBX systems. In some systems, the ring-generation capability is built into the station/telephone interface cards.

Ring Latency In a token-ring network, ring latency is the time required for a transmission packet to go all the way around the ring.

Ring Topology A LAN topology (a MAN topology in SONET) that connects all devices on a network in a ring configuration. The data transmit-

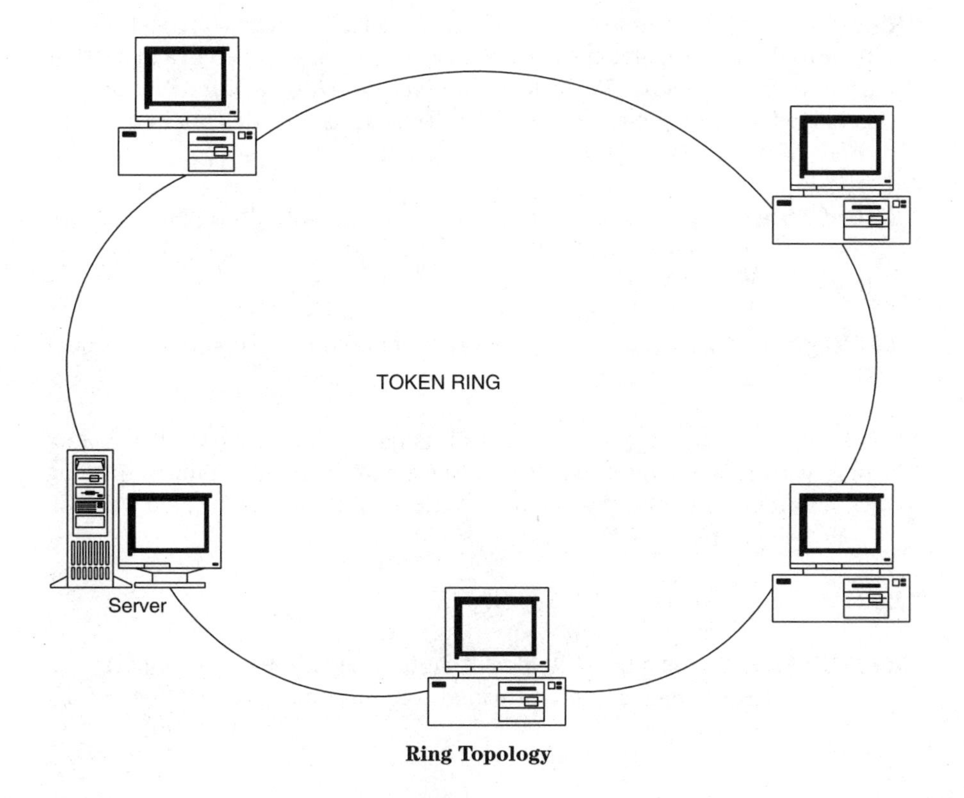

Ring Topology

ted through the network goes through each device. As the devices receive the data, they check to see if it is intended for them. If it is, then they keep it; if it is not, then they pass it along. Unlike ethernet star and bus topologies, the ring topology is not contention based; each device gets a specified turn in sending and receiving data.

Ring Voltage Ring voltage on a POTS telephone line is 90V AC.

Ringer Equivalency Number (REN) A number that references a device's load on a telephone line when the line rings. Telephones, modems, answeringmachines, and other devices connected to telephone lines are required to have this number printed or stamped somewhere on the device, or placed in the device's literature. A telephone line in North America is capable of driving 5 Bells, or 5 devices with a ringer equivalancy number of 1. If a telephone line has more than 5 such devices plugged into it, it may fail to ring when called.

Rip Cord A nylon string that is put in telephone wire and cables when it is manufactured. The string is used by installers to rip the jacket or insulation when it is being installed.

Riser A telephone cable feed inside a building that runs vertically from floor to floor. It is called a *riser* because it is usually placed in a place that architects call *risers*. We call them *elevator shafts*, *plenums*, or *airways*, whichever the riser is used for. Typical riser cables are in the hundreds of pairs in size (100, 200, 300 pair, etc.).

Riser Cable A twisted-pair cable (usually several hundred pairs) distribution system that progresses from the telephone company Demarc or point of entrance in a building to each floor of that building.

RJ (Registration Jack) The prefix to many telephone company connection and interface standards.

RJ11 The telephone jack that most of us have come to know. It has a 6× plug with four conductors. Handset cords are a smaller plug, a 4× plug with four conductors. If you look at the two of them side by side, you will notice the difference. For a photo, see *Jack*.

RJ21 Also known as RJ21X. See *RJ21X*.

RJ21X An RJ21X is a 66M150 block that is designated as the demarcation point for telephone company-provided communications lines. Most RJ21X

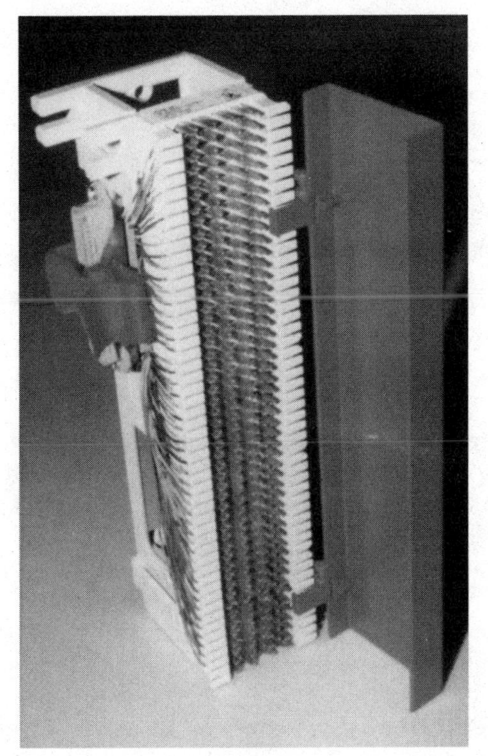

RJ21X

blocks have an orange cover where the telephone numbers of the lines are written.

RJ45 An 8-position, 8-conductor modular jack. RJ45 is used in many computer LAN applications.

RJ48 An 8-position, 8-conductor modular jack. Used to terminate T1 service.

RJ48X An 8-position, 8-conductor modular jack. They are used to terminate T1 service and they have a shorting bar that is built-in for making manual loop-backs.

RMA (Return Material Authorization) See *Return Material Authorization.*

RMS See *Root Mean Square.*

Roaming When a cellular/PCS telephone travels outside of its calling area, it is roaming. When a cellular/PCS telephone roams, it continues to send a signal out that tells any cellular site it can reach (regardless of company) that "I am here." If it can communicate with a cellular site, then the roam indicator is displayed on your phone. If you have a roaming service with your cellular/PCS company, then you can still receive calls even though you are out of your calling area. If you don't pay extra for roaming service, you can still make outgoing calls; however, they cost extra.

Robust A term that is synonymous with fast, flexible, and reliable.

ROH (Receiver Off Hook) See *Receiver Off Hook*.

Robbed Bit Signaling Also called *bit robbing*, but usually known as in-band signaling. The practice of taking a bit here and there in the beginning and end of a digital transmission for use in the overhead of the transmission equipment. Bit robbing is bad news when the signals being multiplexed into the transmission are data. Robbing a bit from a data stream severely corrupts it. Bit robbing is reserved for multiplexing multiple voice circuits onto a T1. Circuits intended to transmit data use out-of-band signaling or clear-channel signaling.

ROH (Ringer Off Hook) A reference to a test result from an *MLT (Mechanized Loop Test)*. This condition means that one of the telephone receivers in a customer's premises has been left off the hook or that another trouble is in the phone line, usually a shorted pair or wet jack.

Rolm (Siemons-Rolm) A telephone equipment manufacturer.

ROM (Read-Only Memory) See *Read-Only Memory*.

Root Mean Square (RMS) A method of calculating the power consumption or power output of an electronic/electrical device. RMS power is ultimately the average of an AC waveform, which is the peak voltage, multiplied by 0.707. The other methods of calculating power include true power, peak power, and transparent power. Most electronic/audio applications use either peak or RMS power. To convert from peak power to RMS power, multiply the peak power rating by 0.707. The result is RMS power. To convert RMS power to peak power, divide the RMS power rating by 0.707. The result is the peak power rating.

Route Nortel's name for a trunk group. See also *Member*.

Router A device used in LAN, MAN, WAN, and GAN networks that routes data traffic to destination addresses by looking at the address portion of

Internet I Router

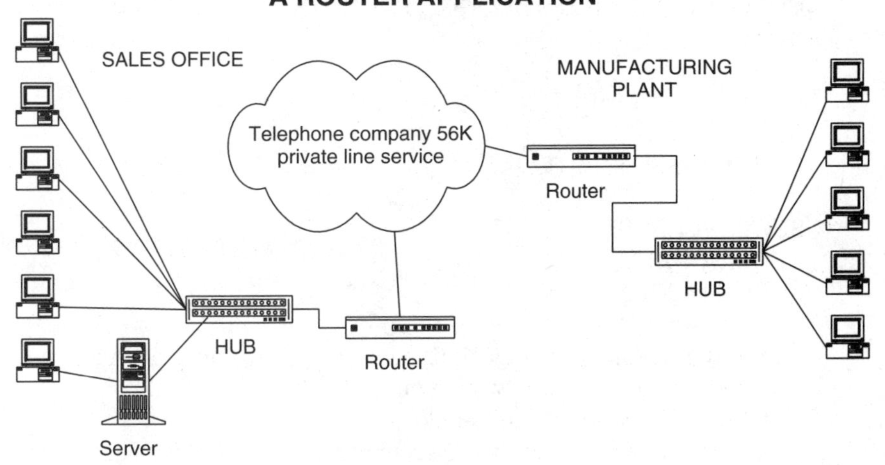

Router

data packets. If you have an ethernet network that is connected to another ethernet network across town via a 56K private line, you want only the data intended for a device on the other network to go across the 56K line to maximize its transmission speed. A router differentiates what data is for which network and routes it accordingly. If the router was not in place, the two networks would act as one large network, and the 56K line would be drastically overwhelmed, slowing the network down to a very low speed. Illustrated is a large-scale Internet domain router, courtesy of Brooks Communications.

Routing Table A reference to call-handling instructions input to an ACD (Automatic Call Distribution) system. The routing table lists each incoming trunk and the steps that the call goes through. The "steps" the call goes through are also called a "call treatment or a script." A routing table could list trunk number 1 and the treatment that calls receive when they come in on that trunk. 2. An incoming digit translation file in a PBX system's memory that instructs the system on which extension to route a call based on DNISI or D10 digits.

Sample Meridian 1 CCR Script

```
GOTO CLOSED IF LOGGED AGENTS QUEUE 1234=0
QUEUE TO 1234
GIVE RAN 105
GIVE MUSIC 100

SECTION LOOP
WAIT 60
GIVE RAN 106
GOTO DIFFICULTIES IF LOGGED AGENTS QUEUE 1234=0
GOTO LOOP

SECTION DIFFICULTIES
GIVE RAN 107
FORCE DISCONNECT

SECTION CLOSED
GIVE RAN 108
FORCE DISCONNECT
```

–RAN 105 Thank you for calling abc company, you call will be answered by the next available agent.

–RAN 106 Please continue to hold

–RAN 107 We are currently experiencing technical difficulties please call alternate number xxx-xxxx

–RAN 108 We are closed, our business hours are . . . please call back during these times.

RS 1. The ASCII control-code abbreviation for record separator. The binary code is 1110001 and the hex is E1. RS Connectors, Protocols 2. *Recommended Standard*, the prefix in RS-232, RS-328, etc.

RS-232C A communications protocol that was developed by the EIA so that data devices could communicate. The standard includes the different functions and signals for two devices to communicate. The signals are physically interfaced to a cable via a 25-pin or 9-pin connector. Each pin having a signal function. In order for an RS-232 connection to work, the cable and pin-outs must match. Even though most modem and DCE/DTE manufacturers use the RS-232 protocol, not all use the same pin-outs. To complete a connection, a cable must be a null cable (null-modem cable), which means that transmit and receive are reversed inside the cable from one end to the other.

RS-328 The first EIA facsimile standard (1966).

RS-366 EIA standard for auto-dialing.

RS-449 EIA standard that is the newer version of RS-232. RS-449 uses a 37-pin connector and each of multiple transmit and receive pairs are balanced. RS-449 is faster and able to transmit longer distances (300 feet) than RS-232 (limited to 50 feet).

RSA (Rural Service Area) The counterpart to *MSA (Metropolitan Service Area)*. A term that refers to the 306 metropolitan areas where the FCC manages cellular and PCS communications. There are also *RSA (Rural Statistical Area)* markets that the FCC determined as separate from each other. 428 RSA markets are in the United States. Each statistical area, 734 in all, has at least two licensed service providers.

RSC (Remote Switching Center) A common term for a long distance carrier's central office, or relay point. Many Long Distance Carriers have offices that are not occupied by personnel, only equipment.

RTE (Registered Terminal Equipment) See *Registered Terminal Equipment*.

RTS (Request to Send) A control signal that has a dedicated wire in the RS-232 protocol. When the far device places a logic "one" or 5-V voltage on this wire, it enables the near modem to initiate a transmission.

RZ (Return to Zero) A transmission format where each positive bit returns or drops to a zero value during its timing period. The drop-to-zero format assists in timing/synchronizing the transmission signal.

SAA (Supplemental Alert Adapter) An AT&T term that is specific to the Merlin telephone system. It is the equivalent to a loud ringer.

Safety Belt Used by communications/power/construction personnel to harness themselves to telephone/power poles or tower structures. Also called a *body belt* and *climbing belt*.

Safety Belt "body belt" "climbing belt"

Sag If an outside plant engineer refers to *sag*, it is the amount that an aerial span dips down between telephone poles. Different cable needs to have a different sag, depending on the climate, the weight of the cable, the type of poles being used, etc.

Satellite An self-sustained electronic device/platform that orbits the earth at an altitude of about 22,000 miles. Communications satellites transmit and receive signals in the microwave range and are used for broadcast TV, telecommunications, global positioning, and many other applications.

Satellite Link A communications path that includes a satellite.

S Band The part of the radio spectrum that ranges from 390 MHz to 1550 MHz.

SBC Communications One of the RBOCs, formerly known as *Southwestern Bell.*

SC Connector A square-shaped snap-on fiber-optic plastic connector. SC connectors come in single or dual. For a photo of a single SC connector, see *Fiber-Optic Connector*.

Scattering Attenuation of light in a fiber optic because of the light changing direction in the fiber.

SCC (Specialized Common Carrier) An old term for an *IXC (interexchange carrier)* other than AT&T.

Schematic A diagram of an electronic circuit. Schematics can be drawn at many different levels from block diagrams to the discrete component level.

Scotchlok A family of splicing connectors that are manufactured by 3M. Scotchloks are used in splicing copper twisted pair in outside and inside plant applications.

SCR (Silicon-Controlled Rectifier) An electronic component that works like a transistor, but much better in power switching applications. An SCR has three leads, one for the anode, gate, and cathode. When a negative bias pulse is applied to the gate, the SCR acts like a switch, turns on and stays on, even though the pulse is gone.

SDI (Serial Data Interface) A connection designed for transmission of data over a media one bit at a time. The other type of transmission is par-

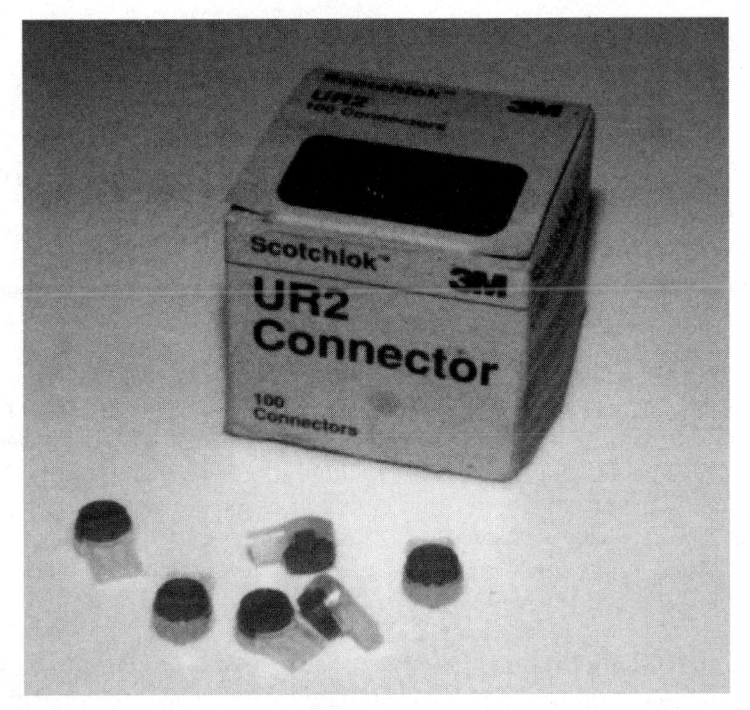

3M Scotchlok

Schematic Symbol for an SCR

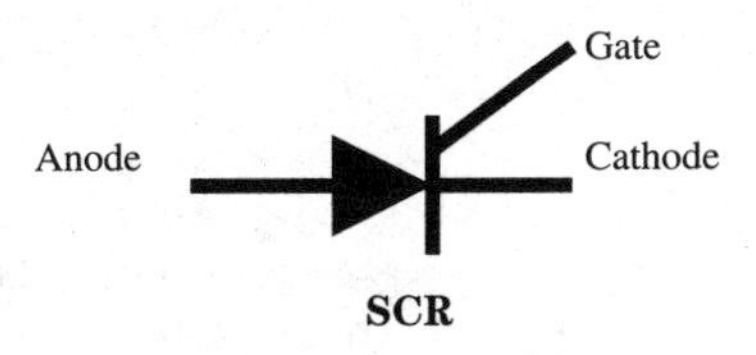

SCR

allel, which sends multiple bits at a time, over multiple wires. Many printers run in a parallel manner, modems work in a serial manner. For a diagram, see *Serial Interface*.

Second Dial Tone The dial tone that you get after dialing 9 on a PBX system. When you first pick-up the handset, you hear a dial tone, which is the PBX internal dial tone, then you dial 9 to get an outside dial tone.

Secondary Winding A reference to the output of an electronic transformer. Transformers have at least one primary and at least one secondary winding. Transformers are made to work with AC voltages. If you connect a transformer to a DC voltage with no filtering electronics, the

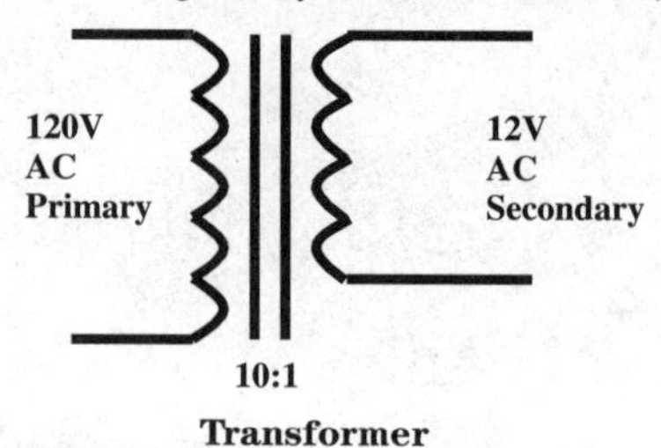

transformer will overheat and be destroyed. Transformers are used to "step-up" or "step-down" AC voltage levels. Transformers are rated with a ratio of the primary to secondary winding. A common transformer is a 10:1. This means that for every 10 windings of wire on the primary side of the transformer, only one winding is on the secondary side. Transformers have the same ratio to voltage as they do windings, so if 120 volts is applied to the primary winding, then 12 volts will be the output on the secondary winding.

Selective Ringing Module (SRM) A device that is attached to each individual network interface for customers that are sharing a party line. The SRM contains electronics that can be configured to recognize different ringing formats using DIP switches. Some different ringing formats that an SRM would differentiate are ring voltage on the ring side, ring voltage on the tip side, ring voltage on the ring side with the tip side grounded, and ring on the tip side with the ring side grounded.

Selectivity The measure in dB of a radio receiver or tuner to select or pass a wanted signal carrier and reject all others. The higher the selectivity in dB, the better the receiver/tuner. Selectivity is sometimes called the *Q* (quality) of a tuner, but the *Q* rating is usually used in reference to specific fixed-frequency filters.

Self Diagnostics A feature of many telecommunications testing and transmitting equipment. PBX switches, microwave radio equipment, SONET equipment, and central-office switches are equipped with troubleshooting aids that indicate where trouble is. They are not always accurate, but they are of great assistance in the diagnosis of faulty equipment or transmission paths.

Self Test The ability of telecommunications test, transmission equipment, or switching equipment to run a test on its hardware components and software. Most telecommunications equipment, as well as personal computer equipment, run a self test when they are first turned on. If there is a problem, then an error of some kind is displayed, which can be cross referenced in a user's or administrator's manual.

Semiconductor Germanium, silicon, and carbon are all semiconductors. They are not great conductors (like copper wire) and they are not insulators (like plastic or rubber). They have special properties that allow a controlled amount of current to flow through them, given a certain amount of voltage, under certain conditions. Transistors, diodes, and other "active" components (devices that require power to do their job) are made from silicon or germanium. Passive devices (such as resistors) are made from carbon.

Sensitivity A reference to the ability of a radio receiver or tuner to receive a tiny electronic signal from an antenna and amplify it. A sensitivity rating is given in microvolts (µV). If a sensitivity rating is better, the sensitivity for the same given output is lower.

SEPT (Signaling End-Point Translator) The part of the SS7 (Signaling System 7) network that receives coded signals from another central office and translates those codes into a number plan or set of codes used in the central-office exchange. For more information, see *Translations*.

Serial Bus A bus that transmits one bit at a time. Serial busses are usually one pair of wires: one used for transmit/receive and the other is ground (or common, for a balanced line). A modem line to your computer can be thought of as a kind of serial bus. See *Parallel Bus*.

Serial Data Interface A connection designed for transmission of data over a media one bit at a time. The other type of transmission is parallel, which sends multiple bits at a time, over multiple wires. Many printers run in a parallel manner; modems work in a serial manner. For a diagram, see *Serial Interface*.

Serial Data Transmission The transmission of data over a media one bit at a time. The other type of transmission is parallel, which sends multiple bits at a time, over multiple wires. Many printers run in a parallel manner, modems work in a serial manner. For a diagram, see *Serial Interface*.

Serial Interface The transmission of data over a media, one bit at a time. The other type of transmission is parallel, which sends multiple bits at a time, over multiple wires. Many printers run in a parallel manner; modems work in a serial manner.

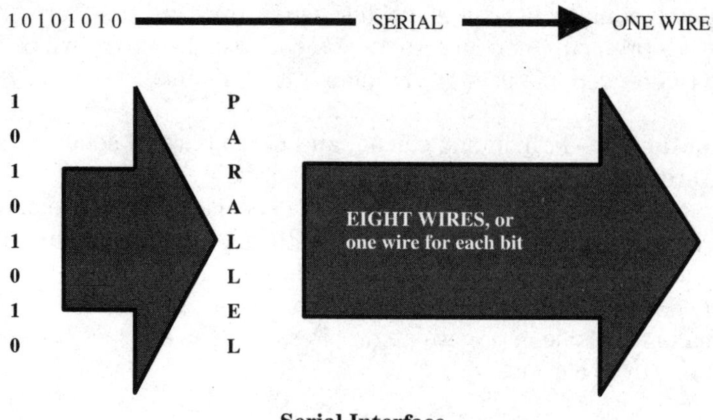

Serial Interface

Serial Port The transmission of data over a media one bit at a time. The other type of transmission is parallel, which sends multiple bits at a time, over multiple wires. Many printers run in a parallel manner, modems work in a serial manner. For a diagram, see *Serial Interface*.

Series Circuit A circuit that has only one path for current through multiple loads. The other type of circuit is a parallel circuit, which has more than one path for current through multiple loads, or devices. For a diagram of a parallel and series circuit, see *Parallel Circuit*.

Server A reference to a client/server environment. A computer that is dedicated to providing services to other computers (PCs). The PCs on the LAN are the clients and the common computer that stores information or programs is the server. File servers store common data for other computers to access via a *LAN (Local-Area Network)*, and there are application servers, which other computers access to run large or complicated tasks. The whole idea behind a LAN, or any other network, is to share information. Servers are computers that act as a place to store shared information.

Service Access Code (SAC) Service access codes are three-digit numbers that are used like an area code, but they are not an area code. These codes are used for special services, such as 800/888 or 900 numbers. Five SACs are in use at the time of this writing: 600, 700, 800, 888, and 900.

Service Affecting A reference to a problem that is critical and is interfering with the operation or ability of a network to provide service.

Service Area The geographic area of a telecommunications service provider. The local service area for USWest is Washington, Oregon, Idaho,

Montana, Wyoming, Utah, Arizona, New Mexico, Colorado, North Dakota, South Dakota, Nebraska, Iowa, and Minnesota.

Service Code A three-digit code or shortened phone number that has a specific purpose, such as 911.

Service Entrance Also called a building entrance. The place where communications cables enter a building.

Session Layer A layer in a communications protocol model. In general, the session layer does the job of establishing and maintaining connection to the communications process of the lower layers. It also controls the direction of the data transfer. The latest model or guideline for communications protocols is the OSI (open systems interconnect). It is the best model so far because all of the layers or functions work independently of each other. For a diagram of the OSI and older proprietary communications models, see *Open Systems Interconnection*. For a functional diagram of the OSI layers, see *OSI Standards*.

SF 1. *Single frequency*, a method of in-band signaling in switched telephone networks. A single 2600-Hz frequency tone is used for signaling. 2. *Superframe*, a framing format for T1. One superframe consists of 12 T1 frames, 193 bits each, transmitted in succession. The superframe format allows for maintenance and monitoring information to be sent along with the 24 DS0 channels. *ESF (Extended Superframe)* is the newer version of T1 framing format.

SFC (Switch Fabric Controller) An interface between the CPU/core and multiple networks of telecommunications switches.

SHARP (Self-Healing Alternate Route Protection) A service offered from Local Telephone Companies over SONET networks. Sharp service is made possible by the SONET ring technology, which incorporates its network on a ring of fiber-optic cable. If the fiber is cut, all traffic is rerouted the other way around the ring. If you are talking on a telephone over a SONET sharp-based service and a fiber is cut, you might hear a very faint "click" sound. Other than that, you would never know there was a problem.

Shift Register An electronic circuit used to temporarily store memory in a serial format. Shift registers are commonly used in the serial-to-parallel conversion for data transmission. Bits are clocked into the register one at a time, then clocked out to their destination when they are needed. Each memory segment of a register is typically an RS flip-flop. For a functional diagram, see *Register*.

Short Circuit A circuit fault. A short is a "short circuit," or an easier path to ground caused by a bad component, water, or other means for electricity to get to where it is not wanted. Many confuse an open with a short. An open is literally a "disconnection" in a circuit. For a diagram of open- and short-circuit faults, see *Open Circuit.*

Short-Haul Modem Also called a "limited-distance modem" or "line driver." Short-haul modems are commonly used to extend the distance of a printer or other *DTE (Data-Termination Equipment)* device from its host. One example is to extend the printer dedicated to printing call-accounting records from a PBX to an accountant's office. For a network diagram and photo of a limited-distance modem, see *Limited-Distance Modem.*

Short-Tone DTMF A reference to a telephone or other dialing equipment that sends a short pulse of touch-tone DTMF (100 to 300 ms), regardless of how long you hold the button down. Some telephony equipment cannot "hear" a tone that is that short, so equipment manufacturers have implemented adjustable short-tone lengths, and options to remove the short tone altogether.

SI The ASCII control-code abbreviation for shift in. The binary code is 1111000 and the hex is F0.

Sideband A sideband is a harmonic radio frequency that is a result of modulation on an AM carrier, and is a transmission characteristic of an FM carrier. In AM radio transmissions, one sideband above the carrier frequency and one sideband below the carrier frequency are created. Each of these sidebands. In FM, the modulation of the carrier itself is an infinite number of sidebands.

Side Tone When you talk on a telephone, you can hear a little bit of your voice being sent back into the earpiece. This is called a *sidetone* and it lets you know that the line is live.

Signal Strength Signal strength is measured in dB, dBrn, or DB.

Signal-to-Noise Ratio Signal-to-noise ratio is the amount of desired signal, in comparison to the amount of unwanted signal, expressed as a ratio.

Signaling System 7 A method of out-of-band inter-office signaling for telephone circuits. Simply stated, *out of band* means that a special separate line is used to carry signaling, such as dialed touch tones, ringing signals, busy tones, (everything, but the actual voices/conversation), etc.

Remember that the two different ways to send signals in telephone transmissions are: in band and out of band. Signals are digits that you dial, dial tone, the phone being off-hook, ringing, etc. An in-band telephone line is like the one in your home; the digits that you dial and the ringing are carried within the channel you talk on. Out-of-band signaling is a method that telephone companies and businesses use for larger PBX applications and data-transfer applications. An out-of-band signaled DS1 has 24 multiplexed channels. The 24th channel carries the signaling for the other 23 channels (phone lines). The advantage of out-of-band signaling is that each channel has an increased capacity to carry data (8Kb/s more) and the 23 channels are not used to find out if a line is busy (both directions, in and out). The off-hook sensing busy signaling and other signaling previously mentioned is done in the 24th channel. If your system receives thousands of calls per day, this can reduce traffic. SS7 makes it easy for long-distance companies to let us dial a phone number, get a busy signal, and not be billed for it because we are not really using a call channel to do this.

Silicon The important thing to know about silicon is that it is an element used to make electronic components. It is used because it has special atomic properties that enable it to conduct or not conduct electricity, depending on the way it is *doped*. Doping is the implantation of impurities into the silicon, additional electrons to be specific. When a transistor, diode, or any other silicon active device is made, at least two types (two types of silicon doped differently) of silicon are used to form a *junction*. The first type of silicon is called a *P-type* (for positive) and the second type is called an *N-type* (for negative). The two (very small) pieces of silicon are placed together to form a P-N junction. A single PN junction is used to make a diode and three pieces of silicon are used to make a PNP or NPN junction transistor. Another element that is not as frequently used, but used in the same manner, is Germanium.

SIM (Single Interface Module) An NEC trademark, this is a smaller PBX in the NEC telephone equipment family. Larger PBXs include the IMG and the MMG.

SIMM (Single-Inline Memory Module) A small circuit board, about 1" by 3", that contains memory components for PCs and other memory-using devices. The edge of the SIMM circuit board has a single row of contacts so that it can be plugged into a socket or slot. The SIMM gets its name from the type of socket it plugs into.

Simplex Communications in one direction. FM radio and broadcast TV are forms of simplex communication. Other methods are half duplex and full duplex.

Single Frequency-Signaling (SF) Mostly referred to as *single frequency*, a method of in-band signaling in switched telephone networks. A single 2600-Hz frequency tone is used for signaling.

Single-Mode Fiber The alternative to multi-mode fiber optic. Single-mode fiber optic has a smaller core, but is capable of longer-distance transmissions. It is used in the public network more often and is the choice for SONET applications. Multi mode has a larger core, and therefore accepts more light and more frequencies of light. Multi mode is used for shorter-distance applications, such as LANs. Multi-mode fiber optic is made with an orange-colored tube or insulation, and single-mode fiber is made with yellow.

Skills Based Routing A method used in customer service call centers where agents log into multiple ACD (Automatic Call Distribution) Queues according to the type of calls they are capable of answering. For example, Johnny is agent (ACD extension) 11777. Johnny can make hotel reservations as well as car reservations, so he logs into Queues 1010 (hotel calls) and 1020 (auto rental calls). The ACD system will then route both of these types of calls to him.

SLC96 Also known as *Slick 96*. A Lucent Technologies "pair-gain" system that multiplexes 96 telephone lines onto eight pairs of twisted-pair wire. It is used extensively in the public telephone network to provide telephone service to areas that do not have enough twisted pairs to meet customer needs. The SLC 96 actually uses four T1 circuits (24 lines per T1) to achieve the 96-line transport. The SLC 96 is configured in a cabinet, one for inside rack-mount central-office use and the other (far end) as an outdoor cabinet. The circuit cards that are incorporated into the SLC 96 design are separate and redundant power cards, battery back-up for the remote end, common equipment (control) cards, and a separate card for every two lines that are multiplexed (48-line cards for a full system).

Slick 96 (SLC96) See *SLC 96*.

Sloppy Floppy Copy To copy data to a floppy disk, then load it onto another computer's hard drive when you can't get the LAN to work right or if you don't have a LAN. Also called *Sneaker Net*.

Slotted Ring A LAN topology that is, for all practical purposes, a switched token ring.

Slots A reference to expansion capability of a PBX system, PC, or other electronic equipment. PBX manufacturers make extra slots for electronic circuit cards to plug into the backplane of a KSU or cabinet for future network expansion.

Smart Card A credit card that not only has a magnetic strip (ROM) like all traditional credit cards, but also has a RAM component. Smart cards are being implemented in places were it is inconvenient to carry cash, like on battleships at sea or in amusement parks. To buy something, you simply place your card into a machine, the machine deducts the balance from your account (which is directly on the card) and the transaction is completed. Your money is actually inside the card, not in a bank account somewhere.

Smart Jack Also known as an *RJ68*. A smart jack is an RJ45 (8-pin modular jack) that has some simple electronic components inside it that enables it to be remotely placed in a loop-back mode for testing purposes.

SMDR (Station Message Detail Reporting) Another term for *call accounting*. A call-accounting system is a computer (usually a dedicated PC) that connects to a PBX switch via a serial data port and monitors the details of every phone call made through that switch. The call details are stored as call records; with the appropriate software, they can be retrieved, sorted, processed, and queried to almost any specific nature that the call-accounting system administrator desires. These systems are used by hotels to track all the calls you make from your room so that they can bill you for them. They are also used by companies to do bill back reports for individual departments within the company.

SMDS (Switched Multimegabit Data Service) A service offered by local telephone companies that is intended for the transport of large amounts of data at high speed from point to point over a switched type of network. Enter in the address or number that you would like your data to be sent and the SMDSU (SMDS Unit) packetizes the data and the SMDS network transports it. SMDS is a packet- or frame-type technology that is available in five transmission rates. Class 1 is 4Mb/s, Class 2 is 10Mb/s, Class 3 is 16Mb/s, Class 4 is 25Mb/s, and Class 5 is 44.7Mb/s.

SMSA (Standard Metropolitan Statistical Area) An area that the FCC manages rights to provide cellular service. Most SMSAs have two cellular service providers.

SNA (Systems Network Architecture) IBM protocol and architecture for mainframe/terminal computing environments.

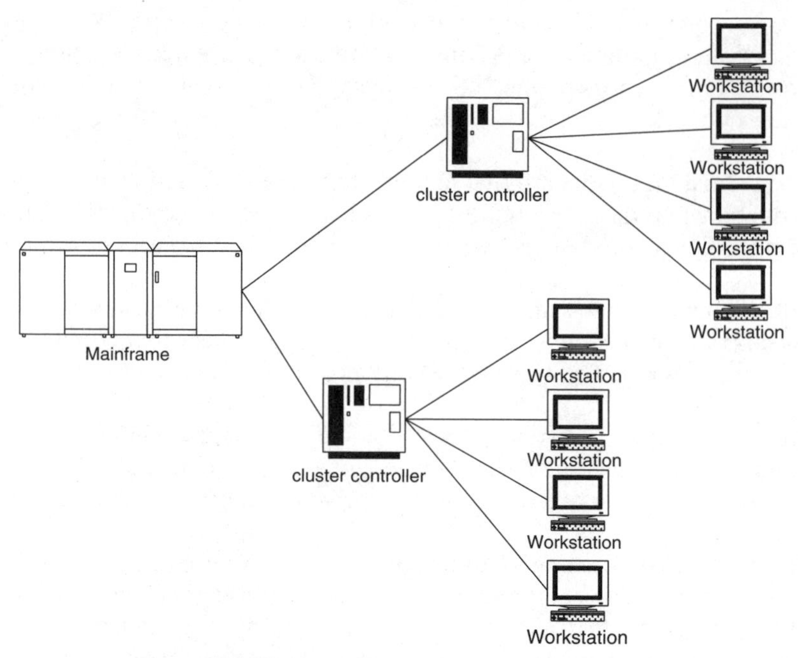

SNA (Systems Network Architecture)

Sneaker Net See *Sloppy Floppy Copy*.

SNI (Standard Network Interface) Also called *TNI (Telephone Network Interface)* and *NI (Network Interface)*. It is a device used to ter-

SNI (Standard Network Interface)

minate telephone service at the customer's location and provide lightning protection. One side of the SNI is for telephone company use only and the other side provides a place for customers to access their telephone lines. For other photos of SNIs, see *Standard Network Interface*, and *Lightning Protector*.

Snips Scissors that telephone cable splicers and cable installation technicians use when installing/splicing cable. Snips have serrated blades, are very sturdy, and are capable of cutting copper that is as thick as a penny. For a photo, see *Cable Knife*.

Snowshoe A device that is used to maintain a minimum bend radius for installed fiber-optic cable. The cable shown in the photo has a slack length to allow for future splicing. The slack is run along the strand and looped around snowshoes.

Snowshoe

SO The ASCII control-code abbreviation for shift out. The binary code is 1110000 and the hex is E0.

Software Also referred to as *application*. Software is the programming instructions that are loaded into a computer's memory that tells it how to function. Microsoft Word is a software application and so is Lotus Notes.

SOH The ASCII control-code abbreviation for start of heading. The binary code is 0001000 and the hex is 10.

Solid State The term *solid state* came about when the transistor was invented. Before the transistor, vacuum-tube amplifiers were in wide use. For a long time, many appliances you bought said "solid state" on them because they had been made with transistors and other silicon devices instead of tubes.

SONET (Synchronous Optical Network) SONET is strictly a broadband transport system. It is implemented over fiber optic and is able to be configured in a ring, which allows it to reroute traffic with no interruption of service if a fiber is cut. CLECs are implementing SONET as the mainstay of their network construction. SONET is based on a hierarchy of *STS (Synchronous Transport Signals)*, which is the electrical version of an *OC-1 (Optical Carrier Level 1)*. An OC 1 has a transmission speed of 51.84 Mb/s. The hierarchy of telephone communications services and their speeds is shown in the table. SONET works similar to switched token ring, except at much higher speeds. SONET permits a virtual tributary to be created from one node to another on a network. Virtual tributaries can be equal to a DS1, DS3, STS-1, or any of the OC

Name/Acronym	Bandwidth	Equivalent DS0	Equivalent DS1	Equivalent DS3	comments
DS0	64Kb/s	1	*	*	one phone line
DS1/T1	1.544Mb/s	24	1	*	popular service
DS1C	3.152Mb/s	48	2	*	equipment
E1/CEPT1	2.048Mb/s	32	1	*	European
DS2	6.312Mb/s	96	4	*	equipment
E2	8.448Mb/s	96	4	*	European
DS3/T3	44.736Mb/s	672	28	1	popular service
E3	34.368Mb/s	512	16	1	European
DS4	139.264Mb/s	2016	80	6	long haul radio
STS-1	51.84Mb/s	672	28	1	electrical OC1
OC-1	51.84Mb/s	672	28	1	SONET
OC-3	255.520Mb/s	2,016	84	3	SONET
OC-12	622.080Mb/s	8,064	336	12	SONET
OC-48	2.488Gb/s	32,256	1,344	48	SONET
OC-192	9.953Gb/s	129,024	5,376	192	SONET

SONET (Synchronous Optical Network)

levels. The important thing to know about a SONET network is that it simply replaces the older telecommunications technology copper twisted-pair outside plant with fiber optic and electronics.

Southwestern Bell Corp Merged with *SBC (Southern Bell Communications)*.

Spade Lug A connector with two flat surfaces shaped like a two-prong fork that crimps onto a wire so that the wire can be mounted with screws.

Span One section of aerial wire.

Spark Gap Two wires, one hot, and one ground that are separated by a gap of air. Spark gaps are used for lightning protection. If the hot lead, or anything connected to it, is struck by lightning, the lighting will arc across the spark gap because it is engineered to be the easiest path to ground.

Speakerphone A feature of telephones. Speaker phone allows a user to talk on the phone as if it were an open intercom system in the room, without using a handset.

Spectrum, Frequency All electromagnetic radiation is categorized by its frequency in Hz. If some of the frequencies were vibrations, rather than

Frequency Range	US designator	ITU designator	use
30Hz to 300Hz	ELF extremely low freq	2	Submarine/power
300Hz to 3KHz	ULF ultra low freq	3	human audio
3KHz to 30KHz	VLF very low freq	4	human audio
30KHz to 300KHz	LF low freq	5	
300KHz to 3MHz	MF medium freq	6	AM radio
3MHz to 30MHz	HF high freq	7	
30MHz to 300MHz	VHF very high freq	8	FM radio Broadcast TV
300MHz to 3GHz	UHF ultra high freq	9	Broadcast TV
3GHz to 30GHz	SHF super high freq	10	terrestrial microwave/satellite
30GHz to 300GHz	EHF extremely high freq	11	terrestrial microwave/satellite
300GHz to 3THz	THF tremendously high freq	12	Heat infrared
3THz to 30THz		13	Infra-red light
30THz to 300THz		14	
300THz to 3PHz		15	Visible light
3PHz to 30PHz		16	Ultra violet light
30PHz to 300PHz		17	
300PHz to 3EHz		18	
3EHz to 30EHz		19	
30EHz to 300EHz		20	
300EHz to 3000EHz		21	

Spectrum, Frequency

electromagnetic waves, people could hear them. Some electromagnetic radiation (theoretically) is visible as light. The full spectrum of electromagnetic radiation is listed in the table, along with a brief description of its use or what types of transmissions are broadcast over those frequencies.

Speed Dial A feature of telephone sets that enables a user to input a frequently dialed telephone number and assign that number a speed-dial code. To initiate the speed dial, the user dials the code instead of the entire number.

Speed of Light The speed of light in a vacuum is 300,000,000 m/s.

Splice The connecting of two wires, cables, coax cables, or cable pairs together. A splice is shown on an engineering diagram as an arrow. The actual splices of twisted-pair telephone cable are done with modular-type splices, plain B wire connectors, or 3M Scotchloks. For photos, see *Plain B Wire Connector*, *Modular Splice Tool*, and *Scotchlok*. Fiber-optic cable is spliced via mechanical or fusion splicing, and coaxial cable is spliced with barrel connectors. For a photo of a barrel connector, see *Barrel Connector*.

Splice Tray A place within a fiber-optic patch panel or other fiber-optic splice closure that holds fusion splices. In the photograph, the fiber optic can be seen below the patch panel. It is wrapped in a circle in the two splice trays.

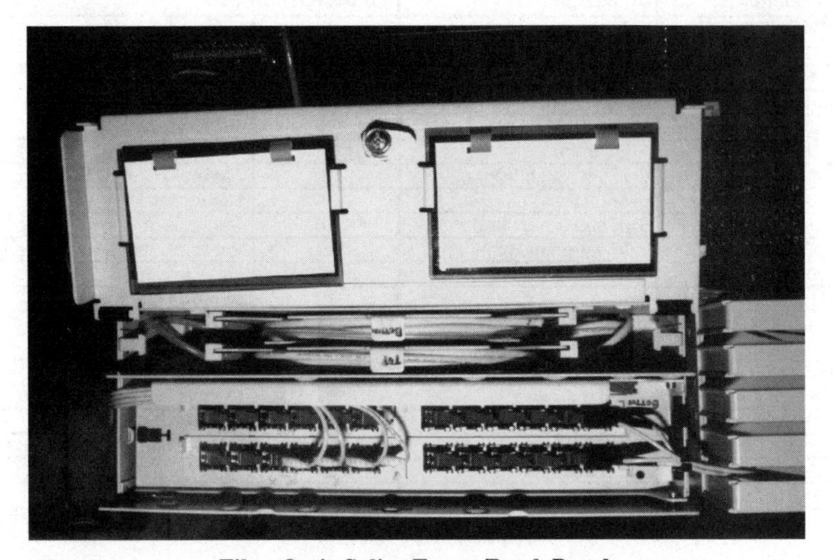

Fiber Optic Splice Tray w/Patch Panel

Split Block Another reference for a 66M150 punch-down wire termination block. The other type of block, or non-split block is a 66M125 block. For a photo, see *66 Block*.

Split Pair The use of one wire from two pairs to make a pair. Sometimes split pairs are done by mistake, but the majority of the time, it is done as a desperate measure to deliver a telephone service to a customer. If there are two bad pairs, but each pair has one good wire in it, the good wire is taken from each pair to make one good pair. Split pairs often cause inductive cross talk and pick up RFI. After RFI filters are placed on the split pair, it will work until new/more telephone wire can be installed or until the old wire can be fixed.

Splitter Used to make a junction point or split a signal so that it will travel down multiple paths over coax. Also called an *RF splitter* and a *UHF/VHF Splitter*. For a photo, see *RF Splitter*.

Spread Spectrum The radio type used in PCS cellular transmissions. Spread-spectrum radio transmits and receives carrier signals over a wide spectrum of frequencies ("channels"). Several technology platforms in wireless communications are considered to be spread spectrum. A spread-spectrum system is any system that occupies more bandwidth than the minimum required for data signal transfer. Two data formats transmitted on spread-spectrum platforms include CDMA (Code-Division Multiple Access) and TDMA (Time-Division Multiple Access)

SPS (Standard Positioning Service) The *GPS (Global-Positioning System)* service that civilians get, but does not correct signal dithering.

Spud A hand shovel especially made for digging holes for telephone poles.

Spudger A device that is shaped and sized like a pencil that telephone technicians use to poke their way through telephone cable with when they are looking for a certain pair of wires.

Spurs What telecommunications and power company personnel wear to climb wooden telephone and power poles. The official name for these devices are *lineman's climbers*. They are also called *climbers*, *hooks*, and *gaffs*. They consist of a steel shank that has straps on it so that it can be strapped to a person's leg. On the inside of the shank is a spike that is used to stab into the pole. For a photo, see *Climbers*.

Squelch An electronic circuit or filter that is incorporated into microphone circuits that makes them have an adjustable sensitivity to the

Spudger

loudness of a sound the microphone will pick up. Squelch circuits are used in speakerphones to cut out background and transient noise.

SRAM (Static Random-Access Memory) *Electronic memory is available in two families,* ROM (Read-Only Memory) and *RAM (Random-Access Memory)*. Memory devices are made from two different technologies: *bipolar (TTL)* and *MOS (Metal-Oxide Semi-conductor)*. Memory is stored by a technique called *writing* and retrieved by a technique called *reading*. ROM devices can only be read, and are programmed during manufacture. *PROM devices (Programmable Read-Only Memory)* can be programmed at a later date by an electronics reseller or electronic assembler for a special application using special equipment. Special ROM devices called *EPROMs (Erasable Programmable Read Only Memory)* can be electronically erased and re-used. RAM has read and write capability. The term *random access* means that any memory address can be read in any order at any time. The two types of RAM are static and dynamic. Static RAM can hold its memory even when power is removed. Dynamic RAM needs constant power to refresh its memory. For a diagram that depicts the types of memory, see *Memory*.

SS7 (Signaling System 7) A method of out-of-band interoffice signaling for telephone circuits. Simply stated, *out of band* means that there is a special separate line used to carry signaling, such as dialed touch tones, ringing signals, busy tones, (everything but the actual voices/conversation) etc.

The two different ways to send signals in telephone transmissions are in band and out of band. Signals are digits that you dial, dial tone, the phone being off-hook, ringing, etc. An in-band telephone line is like the one in your home, the digits that you dial, and the ringing are carried within the channel you talk on. Out-of-band signaling is a method that telephone companies and businesses use for larger PBX applications and data-transfer applications. An out-of-band signaled DS1 has 24 multiplexed channels. The 24th channel carries the signaling for the other 23 channels or phone lines. The advantage of out-of-band signaling is that each channel has an increased capacity to carry data (8Kb/s more) and the 23 channels are not used to find out if a line is busy (both directions, in and out). The off-hook sensing, busy signaling, and other signaling previously mentioned is performed in the 24th channel. If your system receives thousands of calls per day, this can reduce traffic. SS7 makes it easy for long-distance companies to let us dial a phone number, get a busy signal, and not be billed for it because we are not really using a call channel.

ST Connector (Straight Tip Connector) An older type of fiber-optic connector. The newer is the SC connector, which is constructed of plastic instead of metal. For a photo of an ST connector, See *Fiber-Optic Connector*.

Standard Network Interface (SNI) The device used to terminate telephone service at the customer's location and provide lightning protec-

Siecor 6 line Telephone Network Interface

tion. One side of the SNI is for telephone company use only, the other side provides a place for customers to access their telephone lines. For photos of other types of network interfaces, see *Two-Line Network Interface*.

Standby Processor A second (redundant) CPU that takes over if the primary one fails.

Standing Wave In radio transmission, a standing wave occurs when voltage and current form uneven points along a transmitter's antenna or transmission line. This is caused by a mismatch of load (antenna) impedance to transmission-line impedance, and causes an inefficient transmission. The term relating to standing waves is the *SWR (Standing-Wave Ratio)*, which is the ratio of maximum current points on the line to the minimum current points on the line. To picture a standing wave, imagine you make waves in a bathtub so that waves reflecting from the sides of the tub collide in perfect timing with new waves made from your hand. The waves would appear to stand still and thus be called *standing waves*.

Star LAN Network See *Star Topology*.

Star Topology A topology or type of *LAN (Local-Area Network)*. The star topology is used in ethernet applications. Ethernet is one of the oldest communication protocols for personal computers. When A LAN is mentioned, the two things that should immediately come to mind are physical topology and the protocol that the LAN uses to manage communications between devices. Ethernet can be implemented in a bus or star physical topology. The alternative family of LAN protocols is the token-passing type, which is configured as a ring topology (see token ring).

In an ethernet LAN, computers are given a means to communicate with each other called a protocol. A protocol is a set of rules and instructions for communicating. Within the protocol is a "logical topology." Even though a network might be connected as a star, it can still "look" like a bus to the communications equipment because all of the computers/devices are connected to the same wire (in the star diagram, the hub is a device that connects all the wires together). Ethernet works similar to the way that people talk in a group. Instead of using wire to carry the binary coded information as ethernet does, people use air to carry sound information. When there is a silence, then one of the persons in the group is able to speak. When the people speak, they say "Johnny, do you know the answer for 5+5?". Even though all the people in the group hear this message, they know it is for Johnny because the message was "addressed" to him. Only Johnny will respond "10". Then imagine Dawn and

ETHERNET TYPES

PROTOCOL	PHYSICAL TOPOLOGY	WIRING USED
10 BASE 2	BUS	RG 58 COAX (50 ohm)
10 BASE 5	BUS	RG 8 COAX (50 ohm)
10 BASE T	STAR	CAT 4 or 5 UTP/STP*
100 BASE T	STAR	CAT 5 UTP/STP*

* unshielded twisted pair / shielded twisted pair

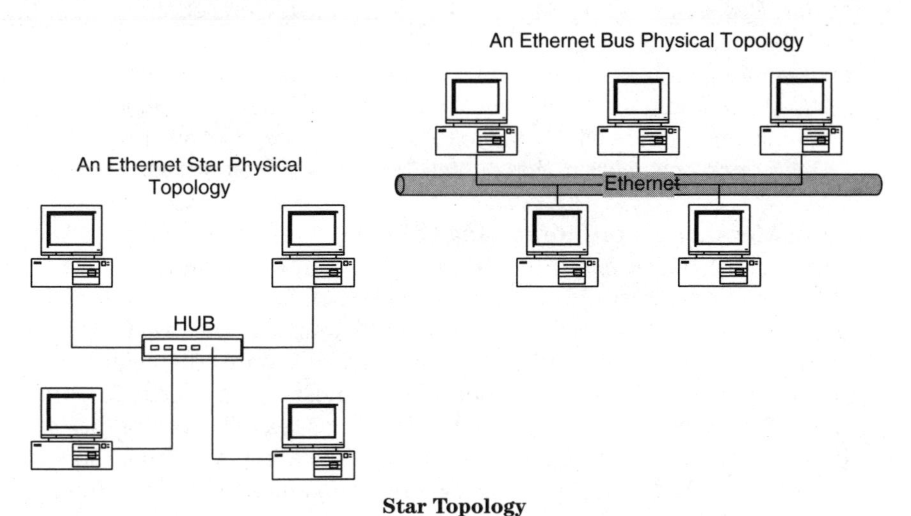

Star Topology

Vicki acknowledges a silence and try to speak at the same time. This is confusing and no one understands the information. Ethernet has the same problem and it is called a *collision*. Collision is the disadvantage of Ethernet. Because of the possibility of collisions (which happen very frequently), ethernet is called a "contention-based" protocol because all of the connected devices are contending for use of the network. Manufacturers have come out with new ways to avoid collisions, called *CSMA/CD* and *CSMA/CA*. Ethernet has many different types of wiring to connect devices and many different *NICs (Network Interface Cards)* to select from that need to be installed in each computer or device on the network. The list shows ethernet protocols and the type of wiring used for each.

Start of Heading (SOH) The ASCII control code for start of heading. The binary code is 0001000 and the hex is 10.

Static RAM (SRAM, Static Random-Access Memory) Electronic memory is available in two families, *ROM (Read-Only Memory)* and *RAM*

(Random-Access Memory). Memory devices are made from two different technologies: *bipolar (TTL)* and *MOS (Metal-Oxide Semi-conductor)*. Memory is stored by a technique called *writing* and retrieved by a technique called *reading*. *ROM devices* can only be read, and are programmed during manufacture. *PROM devices (Program-mable Read-Only Memory)* can be programmed at a later date by an electronics reseller or electronic assembler for a special application using special equipment. Special ROM devices called *EPROMs (Erasable Programmable Read Only Memory)* can be electronically erased and re-used. RAM has read and write capability. The term *random access* means that any memory address can be read in any order at any time. The two types of RAM are static and dynamic. Static RAM can hold its memory even when power is removed. Dynamic RAM needs constant power to refresh its memory. For a diagram that depicts the types of memory, see *Memory*.

Station Message-Detail Reporting (SMDR) Another term for call accounting. A *call-accounting system* is a computer (usually a dedicated PC) that connects to a PBX switch via a serial data port and monitors the details of every phone call made through that switch. The call details are stored as call records. With the appropriate software, they can be retrieved, sorted, processed, and queried to almost any specific nature that the call-accounting system administrator desires. These systems are used by hotels to track all the calls that you make from your room so that you can be billed. They are also used by companies to bill back reports for individual departments within the company.

Statistical Time-Division Multiplexing A multiplexing technology that gives users automatic adjustable bandwidth.

STDM (Statistical Time-Division Multiplexing) A multiplexing technology that gives users automatic adjustable bandwidth.

Step-Down Transformer A transformer that is wired in a fashion to receive an AC voltage on its primary winding and reduce that voltage through electromagnetic induction into the secondary winding. Reversing the way a transformer is wired changes it from a step-down to a step-up and vise-versa.

Step-Up Transformer See *Step-Down Transformer*.

Stepped-Index Fiber Optic A fiber optic that has a core made of glass consisting of one refractive index. Stepped index fiber is available in multi-mode and single-mode. The alternative to stepped-index fiber op-

Schematic Symbol for an iron core step-down Transformer
(iron core is designated by two lines between coils)

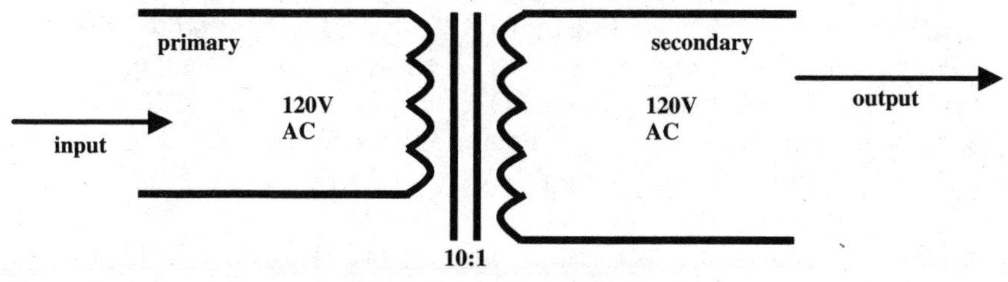

Step-Down Transformer

Schematic Symbol for an iron core step-up Transformer
(iron core is designated by two lines between coils)

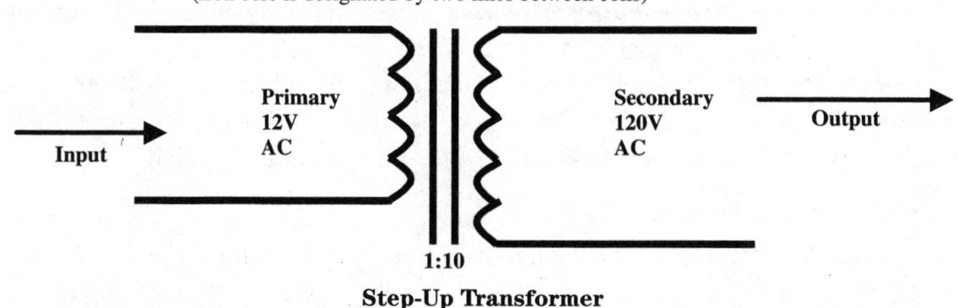

Step-Up Transformer

LIGHT AS IT TRAVERSES THROUGH THE CORE OF A FIBER OPTIC

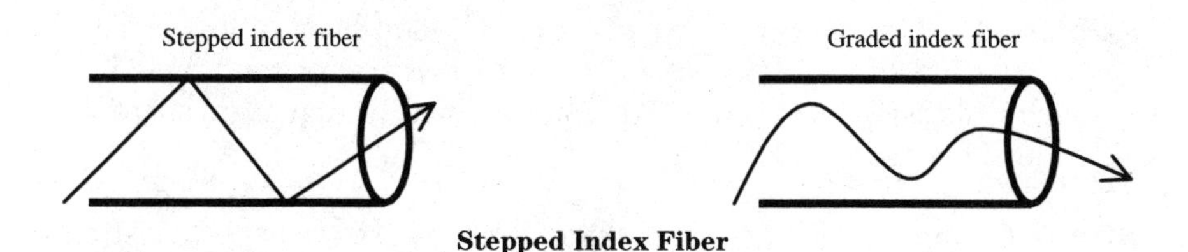

Stepped Index Fiber

tic is graded-index fiber optic. The core of graded-index fiber optic consists of many layers of glass with different refractive indexes that cause the light to gradually bend as it approaches the outside of the fiber. Graded-index fiber (like stepped index fiber) is available in multi-mode or single-mode, and it is more expensive.

Stepper Switch Also called a *crossbar switch*. The old analog telephone switch had mechanical relays that connected telephone calls. This is where the term *switch* comes from. Old central-office switches contained literally thousands of mechanical switches.

STM (Synchronous Transfer Mode) New technologies, such as SONET, *ATM (Asynchronous Transfer Mode)*, and *ISDN (Integrated-Services Digital Network)*, are leading up to STM. It is also referred to as *BISDN (Broadband Integrated-Services Digital Network)*. It will enable the user to have a DS3 private line, a POTS line, or an *ISDN BRI (Basic-Rate Interface)* automatically, depending on the device they use to access the line, a high-definition TV set, a telephone, or a PC.

Stop Bit In serial data transmission, the stop bit is a logical one (1) after the transmission of each character (each character is seven or eight bits long).

STP (Shielded Twisted Pair) The typical wire used in telephone networks and computer LANs. Shielded twisted pair is not as commonly used as *UTP (Unshielded Twisted Pair)*. The physical make up of twisted-pair wire is solid 24 AWG wires, twisted into pairs. Four pairs (two wires make one pair) is the most commonly used for new installations these days because it is only a penny or two more per foot than three pair. Twisted-pair wire also comes in different categories:

- Cat 3 for telephone
- Cat 4 for 10 base T LAN (local-area network) or telephone
- Cat 5 for 100 base T ethernet or token ring LAN
- Cat 7, (the newest) which is a little better than cat 5 for data, but a lot more expensive.

Straight-Tip Connector A type of round, metal fiber-optic connector, usually called an *ST connector*. The newer connector (and rapidly becoming more popular) is the *SC connector*, which is square in shape and made of plastic.

Strand Clamp A pole-attachment device that is used to hold steel strand to utility poles.

Strand Clamps and 14' Mounting Bolts

Stranded Copper See *Stranded Wire*.

Stranded Wire A wire that is made up of many small wires, rather than one big solid one. Stranded wire is not used in telephony applications because it doesn't stay connected to 66M150 or AT&T 110 termination blocks.

Strand, Steel The support for telephone cable, cable-TV coax, and fiber-optic cable when installed in aerial applications. See also *Pole Attachment*.

Strand, Steel

Strap An electrical connection (usually a wire or metal jumper) from one point to another. *Strap* is a common term used in reference to the configuration of rectifiers, power supplies, and circuit cards.

STS (Synchronous Transport Signal) In SONET networks, the optical signal must be converted to electricity at one time so it can be de-multiplexed and further processed. This electrical version of the SONET OC-1 level signal is called a *synchronous transport signal 1 (STS-1)*, and is transported from node to node or node to digital cross-connect system, or to DSX cross-connect panels via 50-ohm coax.

STS 1 (Synchronous Transport Signal 1) See *STS*.

STS 1

STS 3 (Synchronous Transport Signal 3) Referred to by some as a level 2. The electrical equivalent of an *OC-3 (Optical Carrier Level 3)*. 155Mbps.

Studio Transmitter Link (STL) A studio transmitter link is basically a remote control for a distant transmitter. It is a separate radio channel from the broadcast radio station itself and it is used to adjust the amount of power that the transmitter emits, performance monitoring, and provide remote access for testing and troubleshooting.

STX The ASCII control-code abbreviation for start text. The binary code is 0010000 and the hex is 20.

SUB The ASCII control-code abbreviation for substitute. The binary code is 1010001 and the hex is A1.

Subscriber A telecommunications customer. This includes telephone, cable TV, and cellular (PCS and wireless).

Subscriber Loop The pair of wires that runs from a telephone company central office (or from extended transmission equipment) to the customer's network interface. A loop is a pair of wires.

Subscriber Loop Carrier (Another term for SLC 96) See *SLC 96.*

Super Server A server that has multiple microprocessors.

Superframe Format A framing format for T1 that consists of 12 T1 frames, 193 bits each, transmitted in succession. The superframe format allows for maintenance and monitoring information to be sent along with the 24 DS0 channels. *ESF (Extended Super Frame)* is the newer version of T1 framing format.

Supervisory Signal A way that telephone electronics communicates with each other to initiate a command. If you have call waiting on your home telephone, when you get another call (hear the beep or click indicating another call), you momentarily press the hook switch (flash button) to signal the central office to give you the other line. This is an example of a supervisory signal. It is called a supervisory signal because it relates to the connect and disconnect of a phone line. The general term for the ability for a central office or PBX switch to recognize that a telephone conversation has ended is *disconnect supervision.* When the telephone is "hung-up," the central office or PBX recognizes the decrease in current flow and disconnects the call. *PBX (Private Branch Exchange)* phone systems have disconnect supervision, which means that when a call is ended, it recognizes a "hook-flash" from the central office and disconnects the call, or vice-versa. If the PBX did not have this feature, it would not release (hang-up) telephone calls.

Surface Mount A reference to a modular jack that is shaped like a box, and can be fastened to the surface of a wall, baseboard, or anywhere else you do not have prewired outlets. Surface-mount jacks are also called *biscuit jacks* and *baseboard jacks.*

Surge Protector A power-filtering device. Most surge protectors come in the form of an extension cord with six to eight outlets box attached to the end. Not all surge protectors are created equal. Most surge protectors are more useful as an extension cord than any kind of protection from a voltage surge or spike. Some good surge protectors use fast-switching components to sense overvoltage and spikes on the power source. These good surge protectors cost about $50. If you are truly concerned about protection from all of the evil electrophysical characteristics of public power, a small UPS system is the best protection. A good *UPS (Uninterrupted Power Supply)* costs about $150.

SVC (Switched Virtual Circuit) Any circuit that can be connected for a temporary amount of time with the use of electronic circuit switching equipment. This includes a plain-old telephone call. When the phone is on the hook and no one is using it, the telephone wire runs to the central office and ends. When you pick up the receiver and dial a number, the central office makes a connection through its electronics from one phone line to another that lasts only as long as the receivers are off hook. It is a "switched" circuit because it can be switched to any telephone with a number that you dial. It is virtual because the actual path through the electronics is multiplexed, as opposed to a physical pair of wires.

SVGA (Super Video Graphics Array) The suggested monitor for the PC that is used for call-accounting applications, and other computer-telephony integration applications. The SVGA monitor is capable of resolution to 1024 by 768 pixels (dots of light on the computer screen) per inch. SVGA is the newer version of *VGA (Variable Graphics Array).*

Switch Another name for a PBX (phone system) or central office.

Switch Room A room that is dedicated for switching equipment. Switch rooms are usually kept at a temperature of 65 degrees F and a humidity of 50% to reduce ESD.

Switched 56 A service offered by local and long-distance telephone companies that works like a regular telephone line except, that it is intended for data/modem use. Switched 56K lines have 7-digit telephone numbers (plus area code) and they are available in digital or analog. When the telephone company installs the line, they condition the copper pair (remove bridge taps and coils) and install a *DSU (Digital Service Unit)*, which is a line/signal amplifier. Connect it to your 56K modem; it will transmit at a rate of 56,000 bits per second as long as you are talking to another 56K modem on another 56K line.

Switched Ethernet An ethernet protocol that gives each computer or device connected to the network its own channel to communicate with. In plain old ethernet, all of the devices communicate on the same wire (channel). In switched ethernet, the physical star (or bus) is multiplexed into a variable number of channels, one for each device that is

Switched Ethernet

communicating. The speed of each channel depends on the number of devices communicating at any one time. Collision is eliminated in switched ethernet, which was the largest inefficiency with the previous ethernet protocols (CSMA/CD and CSMA/CA). This newer protocol transfers data at a much faster rate because of the newfound efficiency. Switched ethernet requires its own special NICs *(Network Interface Cards)* and other hardware. It also requires its own software drivers.

Switched Multimegabit Digital Service (SMDS) A service offered by local telephone companies that is intended for the transport of large amounts of data at high speed from point to point over a switched type of network. You enter in the address or number that you would like your data to be sent and the SMDSU (SMDS Unit) packetizes the data and the SMDS network transports it. SMDS is a packet- or frame-type technology that is available in five transmission rates. Class 1 is 4Mb/s, Class 2 is 10Mb/s, Class 3 is 16Mb/s, Class 4 is 25Mb/s, and Class 5 is 44.7Mb/s.

Switched Private Line A reference to switched 56K service.

Switched Service Basic voice telephone lines (POTS lines) are switched services.

Switched Virtual Circuit A telephone line is a switched virtual circuit. The circuit only exists while the conversation is happening. After the conversation is over, the circuit, which acts like a pair of wires from point A to point B, is gone. Another form of a switched virtual circuit is like those in switched ethernet or switched token-ring technology.

Switching Center Another name for a telecommunications company's central office. A location for switching equipment/electronics and transport equipment/electronics.

Switching Hub Also called a concentrator. A *LAN (Local-Area Network)* element that links a device to a network with a specific amount of bandwidth or exchange rate of data, regardless of the number of users on the network. A switching hub performs the same function as a nonswitching hub, except that the nonswitching hub only connects many users into the same channel, where they all share the same bandwidth. For a photo of a switched ethernet hub, see *Switched Ethernet*.

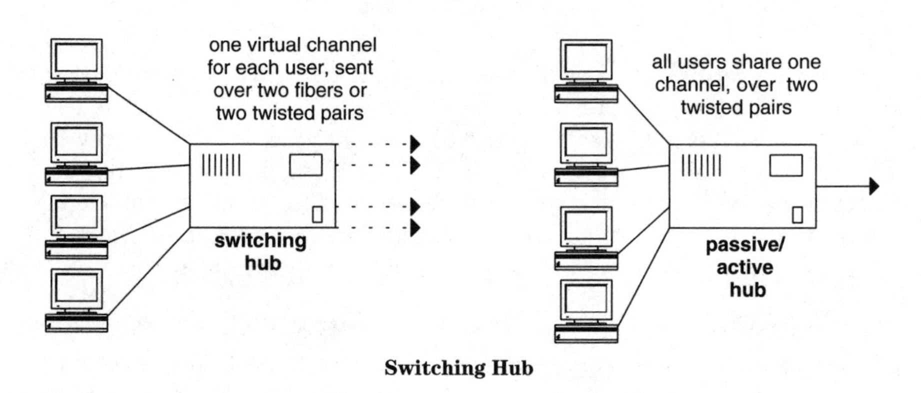

Switching Hub

SYN The ASCII control code abbreviation for synchronous idle. The binary code is 0110001 and the hex is 61.

Synchro Daemon A background program or subroutine that is used for a timing reference in telecommunications platforms.

Synchronize See *Synchronous.*

Synchronous "With time." A reference to communications equipment that is timed from a common timing source, such as a bits clock. SONET networks are synchronous and modem transmissions are asynchronous. *Asynchronous* means "without timing." People talk asynchronously. Even though one person talks very fast and another very slow, they both exchange information and process it in their brains.

Synchronous Data Link Control An IBM term. The protocol used in IBM *SNA (System Network Architecture)* computer networks.

Synchronous Transfer Mode (STM) See *STM.*

Synchronous Transport Signal (STS 1) See *STS 1.*

System Speed Dial A feature of *PBX (Private Branch Exchange)* and key telephone systems. Unlike standard speed dial, where each individual user programs a speed dial under a button on their own phone, system speed dial programs a speed-dial number under a code. The telephone number programmed in the system speed dial can be dialed by entering the code from any telephone on the system (that has the feature). The Northern Telecom Meridian 1 PBX system is capable of having 1000 system speed-dial numbers.

Systems Network Architecture (SNA) See *SNA.*

T

T1 A T1 ("T" one) is a standard 1.544Mb/s carrier system used to transport 24 telephone lines or various broadband services from one point to another (also called a *DS1*, but a DS1 is the service given to a customer without the –135-V carrier). T1 is the standard carrier for the United States, Canada, Japan, and Singapore. All other countries use the E1 standard (30 channels on four wires). The T1 is a four-wire circuit, two wires for transmit and two wires for receive. The T1 line voltage is –135 V. The T1 circuit can carry voice or data. Its use determines the variables of T1 service, framing format, and line format. See also *DS1*.

T1 Test Set Many instruments are manufactured for testing T1 transmission circuits.

T1C A 3.152 Mb/s multiplexed signal that carries the equivalent of two T1 signals. The T1C signal is not a widely embraced form of transport; however, it does exist within the stages of some multiplexing equipment.

TA (Terminal Adapter) A device that converts an *ISDN (Integrated Services Digital Network)* line into regular *POT (Plain-Old Telephone)* service so that you can connect a standard telephone and/or modem to an ISDN line. For a diagram, see *Terminal Adapter*.

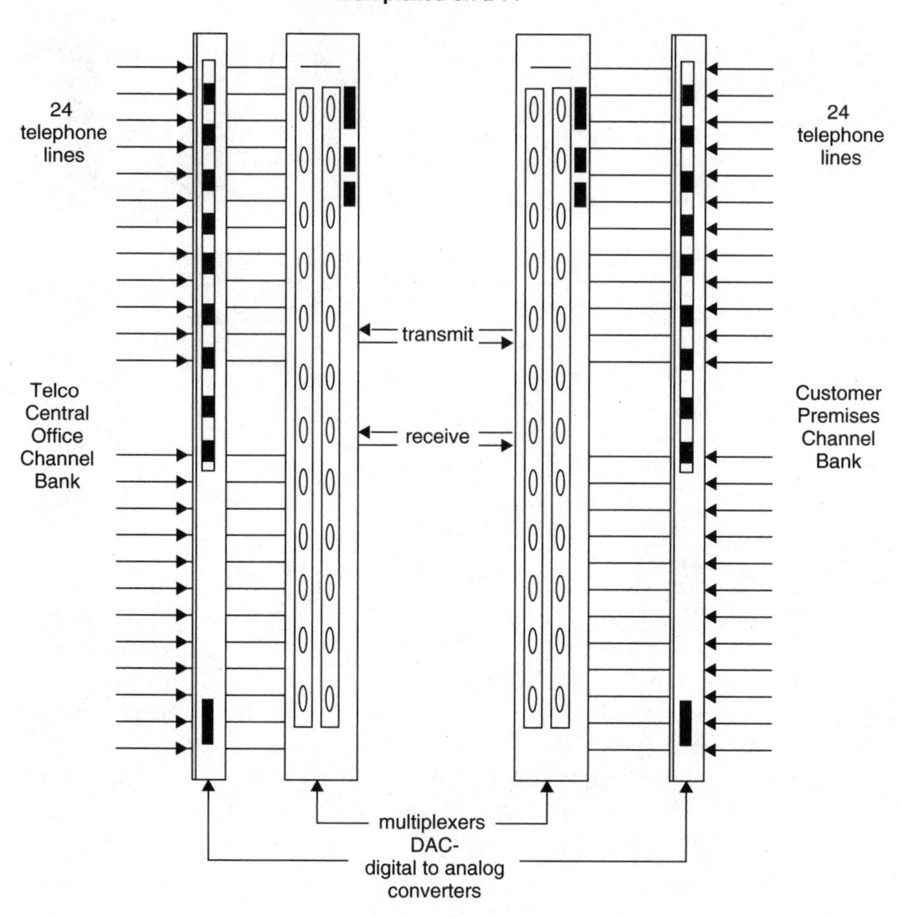
24 DSO voice channels
multiplexed on a T1
24
telephone
lines
Telco
Central
Office
Channel
Bank
transmit
receive
24
telephone
lines
Customer
Premises
Channel
Bank
multiplexers
DAC-
digital to analog
converters

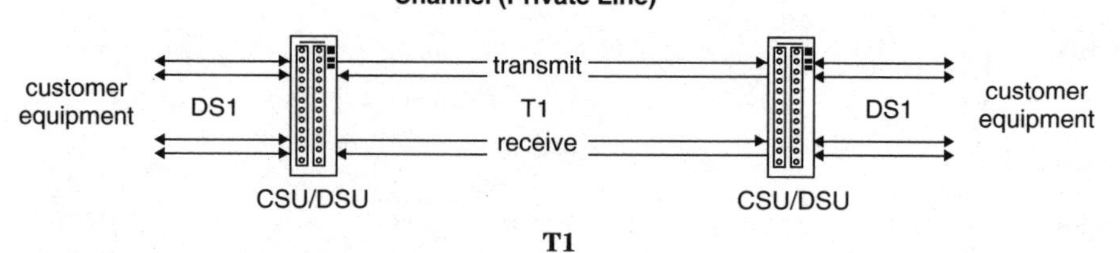
A T1 as a Data Communications
Channel (Private Line)
customer
equipment
DS1
transmit
T1
receive
DS1
customer
equipment
CSU/DSU
CSU/DSU

T1

DS1/T1circuit/line types and applications

Line format/coding	framing format	signaling	Application
AMI	SF/D4	in-band	24 voice/modem channels
AMI	ESF	in-band	24 voice/modem channels
AMI	ESF	out-of-band	23 voice/modem or digital/data channels
B8ZS	SF/D4	in-band	24 voice/modem channels
B8ZS	ESF	in-band	24 voice/modem channels
B8ZS	ESF	out-of-band	23 voice/modem or digital/data channels

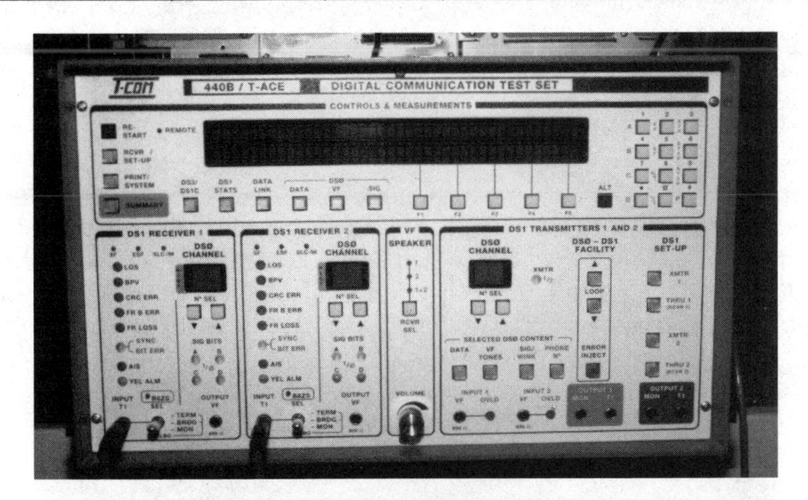

T1 Test Sets

Talk Battery Typically –10 V on the public telephone network. When a telephone line is idle, a –52-V central-office battery is on the line. When the receiver is picked up, the voltage drops to 10 V (talk battery). The ringing voltage on a telephone line is 90V AC.

Talk Path Two wires with a battery voltage that can be used to connect two telephones to provide a simple talk path. Sometimes when telephone company employees are working on telephone lines, they connect a battery to a pair of wires, connect a test set (butt-set/goat/test telephone) to each end, and talk to each other. They call this arrangement a *talk path*.

Tandem 1. A classification of telephone company central office or node that contains a switch in which all inter and outer area-code traffic is handled. The main LEC central office is in an area code where the hand-off for long-distance service happens. For a diagram, see *Access Tandem*. 2. A central office that carries a call, but does not connect it with the end customer, it switches ("sends") the call to the central office from which the called customer is served.

Tandem Office See *Tandem*. .

Tandem PBX A *PBX (Private Branch Exchange)* switch that carries telephone calls, but does not terminate them to a telephone. It switches the calls to another PBX. See also *Tie Trunks*.

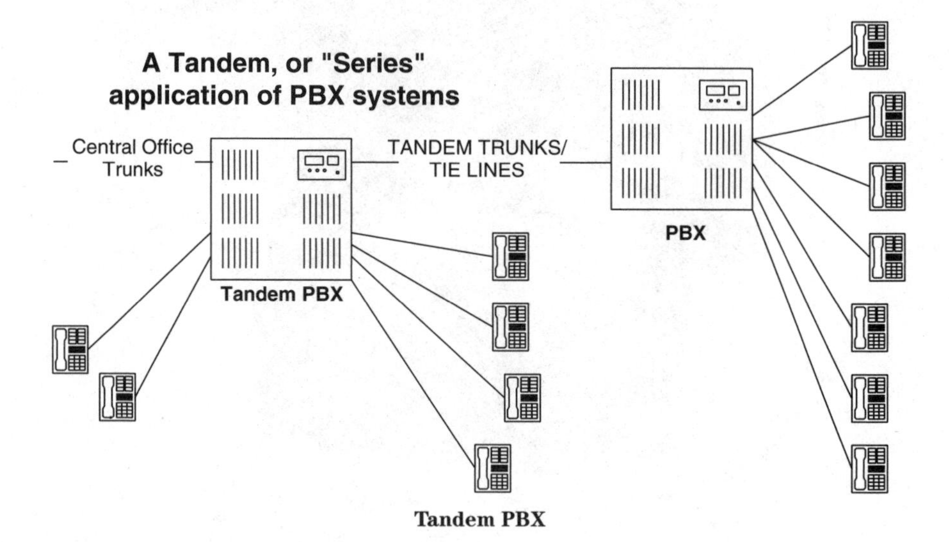

Tandem PBX

Tandem Switch A central office that carries ("links") a call, but does not connect it with the end customer, it switches ("sends") the call to the central office from which the customer is fed. See also *Tandem.*

Tandem Trunks Trunk lines that connect central offices/switching equipment. See *Tandem PBX.*

Tap A device used to monitor telephone lines. With newer technology, the telephone company is capable of tapping (monitoring) a telephone line with a stroke of a few keys. They can even set the telephone line to be monitored by a different telephone line, anywhere they want. Telephone companies (especially the Bell Companies) have extremely strict security guidelines regarding the monitoring of telephone conversations. They will not set up a tap or a monitor service without legal procedures being followed according to the laws of the area where they are operating. Watch dogs are in place to ensure that telephone company employees do not monitor telephone lines when they are not supposed to. Some "spy" shops sell telephone line tapping and recording devices, but their use is not recommended because of strict laws regarding telephone privacy.

Tap Button Key A button on some *PBX (Private Branch Exchange)* telephones. Most PBX systems are purchased with electronic/digital telephone sets. Those that are not, are purchased with standard single-line telephones that have an extra button, called a *flash* or *tap key* (most PBX systems can be configured for either). The flash/tap key sends a hook-flash signal to the PBX system to tell it that it is going to receive a command (e.g., more dialed digits) to activate a feature, such as transfer, hold, or conference.

Tape Drive A RAM storage device that utilizes magnetic tape for data storage. The magnetic tape is similar to that of the magnetic tape used in a typical audio cassette. The disadvantage of tape-drive memory storage is the long amounts of time required to retrieve data.

TAPI (Telephone Application Programming Interface) A form of *Computer Telephony Integration (CTI)* that is developed and made by Microsoft. It brings telephony applications, such as modem set-up, to the *GUI (Graphical User Interface)* level.

Tariff A pricing structure of telecommunications services that is offered by a communications services company and accepted by the *Public Utilities Commission (PUC).*

TBM (Transport Bandwidth Manager) Northern Telecom's name for their SONET-operating platform.

TBOS (Telemetry Byte-Oriented Serial) An alarming or maintenance protocol for communicating with distant or remote equipment.

T Carrier A reference to T1 or T3 transmission systems. A T1 has 24 voice channels (standard telephone lines) on one 4-wire circuit, and a T3 has 28 T1 channels (also called *tributaries*) on one 4-wire circuit. For a diagram of a T1 and a chart showing T1 signaling and framing applications, see *T1*.

T Connector A coax connector used in *LAN (Local-Area Networks)* that has a male BNC on one end and two female BNC connectors on the other. It's shaped like a "T."

TCP (Transmission Control Protocol) A data-packet protocol (rules for communications between devices) for data exchange between computers, printers, servers, or anything else that is capable of being loaded with the protocol driver. The protocol driver can be loaded in software or embedded in hardware (PROM). It was developed by *ARPA/DARPA (Advanced Research Projects Agency/Defense Advanced Research Projects Agency)* of the U.S. federal government.

TCP/IP (Transmission Control Protocol/Internet Protocol) It was initially called TCP, but then it became the standard for the Internet, and the "/IP" was added to the end. See *TCP*.

TCR (Transaction Confirmation Report) The report that a fax machine prints. It details all of the faxes that were sent and received for a specified duration of time.

TDD (Telecommunications Device for the Deaf) A reference to devices that attach to telephone handsets and allow deaf people to type messages to each other. Most of the devices look like Wyse terminals with an acoustic coupler attached. (An acoustic coupler is a pad that a telephone handset rests on so that data transmitted over the line can be received through the mouthpiece and earpiece. The transmission rate is a very slow 300 baud)

TDM (Time-Division Multiplex) For an explanation and diagram, see *Multiplex*.

TDMA (Time-Division Multiple Access) Also called *FDMA (Frequency-Division Multiple Access)*. Another name for time-division multiplexing. Only TDMA and FDMA band can multiplexed signals together over a transmission frequency. Think of TDMA as a "whole-lotta multiplexing." It is used in spread-spectrum cellular radios (transmitters) that the cellular telephone industry uses.

TDR (Time-Domain Reflectometer) A testing device that measures the distance of a copper twisted pair. The Dynatel 965T is a popular TDR. A TDR works by transmitting a signal down a copper twisted pair, then it waits for a reflection to come back. When the reflection returns to the device, the time difference is used to calculate the distance that the signal traveled. For a photo of a TDR, see *965T*.

TE (Terminal Equipment) The equipment at the end (customer side) of an ISDN line. It is classified in two categories: Type 1 (TE1) for equipment that is directly ISDN compatible) and Type 2 (TE2) for equipment that requires a converter to convert the ISDN *BRI (Basic Rate Interface)* into two separate phone lines so that analog modems and telephones can be used.

TE 1 (Terminal Equipment Type 1) Equipment that is directly ISDN compatible or is capable of plugging directly into an ISDN line.

TE 2 (Terminal Equipment Type 2) Regular telephones and terminal adapters that convert the digital ISDN signal into an analog signal so that normal POTS (Plain Old Telephone service) equipment can be used.

Telco Abbreviation for telephone company.

Telecommunications To communicate across a distance.

Telemetry A reference to the remote monitoring of communications equipment, and its environment.

Telephone A device that consists of six major parts. A switch-hook, a dialing circuit (DTMF tone or Rotary), a ringer, a microphone, a speaker (for the ear-piece), and a hybrid coil.

Telephony Server Also known as a *FEP (Front-End Processor)*. A communications "front end" device that can be loaded with a "firewall" to prevent unwanted users from accessing the communications network. An FEP can also perform routing, and differentiate between different communications protocols, depending on the software that runs on it. For a diagram, see *Front End Processor*.

Teletext Data that comes over a television transmission in the form of text at the bottom of the screen.

Teletypewriter (TTY) A device that used to work like a telegraph (its almost as old), except that it would send typewritten messages over phone lines instead of Morse Code. It had a keyboard and a printer.

Television (TV) The current name for video broadcasting and the receiver used to pick it up. Currently, the standard for terrestrial broadcast television in North America is NTSC, which offers a video resolution of 525 × 495 lines and FM stereo audio. Different standards, such as PAL and SECAM, are used in other parts of the world. The FCC is now implementing new HDTV standards, which should offer a higher-resolution signal via a digital transmission process. The following diagram shows the different blocks of a television receiver.

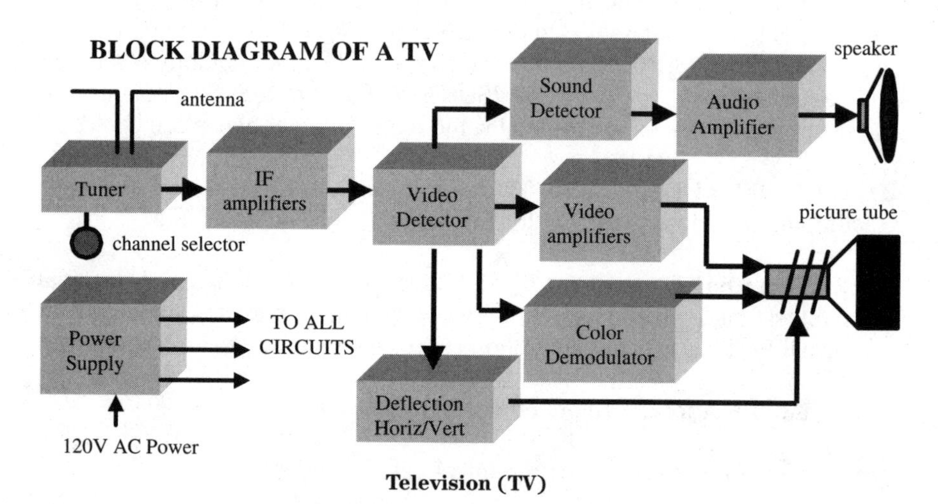

Television (TV)

Telex An older international switched message service that is being replaced by fax machines and the Internet/public packet-switched network.

Tera The prefix for Trillion. Five THz means 5,000,000,000,000 hertz.

Terabyte (Tb) 1,000,000,000,000 bytes.

Terminal A closure where a telephone cable is terminated. It is usually a green box if it terminates buried cable or a silver box if pole mounted. 2. A video I/O device with a keyboard that is used to enter and retrieve information (data) from computers.

Terminal Adapter A device that converts an *ISDN (Integrated Services Digital Network)* line into regular *POT (Plain Old Telephone) service* so that you can connect a standard telephone or modem to an ISDN line.

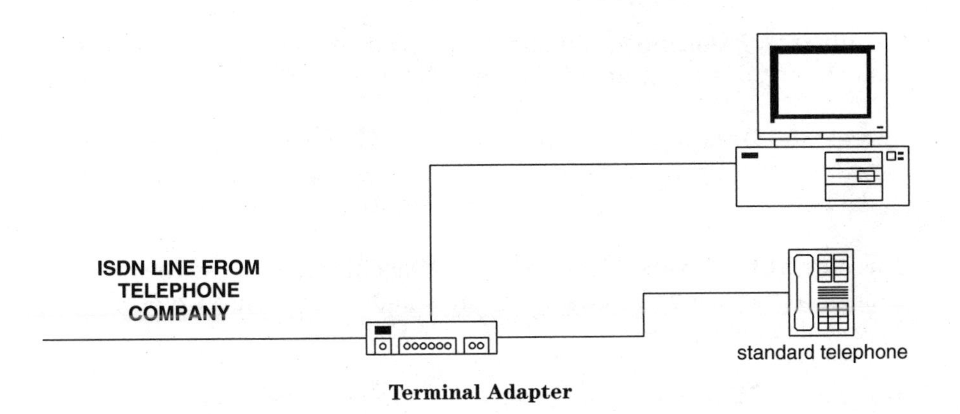

Terminal Adapter

Terminal Block A reference to a 66M150, AT&T 110, Crone, or some other block that is designed to terminate (permanently affix) wire.

Terminal Emulation The use of a PC to act or communicate as a dumb I/O terminal. To use a PC in this application (such as to plug into a microwave link and boost its power), it must be equipped with terminal-emulation software. The microwave-link device has its own microprocessor and only needs a device (usually a VT100 terminal) to communicate with. Terminal emulation software allows your PC to "look like" a VT100 terminal to the microwave radio equipment.

Terminal Equipment The equipment at the end of a communications circuit or the equipment that is on the "receive" end of the transmission, such as a printer or telephone.

Terminals 1. A workstation, as in an IBM SNA network. 2. A reference used to identify where copper twisted pairs are terminated in access points, cross boxes, and terminals. Physically, a binding post is a pair of teeth on a 66M150 block or a pair of 9⁄16" lugs. Each binding post has a number. When a technician looks for a specific pair in a cable (called a *cable pair*), they refer to documents that list the pairs and binding posts on which they are spliced. 3. A box on a telephone pole or on the ground where the telephone cable is terminated. From the terminal, the service wire is run to the house or building. For a photo of a buried cable terminal, see *Pedestal*.

Terminate 1. To complete a communications path. 2. To permanently fasten wire to some form of connectivity, such as an RJ45 jack, 66M150 block, AT&T 110 block, etc.

Terminator, Ethernet A 50-ohm BNC "dead end" connector that is used at the end of a coax run in ethernet *LANs (Local-Area Networks)*.

Terrestrial Data Service A reference to data signal transport service via terrestrial microwave. Terrestrial Microwave can be used for voice or data transport. See also *Terrestrial Microwave*.

Terrestrial Facilities Communications facilities that are on the ground, such as transmitters, telephone lines and poles, central offices, etc. Anything but satellites.

Terrestrial Microwave Microwave radio has become a very economical way to bypass construction costs of broadband private-line services.

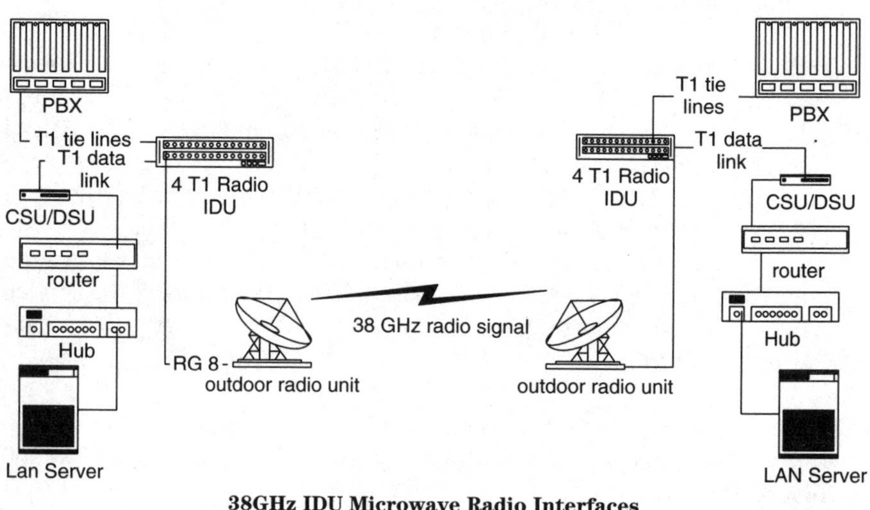

38GHz IDU Microwave Radio Interfaces

Many *CAPs (Competitive-Access Providers)* have access to microwave radio resources, such as licensing, equipment, and installation. Digital microwave is also called an *eyeball shot*, *38 Gig*, or just *radio*. Most of the microwave being installed for private line service today is in the 33- to 39-GHz frequency range. These microwave units use an FM-FSK over two sidebands for transmitting at full duplex. They are available in T1, DS3, and STS-1 (which is a DS3 formatted for SONET). 38-GHz microwave has a range that depends on the size of the antenna (dish) placed on the outdoor radio unit. The choices in antenna size are one or two feet in diameter. The one-foot antenna has a maximum range of one to three miles, depending on the regional weather conditions (rainfall, snow, and especially fog drastically attenuate microwave transmissions). The two-foot dish has a range of two to seven miles, also depending on the weather in the region. Below is a microwave application. For a photo of a microwave antenna, see *Microwave*. The photo illustrates an *IDU* (*Indoor Unit*. The dish is referred to as the *ODU*, *Outdoor Unit*), as pictured in the diagram.

Test Set Also called a "goat" or "butt set." A test telephone set used by telephone installation and repair personnel. Instead of a plug on the end of the cord, it has a pair of alligator/bed-of-nails clips.

Test Shoe A testing device aid that connects to distribution frame blocks or other cable-terminating blocks that helps prevent shorts and crosses while technicians work on or test the cable.

Test Tone Also called an installer's tone. A small box that runs on batteries and is used to put an *RF (Radio Frequency)* tone on a pair of wires. If a telephone technician can't find a pair of wires by color or binding post, they attach a tone to one end, then go to the other end and use an inductive amplifier (also called a *banana* or *probe*) to find the beeping tone.

TFT (Thin-Film Transistor) A transistor used in PC laptop displays. Each pixel is controlled by an individual transistor.

Thermal Noise Also referred to as *ambient current*. The result of heat that causes random movement of electrons in a circuit when the power is off. The current/movement of electrons causes ambient voltage. Ambient current/voltage is why oscillator circuits start oscillating when the power is turned on. The natural oscillations of the electrons become filtered and amplified when the power is applied to the circuit. As electronic circuits become warmer, thermal noise becomes more prevalent. If it is amplified so that you can hear it, it sounds like white noise or a radio that is not tuned to a station.

THF (Tremendously High Frequency) The American standard name for frequencies within the spectrum of 300 to 3000 GHz. For more information, see *Spectrum, Frequency*.

Thick Ethernet (Thicknet) An earlier ethernet *LAN (Local-Area Network)* on 50-ohm coax cable. Thick ethernet was given its name after a thinner version of the RG-58 coaxial cable started being used. Thick ethernet is slightly better than its thin counterpart for transmission distance, but thin ethernet coax costs less and is much easier to work with. Whether you are using RG-58 thick or thin coax is not so much of an issue because the network interface cards inside the computers don't know the difference. For a diagram, see *Thin Ethernet*.

Thin Ethernet (Thin-net) A popular ethernet *LAN (Local-Area Network)* that utilizes thin RG-58 coax. The less-popular thick ethernet cable is slightly better for distance, but harder to work with and install. Whether you are using RG-58 thick or thin coax is not so much of an issue because the network interface cards inside the computers don't know the difference.

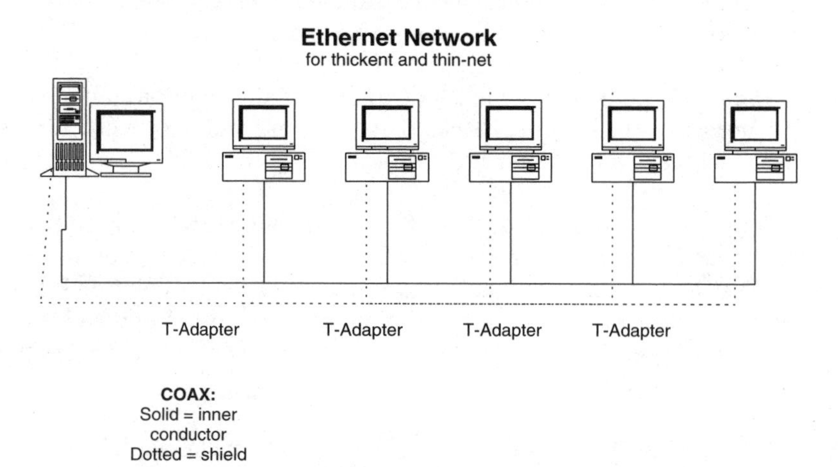

Thin Ethernet (Thin-net)

Third-Party Call Not a direct call and not a collect call. A third-party telephone call is charged to a telephone number other than the one that is being dialed or used to make the call. Calling cards have been offered by many phone companies to make third-party calls easier. Without a calling card, you need an operator to dial a third-party call. Typical charges for third-party calls are about 30 cents per minute. The expensive rate has brought on competition from prepaid calling-card companies.

THz (Terahertz) Tera is the prefix for trillion. For example, 5 THz means 5,000,000,000,000 Hertz.

TIA (Telecommunications Industry Association) An organization that hosts great communications trade/product shows. For more information, see *http://www.tiaonline.com.*

Tie Communications (Telephone Interconnect Equipment Communications) An older telecommunications equipment manufacturer/distributor that had their name stamped on *PBX (Private Branch Exchange)* and key telephone systems, like the BK2464, Delphi 6/16, and the *VDS (Visual Display System).* Tie Communications' old products can still be purchased in catalogs and from telephone equipment companies, but it is now Nitsuko.

Tie Line A tie trunk that is dedicated to one phone. Tie trunks are telephone lines that connect PBX systems together.

Tie Trunk A telephone line that connects two *PBX (Private Branch Exchange)* telephone systems together so that calls can be transferred between them.

Tight Jacket Buffer The alternative to loose tube buffer in fiber-optic cable manufacturing. Tight jacket buffer cable has insulation around the cable that is tight against the fiber optic. Loose tube buffer allows the fiber to move freely within the cable.

T Interface An ISDN-compatible digital interface. For example, a T interface fits between an ISDN telephone and an ISDN line.

Time-Division Multiple Access (TDMA) See *TDMA.*

Time-Division Multiplex (TDM) The process of encoding two or more digital signals or channels onto one through timesharing the media (wire,

Individual 4 channels to multiplexed. The dotted vertical lines are timing segments.This multiplexed output signal is four times faster than a single input signal.

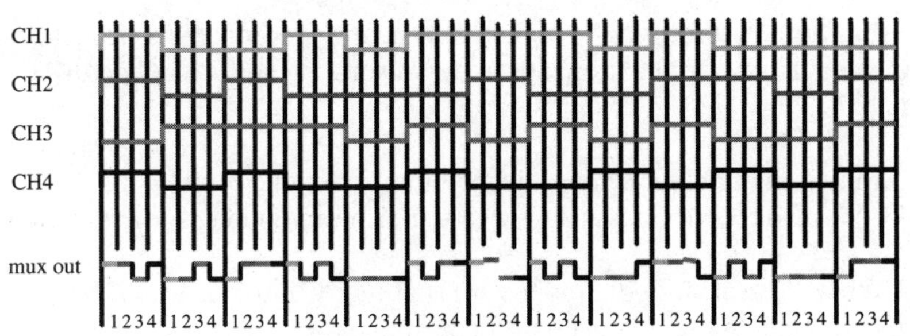

The 4 channel multiplexed output.

Time-Division Reflectometer (TDR)

air, fiber, etc.). The reason that we multiplex channels together in communications is because it saves money. When we use all of the wires in a cable and need more, it costs less to add electronics on the ends of a cable than to install a new one (imagine the expense from LA to NY). A T1 encodes 24 channels into one by using frequency-division multiplexing. In a simpler explanation, a T1 makes it possible to place 24 lines that once needed 24 lines on only two pairs. When a group of signals are multiplexed together, they are all sampled at a high rate of speed, faster than the combined speed of all the channels being multiplexed. The following diagram illustrates the concept of *Frequency-Division Multiplexing (FDM)*.

Time-Domain Reflectometer (TDR) A testing device that measures the distance of a copper twisted pair. The Dynatel 965T is a popular TDR. The TDR works by transmitting a signal down a copper twisted pair, then it waits for a reflection to come back. When the reflection returns to the device, the time difference is used to calculate the distance that the signal traveled. For a photo of a TDR, see *965T*.

Time Out 1. Refers to a "system time-out" test result on an *MLT (Mechanized Loop Test)* from a telephone company central office. When the MLT system is busy, it holds the test request in queue for several minutes until the testing system is available. If it does not become available, then the MLT system returns a "system time-out" result to the user/tester, meaning that the line was not tested. 2. A condition of a central-office line or trunk that occurs when a telephone is left off hook or the pair (facility), it is on is shorted. The central office sends a message that says "please hang-up and try your call again" followed by some harsh, loud beeping tones (also called an *off-hook indicator/warning tone*). If the line is not returned to a "on-hook" state, the central-office equipment will automatically time-out the line. When the line is timed-out, the –52-V battery is disconnected to save power. The central-office equipment will restore the –52-V battery voltage when the ringer is placed back on hook or the short is repaired.

Time Sharing A reference to the shared use of a computer's processing power by many applications or many users on a *LAN (Local-Area Network)*.

Time Slot A channel on a time-division multiplexed circuit (a T1, for example) is also referred to as a *time slot*. For more details, see *Multiplex*.

Tinsel Wire The stranded copper wire that many base cords and handset cords are made with. It is used in these applications because of its ability to withstand lots of bending without breaking.

Tip 1. The positive side of a two-wire telephone circuit, which is supposed to be terminated to black (top position). The other side of a two-wire telephone circuit (or the other wire) is called *ring* and it is designated red (bottom position) when terminated (connected to something). 2. The "tip" of a phono-type plug, which is used for signal patching and for consumer headphones. The other part of a phono plug is the "ring," which is the lower part of the phono plug.

Tip and Ring The more electrically positive side of a *POTS (Plain Old Telephone Service)* telephone line (0 V) is tip. It is designated internationally as black, but in the U.S., it is often designated green. It's counterpart is ring (the more negative side, 52 V), which is designated red internationally and in the U.S. When tip and ring are terminated on a connecting block, tip usually goes on top (left), and ring usually goes on the right (bottom).

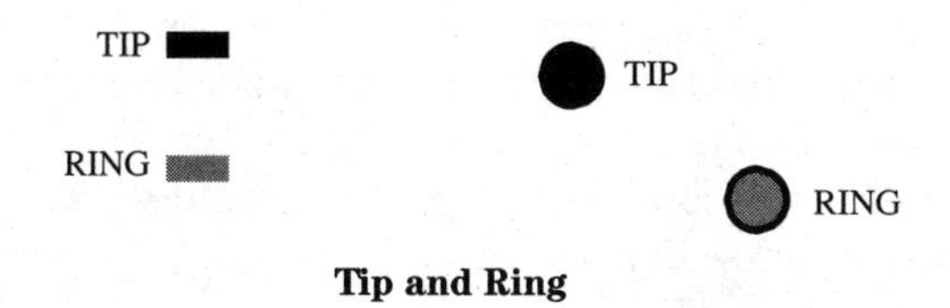

Tip and Ring

Token The bit message in a token-ring network that is passed from one node (computer or other device) to the next that grants the right to transmit to the other nodes on the ring architecture.

Token Bus The physical ring of wire or fiber optic that interconnects computers and other devices (printers) on a token-ring physical *LAN (Local Area Network)* topology. The "physical topology" is the way that the LAN is wired (ring or star) and the "logical topology" is the way that the devices communicate (the protocol).

Token Passing A reference to a token-ring logical topology. The token-ring topology is a type of *LAN (Local-Area Network)*. What makes a token-ring network unique from other topologies is that each device on the network takes turns using the ring (the wiring, or fiber) in an organized, systematic way. The organization is achieved by having a "virtual token" that the devices hand off to each other. While a device possesses the token, it is allowed to transmit over the network. The other popular type of network is called *ethernet*, which includes 10-BaseT, Thicknet, Thinnet, and 100-Base-T (the newer). In ethernet protocols, there is no "token" and no taking turns. Devices simply listen to the network; if it is clear, they transmit. The problem with this is that many devices attempt to communicate at close to the same time. The transmitted data collides on the network, becomes corrupted, and must be retransmitted. Ethernet is referred to as a *contention-based protocol* because the devices (computers) are always contending for the use of the network.

Token Ring A physical and logical *LAN (Local-Area Network)* topology. The physical topology is the way that the computers and other devices are wired together. For a token-ring physical topology, the devices are wired in a ring. The logical topology is the protocol, the rules for communication. In a ring logical topology, each device listens to everything transmitted on the network and only processes what is intended for it to receive (devices know what is meant for them by reading the address information). The devices connected to the network only transmit when they have possession of the *token*. The token is a message that each of the computers pass to each other that allows them to transmit data. This organized method of network utilization prevents more than one device from communicating at one time. If two devices try to communicate at the same time, the data that they both send becomes corrupted. We can't talk to two people at the same time because the words we hear become mixed up and we become confused. The same goes for computers on Local Area Networks. For a diagram, see *Ring*. For more information, see *Token Passing*.

Toll Restriction A *PBX (Private Branch Exchange)* and key telephone system feature that enables an administrator to make a telephone extension unable to make long-distance (toll) telephone calls. Toll restriction can be made to restrict the dialing of a specific area code, a specific group of area codes, or all area codes.

Tone Generator The part of a *PBX (Private Branch Exchange)* system that creates dial tone. The tone generator in a telephone system is usually an individual circuit card, with its own slot in the system cabinet.

Tone Probe Also called a *banana*. Telephone line installers use test tones (also called *installer's tones*) to place an *RF (Radio Frequency)* signal on a pair of wires so that they can locate that pair on the other end of a feed.

Topology The two types of topology are physical and logical. The logical topology defines the way that a LAN communicates. The physical topology defines the way that a LAN is physically wired. For example, even though an ethernet network might be physically wired into the formation of a star, it really works as though it were a bus. The wire is just physically laid out and connected differently, and the electronics are a little different.

Touch Tone (DTMF, Dual-Tone Multiple Frequency The tones that you hear when you dial a single-line push-button phone. The tones are a mixture of two frequencies. For a diagram of DTMF frequencies, see *Dual-Tone Multiple Frequency*.

T PAD (Terminal Packet Assembler/Dissembler) A PAD that is specifically located on the host end of a communications link. Even though PADs are the same on each end, sometimes technicians refer to them in a specific manner. The T PAD is a device that is located at the terminal or user end (as opposed to the host end) of a virtual communications link in a frame protocol environment that reassembles and disassembles large files of data. The *HPAD (Host Packet Assembler Dissembler)* also adds and removes address, envelope, and HDLC information.

Trace Also called a *trap*. When the telephone company tracks the calls made to a customer's telephone number to catch a malicious caller. The standard for malicious call trace is to dial *57 after receiving a malicious call. The telephone company will charge anywhere from $1 to $4 to investigate the call. By using the service, you agree to press charges. The only disturbing thing is that you never find out who made the calls until the court hearing.

Trac Splice Closure An aerial splice closure designed to resist condensation. Trac Splice Closure is a registered trademark of RayChem.

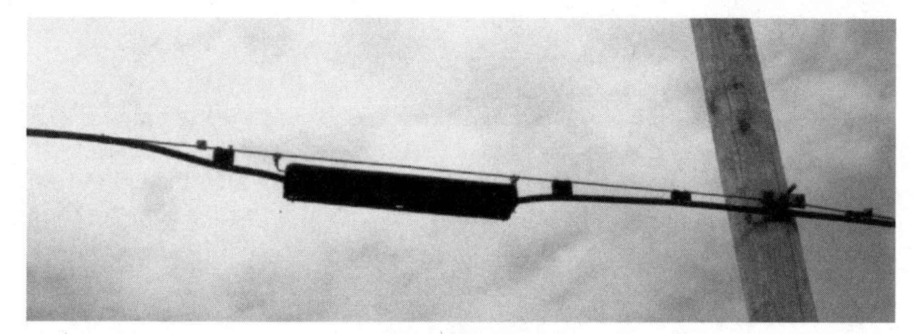

Trac Splice Closure

Traffic A measure of the amount of call attempts and active calls on a telephone switch. Traffic is measured in centum call seconds (*CCS*, one phone call for one second) or Erlangs. Many larger *PBX (Private Branch Exchange)* telephone systems and central-office switches now have *CTI (Computer Telephony Integration)* applications that will calculate traffic, CPU % utilization, busy hours, and other useful information.

Traffic Engineering The process of figuring out how much equipment and what equipment will be needed and how to allocate the resources of that equipment to prevent call blocking, or keep call blocking to a minimum. Telephone switches and *PBX (Private Branch Exchange)* systems are engineered according to the "busy hour" of the network, which is the time when the network has the most traffic. The busy hour can be for the day, month, or year. Traffic is measured in *centum call seconds* (*CCS*, one phone call for one second), or Erlangs. Many larger *PBX (Private Branch Exchange)* telephone systems and central-office switches now have *CTI (Computer Telephony Integration)* applications that will calculate traffic, CPU % utilization, busy hours, and other useful information.

Transceiver A transmitter and receiver built into the same device. A CB radio is a type of transceiver.

Transducer An electronic component that converts one form of energy to electrical energy or vise-versa. Some examples of transducers are:

• Photocell	Light to electricity
• Speaker	Electricity to sound
• Microphone	Sound to electricity
• Light bulb	Electricity to light
• Coil winding/motor	Electricity to motion
• Coil winding/generator	Motion to electricity

Transfer A feature of *PBX (Private Branch Exchange)* telephone systems and key telephone Systems. The transfer feature allows users to send a phone call to another extension by pressing a "transfer" button while on the call, entering the extension they want to transfer to, then pressing the transfer key again (this is true for Northern Telecom Systems).

Transfer Rate How fast a data transmission can send data in *Bits Per Second (bps, b/s)*, e.g., T1 has a transfer rate of 1.544Mb/s including payload and overhead.

Transformer An electronic or electrical component used to "step-up" or "step-down" AC voltage. A transformer wired in a fashion to receive an

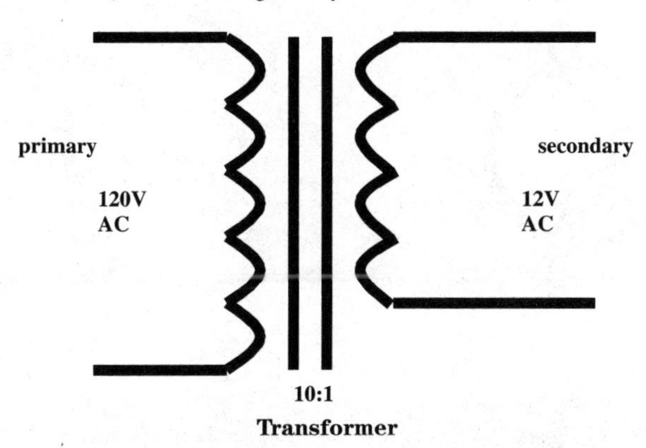

AC voltage on its primary winding and increase that voltage through electromagnetic induction into the secondary winding. Increasing voltage through a transformer does the opposite for current. Reversing the way that a transformer is wired changes it from a step-up to a step-down transformer and vice-versa. The amount the voltage is stepped up or down is equal to the ratio of wire windings of the primary and secondary. The diagram shows a step-down transformer that has a 10:1 winding ratio.

Transit Delay The time required for a transmission to get from one point on a network to another.

Transistor A device that is used as a switch (for logic) or amplifier (signal). The great thing about transistors is that they use far less power than vacuum tubes, they are much smaller, they are much faster, and they cost less. The only disadvantage of transistors to tubes is that transistors cannot amplify odd-order harmonics (natural sounds/music) and they are much more susceptible to cosmic radiation (outer-space/nuclear-fallout applications).

Translating Bridge Also called a *protocol converter*. A device that is usually in the form of a PC that is loaded with software that converts protocols, such as ethernet and token ring. The PC has a network card installed for each of the different network connections.

Translator A device that receives a transmission, converts it, and retransmits it on a different format or frequency. Translators are used to rebroadcast radio programming. Translators in data networks are called *translating bridges*.

SCHEMATIC SYMBOLS FOR POPULAR TRANSISTORS

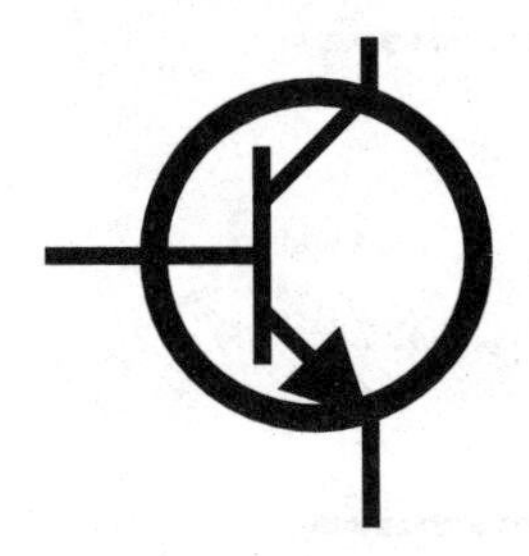

Bipolar NPN

DE MOSFET (CMOS) N-Channel

Transistor

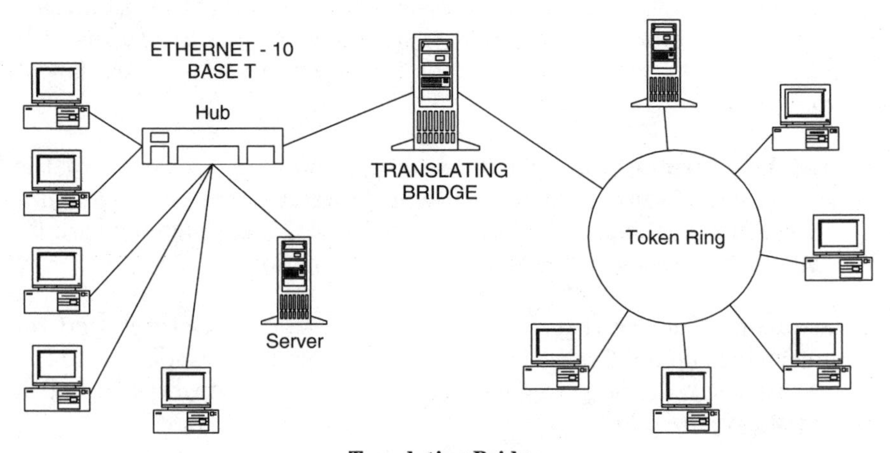

Translating Bridge

Translations The part of switching or switch set-up that converts signaling messages from other switches. Translations are very important and are probably the most crucial part of the switch set up. When a telephone call is handed from one switch to another or from one central office to another, digits (addressing information) are sent with it. Each switch has its own definitions for number set up and configuration. The switch converts these numbers for every call processed and the process is called *translating*. If you are very interested in translations, several good books are published by McGraw-Hill on Signaling System 7, and some good classes are offered by Northern Telecom.

Transmission The sending of a signal, analog, digital, or light-wave, across a media.

Transmission Control Protocol (TCP) A data packet protocol (rules for communications between devices) for data exchange between computers, printers, servers, or anything else that is capable of being loaded with the protocol driver. The protocol driver can be loaded in software or embedded in hardware (PROM). It was developed by *ARPA/DARPA (Advanced Research Projects Agency/Defense Advanced Research Projects Agency)* of the U.S. federal government.

Transmission Protocol The organized processes and rules that communications equipment use to transfer bits and bytes (data), or just an analog signal. The many communications protocols and layers of protocols that carry other protocols (called *protocol stacks*), include ISDN, ethernet, token ring, POTS signaling, DS1, ATM, frame relay, and SONET. *AM (Amplitude Modulation)* and *FM (Frequency Modulation)* could be thought of as analog protocols.

Transmitter, Radio A radio transmitter emits an electromagnetic field. The electromagnetic radiation is simply an antenna being made to change its magnetic field at the rate of a carrier frequency. The magnetic field then traverses through the farthest reach of the magnetic field, which could be many miles. The distance that the signal will reach is determined by the transmitting power, weather, sunspot cycle, time of day, ionospheric conditions, and surrounding terrain. What makes the radio signal carry a voice or music is called *modulation*.

Transparent Network A network connection that is made through more than one network, but appears to be only one network to the end user. Many *Local-Area Networks (LANs)* within companies are connected to the Internet so that users can access the Internet or Internet e-mail by clicking on a desktop icon. This is a form of transparent networking. A nontransparent version of this would be to open your communications application, dial out on your modem, enter your e-mail address, then read your e-mail.

Transparent Routing To send data across more than one network without any additional coaxing from the user. The network does all of the protocol conversion and gives the illusion that there is only one network.

Transponder An electronic device that receives a radio signal, then emits a response to the signal that it has received.

Transport Layer A layer in a communications protocol model. In general, the transport layer performs the function of error correction and the direction of data flow (transmit/receive). The latest guideline for communications protocols is the *OSI (Open Systems Interconnect)*. It is the best model so far because all of the layers (functions) work independently of each other. For a diagram of the OSI, DNA, and SNA function layers, see *Open Systems Interconnection*. For a functional/conceptual diagram of the OSI layers, see *OSI Standards*.

Transport Medium A reference to the physical component of what a transmission signal is being sent over; it can be fiber optic, coax, copper twisted pair, or air.

Transport Protocol A reference to how data should be presented to a device or layer in a transmission process. In general telecommunications transport, protocols include SONET, T1, STS-1, and OC-N. Transport protocols can become very complicated, but if you understand the four listed, you are in good shape.

Trap See *Trace*.

Trellis Coding An error detection method used in high-speed modems.

Tremendously High Frequency (THF) The American standard name for frequencies within the spectrum of 300 to 3000 GHz.

Tributary Circuit A lower-level multiplexed channel that exists within a larger multiplexed channel. For example, a T1 channel riding inside or on an OC-3 SONET (an OC-3 can carry 84 T1 circuits) carrier is considered to be a tributary. This term is frequently used within the realm of SONET networking.

Triode An electronic vacuum tube that works in similar applications as a transistor. Like a diode vacuum tube, it has a heater filament, anode, and cathode; the triode has a third added element, called a *grid*.

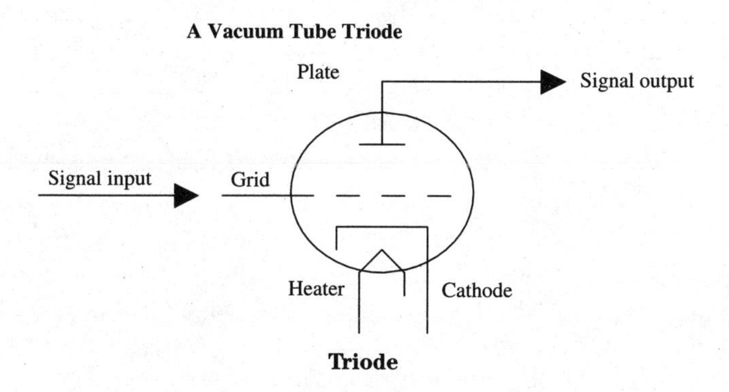

Triode

TRL (Transistor/Resistor Logic) Logic circuits that consist of bipolar transistors, rather than *CMOS (Complimentary Metal-Oxide Semiconductor)* transistors. TRL is used in environments with lots of static electricity.

Trunk A trunk is a line that comes from a central office. The type of trunk determines which type of signaling the line requires to work. A loop-start trunk is a two-wire central-office trunk or dial-tone line that recognizes an "off hook" situation when a telephone switch hook puts a 1000-Ω short across tip and ring when the handset is lifted. This is the most common type of line. It is also called a *POTS line* or *plain-service line*. Other types of trunks are ground start and *E&M Trunks (Ear and Mouth)*, T1 with in-band signaling, ISDN PRI, and ISDN BRI.

Trunk Group A group of telephone lines that connect a *PBX (Private Branch Exchange)* or key telephone system to the phone company, and are used for a specific applications, such as incoming customer-service lines, sales lines, or information lines. A specific group of trunks can also be configured for outgoing calls only.

Truth Table A diagram used to portray a logic statement. Logic is a mathematical process first developed by the Irish mathematician George Boole in the 1850s. The premises of logic is to tell if a certain statement is true or false. An example is "the light is ON." This statement can only be true or false. Logic couples this statement with others like "the switch

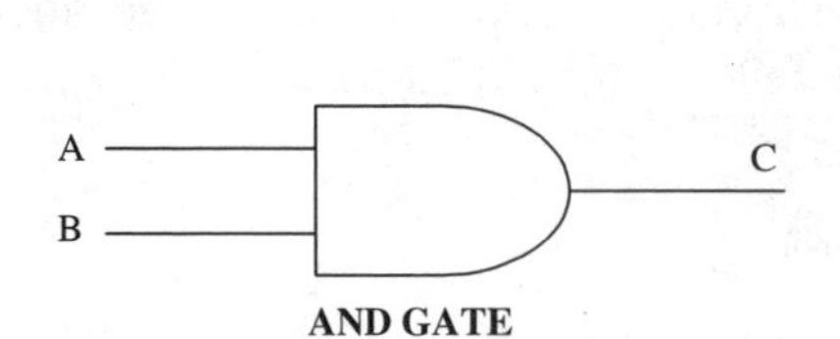

TRUTH TABLE

A	B	C
0	0	0
0	1	0
1	0	0
1	1	1

Truth Table

is ON, the power is ON; therefore, the light must be ON." If the switch is OFF and the power is ON, then the light is OFF." "If the switch is ON and the power is OFF, then the light is OFF. These statements depict the truth table for an AND electronic logic gate, which is a primary building block of microprocessors. In the table depicted, the light switch would be A, the power would be B, and the light bulb would be C. Truth tables are written in ones and zeros, rather than ONs and OFFs. The science of this math is called *Boolean Algebra*. It is a book in itself and is usually covered well in textbooks that cover digital electronics.

T Span Another name for a T1 or T3. See *T1*.

TTL (Transistor-Transistor Logic) Digital devices that consist of many bipolar transistors (the original transistor as we know it). TTL is not as static sensitive as its newer and more commonly used competitor, *CMOS (Complementary Metal-Oxide Semiconductor)* transistor, but it uses much more power.

TTR (Touch-Tone Receiver) An electronic device that converts *DTMF (Dual Tone Multi-Frequency)* tones dialed by telephones and other communications equipment to digits.

TTY (Teletypewriter) See *Teletypewriter*.

Tube (Vacuum Tube) An active electronic device that is literally a tube, with no air or other gasses in it. Tubes were the predecessors to transistors and other "solid-state" devices. Tubes have four main parts: the plate, the cathode, the filament (or heater), and the grid. The tube works when electricity flows through the tube, from the cathode to the plate. The electricity flows when the filament is heated, which is actually the source of the electron flow. The signal to be amplified is fed to the grid, which manipulates the amount of current flowing through the tube. The resultant is an amplified output on the plate. Tubes are still used in many applications. Today's vacuum tube applications include high-power applications (such as radio transmitters), high-end consumer audio equipment (where it is necessary to have odd-order harmonics accurately reproduced, transistors do not amplify odd-order harmonics, tubes do), and high radiation-risk applications (such as outer space because transistors are sensitive to cosmic radiation). For a schematic symbol of a vacuum tube, see *Triode*.

Turnkey A method of network installation. Also referred to as *EF&I (Engineer, Furnish, and Install)*. Many companies have equipment or entire networks built in this manner, so that when it is ready, all they need to do is use it.

TV (Television) See Television.

Twisted Pair Communications wiring consists of 19 to 26-AWG solid and insulated wires. Twisted-pair wire consists of "pairs" of color-coded wires. Common sizes of twisted-pair wire are 2 pair, 3 pair, 4 pair, 25 pair, 50 pair, and 100 pair. Twisted-pair wire is commonly used for telephone and computer networks. It comes in ratings of CAT 3 (for voice), CAT 4 (voice and 10-base-T), and CAT 5 (for 100-base-T and token ring).

Two-Wire Circuit A circuit that utilizes two wires to work. A plain telephone line like, such as the one in your house, is a two-wire circuit, and the two wires are called a *subscriber loop*.

Type 1 Cable 3 IBM two-pair 22 AWG shielded data cable designed for switched token ring. Characteristic impedance is 150 ohms. It is usually terminated to genderless IBM data connectors.

Type 2 Cable 3 IBM two-pair 22 AWG shielded data cable designed for switched token ring. Characteristic impedance is 150 ohms. It is usually terminated to genderless IBM data connectors.

Type 3 Cable 3 IBM 24-AWG nonshielded four-pair data cable designed for switched token ring. Characteristic impedance is 150 ohms. It is usually terminated to genderless IBM data connectors.

Type 4 Cable Presently, no type-4 cable exists.

Type 5 Cable 3 IBM 2 optical fiber-data cable designed for switched token ring. Characteristic impedance is 850 ohm. It is usually terminated to genderless IBM data connectors.

Type 6 Cable 3 IBM two-pair 26-AWG shielded data cable designed for switched token ring. Characteristic impedance is 105 ohms. It is usually terminated to genderless IBM data connectors.

Type 7 Cable Presently, no type-7 cable exists.

Type 8 Cable 3 IBM 2-pair flat 23-AWG shielded data cable designed for switched token ring. Characteristic impedance is 150 ohms. It is usually terminated to genderless IBM data connectors.

Type 9 Cable 3 IBM 2-pair 26-AWG shielded data cable designed for switched token ring. Characteristic impedance is 150 ohms. It is usually terminated to genderless IBM data connectors.

U Interface A two-wire (one pair) *ISDN (Integrated Services Digital Network) BRI (Basic Rate Interface)* telephone line. This popular type of digital telephone line has two B (bearer channels) and one D channel. For more information, see *ISDN*.

UART (Universal Asynchronous Receiver Transmitter) The part of a computer's serial communications port (most PCs have two) that converts the parallel data from the data bus into serial data to be sent to a device connected to the port, like a modem.

UCD (Uniform Call Distributor) A less-expensive and less-smart version of an ACD (Automatic Call Distributor). The UCD receives incoming calls and equally distributes them among agents in a call center. For more details, see *ACD*.

UHF (Ultra-High Frequency) The part of the radio-frequency spectrum that ranges from 300 MHz to 3 GHz. It is used for broadcast TV and other radio communications.

UHF/VHF Splitter Used to make a junction point or split a signal so that it will travel down multiple paths over coax. It is also called an *RF splitter* or *splitter*. For a photo, see *RF splitter*.

UL (Underwriters Laboratories) A private company that certifies manufacturer's products for safety.

Ultra High Frequency (UHF) The part of the radio-frequency spectrum that ranges from 300 MHz to 3 GHz. It is used for broadcast TV and other radio communications.

Under-Floor Raceways Wire ducts that are designed to be used under raised-floor systems, like those in computer rooms.

UNI (User Network Interface) The device that combines different types of communications protocols onto one broadband channel. A UNI can be in the form of a front-end processor or a communications server.

Uniform Call Distributor (UCD) A less expensive and less smart version of an *ACD (Automatic Call Distributor)*. The UCD receives incoming calls and equally distributes them among agents in a call center. For more details, see *ACD*.

Uninteruptable Power Supply (UPS) A battery back-up system. When the power goes out, the UPS converts the DC battery power to AC power to run the system.

Unity Gain Gain (another word for amplification) in an electronic circuit that is equal to 1. In oscillator circuits, and some other radio receiver circuits that have a feed back (generate their own input), the amplification between the output and the input must be one. If the gain is not one, the signal will either diminish to nothing or be amplified beyond the saturation point of the active component (e.g., transistor) in the circuit.

Universal Asynchronous Receiver Transmitter (UART) The part of a computer's serial communications port (most PCs have two) that converts the parallel data from the data bus into serial data to be sent to a device connected to the port, like a modem.

UNIX An operating system similar to MS Windows, only it is designed to operate on *RISC (Reduced Instruction Set Computers)*, like those made by Sun Microsystems (It can work on Intel Pentium and other 586-based PCs, too). UNIX performs all the functions that MS Windows does, only the icons look different and have different names. UNIX and RISC computers are used in conjunction with *CTI (Computer Telephony Integration)* applications, such as front-end processors and integrated voice-response systems, because of their ability to execute small instructions and tasks very quickly, and on a real-time basis.

Unlisted Number A phone number that is not listed in a telephone book, but can be found by calling directory assistance and giving the person's name. The other type of private listing is an unpublished number, which is not listed in a telephone book and cannot be found by calling directory assistance.

Unpublished Number See *Unlisted Number*.

Unshielded Twisted Pair (UTP) Twisted-pair wiring that is unshielded, meaning it does not have a foil wrapping around the group of conductors within the jacket. Unshielded twisted pair is the most commonly used wiring for voice and data networks. Twisted-pair wire consists of pairs of color-coded wires. Common sizes of twisted-pair wire are 2 pair, 3 pair, 4 pair, 25 pair, 50 pair, and 100 pair. Twisted-pair wire is commonly used for telephone and computer networks. It comes in ratings of CAT 3 (for voice), CAT 4 (voice and 10-base-T), and CAT 5 (for 100-base-T and token ring). See also *Plenum* and *PVC*.

UPS (Uninteruptable Power Supply) A battery back-up system. When the power goes out, the UPS converts the DC battery power to AC power to run the system.

US The ASCII control-code abbreviation for unit separator. The binary code is 1111001 and the hex is F1.

USART (Universal Synchronous/Asynchronous Receiver/Transmitter) The part of a computer's serial communications port (most PCs have two) that converts the parallel data from the data bus into serial data to be sent to a device connected to the port, like a modem.

User Network Interface The device that combines different types of communications protocols onto one broadband channel. A UNI can come in the form of a front-end processor or a communications server.

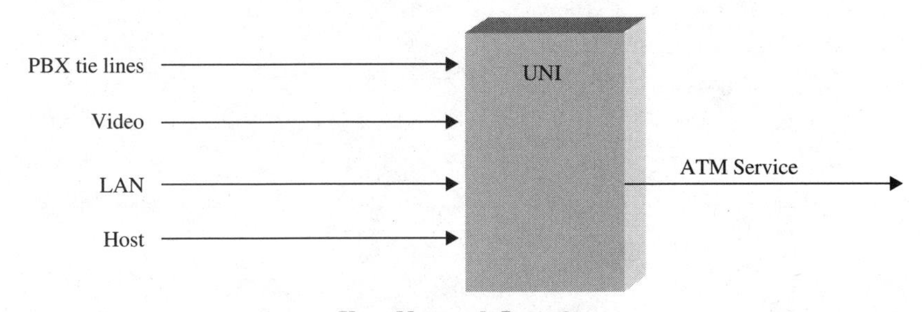

User Network Interface

USOC (Universal Service Order Code) A code that defines different equipment and services within Regional Bell Operating Companies

(RBOCs). Different RBOCs have different meanings for each code, as far as services, but the USOC codes for equipment has remained the same since the old Bell System (pre-1984).

USRT (Universal Synchronous Receiver Transmitter) The part of a computer's serial communications port (most PCs have two) that converts the parallel data from the data bus into serial data to be sent to a device connected to the port, like a modem.

USWest A *Regional Bell Operating Company (RBOC)*. Their territory includes the States of Washington, Oregon, Idaho, Montana, Wyoming, North Dakota, South Dakota, Minnesota, Iowa, Nebraska, Utah, Colorado, Arizona, and New Mexico.

UTP (Unshielded Twisted Pair) Twisted-pair wiring that is unshielded, meaning it does not have a foil wrapping around the group of conductors within the jacket. Unshielded twisted pair is the most commonly used wiring for voice and data networks. Twisted-pair wire consists of "pairs" of color-coded wires. Common sizes of twisted-pair wire are 2 pair, 3 pair, 4 pair, 25 pair, 50 pair, and 100 pair. Twisted-pair wire is commonly used for telephone and computer networks. It comes in ratings of CAT 3 (for voice), CAT 4 (voice and 10-base-T), and CAT 5 (for 100-base-T and token ring). See also *Plenum* and *PVC*.

V

V.13 An older modem standard that enabled full-duplex modems to act as half-duplex modems when necessary.

V.14 An older modem protocol that enables modems equipped with error correction to talk to modems that don't have error correction.

V.17 A data-transfer protocol for one-way facsimile (fax machines).

V.21 A modem protocol that is now used more outside the U.S. and Canada than within. Modems that are made in the U.S. and Canada are capable of receiving or sending transmissions to these modems.

V.22 Bell 212A An older modem protocol for transfer rates of 1200 bp/s.

V.22 bis An older modem protocol that allows for automatic increase/decrease in speed of the V.22 protocol to 2400 bp/s. The "bis" is an indicator of the ITU-T that means "second edition, or second in family)".

V.23 An older modem protocol that provides up to 1200 bp/s in a forward channel and 75 bp/s in a reverse or backward channel.

V.24 A popular serial data protocol used to transfer data between electronic equipment. V.24 incorporates separate conductors for transmit and receive.

V.26 An older four-wire circuit protocol for 1200 baud modems. The two pairs (four wires) were transmit and receive.

V.27 An older modem standard that was capable of being configured for two-wire half-duplex operation or four-wire full-duplex operation.

V.29 A modem standard for transmission rates of 9.6Kbp/s on private lines (no dialing, just wire).

V.32 A newer modem standard that has a line rate of 9600 baud, and a compressed data-transfer rate of 14.4 Kb/s.

V.34 A modem standard for data transmission that compresses data and transfers it at speeds up to 19.2 Kp/s. It is also referred to as *Vfast*.

V.34 bis A newer version of the V.34 standard that incorporates leaner compression techniques to achieve speeds that extend beyond the original 19.2 Kb/s to 31.2 Kb/s and 33.6 Kb/s.

V.35 A data-transfer protocol that is no longer published by the ITU-T.

V.42 A modem protocol that includes many revisions as a family. V.42 modems are backward compatible with other previous V.XX standards. The V.42 protocol and compression standards combined can compress and transfer data at speeds that range from 1200 bp/s to 57.6Kb/s and up.

V.54 The ITU-T standard for loop-back test capability in modems. A loop-back test is when you send data to a far-end modem, and the modem simply sends your own data back to you. If you receive your own data in a loop back, then you know that your modem and the line are OK.

V.110 *ISDN (Integrated Services Digital Network)* two-wire *BRI (Basic Rate Interface)* protocol standard for data transfer over the B (Bearer) channel.

V.120 *ISDN (Integrated Services Digital Network)* two-wire *BRI (Basic Rate Interface)* protocol standard for data transfer over the B (Bearer) channel.

V&H Coordinates A method for calculating airline mileage between cities. For a listing of cities and their respective V&H coordinates, see *Airline Mileage*.

V (Volt) The basic unit of electric force, pressure, or *EMF (Electromotive Force)*. In Ohm's Law formulas, EMF is used as the designator for voltage, *E*. The two main components of electricity are current (amperage) and voltage. Voltage is also referred to as the potential energy between two points in a circuit.

V Fast Also called V.34. A modem standard/compression protocol that transfers data at speeds up to 19.2Kp/s.

V. Standards Standards for telecommunications that are recommended by the *ITU-T (International Telecommunications Union Telecommunications)*. The ITU-T is an international organization comprised of members of almost every country in the world. The standards are recommended so that manufacturers of communications equipment have guidelines to make equipment that is compatible.

Vacant Code A central-office prefix (the first three digits in a seven-digit phone number) or area code that is not in use or not assigned yet.

Vacuum Tube An active electronic device that is literally a tube, with no air or other gasses in it. Tubes were the predecessors to transistors and other solid-state devices. Tubes have four main parts: the plate, the cathode, the filament (or heater), and the grid. The way that the tube works is that electricity flows through the tube from the cathode to the plate. The electricity flows when the filament is heated. This is actually the source of the electron flow. The signal to be amplified is fed to the grid, which manipulates the amount of current flowing through the tube. The resultant is an amplified output on the plate. Tubes are still used in many applications. Today's vacuum tube applications include: high-power applications (such as radio transmitters), high-end consumer audio equipment where it is necessary to have odd-order harmonics accurately reproduced (transistors do not amplify odd-order harmonics, but tubes do), and in applications in outer space where radiation damage is a high risk (transistors are sensitive to cosmic radiation). For a schematic symbol of a triode vacuum tube, see *Triode*.

Vampire Tap A *LAN (Local Area Network)* connector that crimps onto the outside of a coax cable. The tap connector is equipped with teeth that penetrate through the jacket, shield and dielectric of the coax to make a solid connection with the inner conductor. The inner conductor of a coax is used as the bus in ethernet networks.

Variable Resistor Also called a *potentiometer (pot)*. A resistor that is usually made of carbon film and has a control knob or slide connected to it. Many electronic control knobs are connected to variable resistors.

Schematic Symbol for a Variable Resistor/Potentiometer

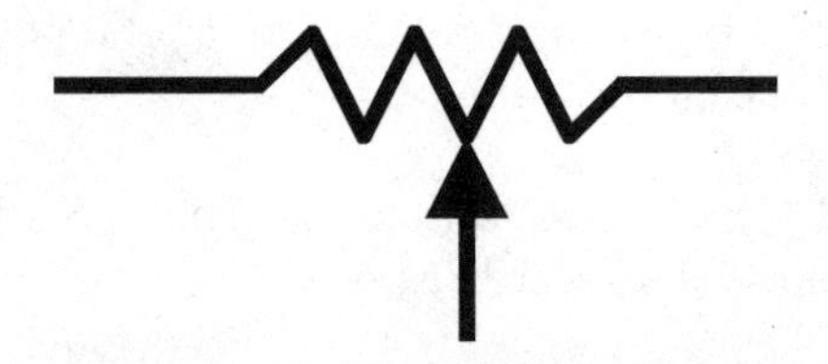

Variable Resistor

VDT (Video Display Terminal) Another name for a monochrome or color monitor used as an output device for computers.

Vertical and Horizontal Coordinates See *Airline Mileage*.

Vertical Redundancy Check (VRC) A part of *Longitudinal Redundancy Checking (LRC)*. A method of checking for errors in communications/modem transmissions by combining vertical error checking and longitudinal error checking. A transmission device sends data in bytes, which are logically stacked on top of each other. The stack forms a block. The last bit of each line is used to form a check sequence. LRC is about 85% accurate in detecting and re-transmitting blocks that contain errors. The newer method of error checking is *CRC (Cyclic Redundancy Checking)*.

LRC BLOCK AND BITSTREAM DIAGRAM

DATA BLOCK

Even parity

byte 1	1	0	1	1	0	1	1	1	0
byte 2	1	0	0	1	1	0	1	0	0
byte 3	0	1	1	0	0	0	0	1	1
byte 4	1	1	1	1	1	1	1	1	0
byte 5	0	0	0	0	0	0	0	0	0
byte 6	1	1	1	1	0	0	0	0	0
byte 7	0	0	1	1	0	0	1	1	0
byte 8	1	0	1	0	1	0	1	1	1
byte 10	1	1	0	0	1	0	1	1	

Longitudinal Parity Sequence (byte 9) SHADED
Vertical Parity sequence (byte 10) bottom row

LRC BIT STREAM

byte 1	Byte 2	byte 3	byte 4	byte 5	byte 6	byte 7	byte 8	byte 9	Byte 10
10110111	10011010	01100001	11111111	00000000	11110000	00110011	10101011	100001	11001011

Vertical Redundancy Check (VRC)

Very High Frequency (VHF) Radio frequencies in the range of 30 to 300 MHz. VHF is utilized for broadcast television and FM radio.

Very Large Scale Integration (VLSI) A classification of *Integrated Circuit (IC)*. Microchip/ICs are classified as *SSI (Small-Scale Integration)*, *MSI (Medium-Scale Integration)*, and *LSI (Large-Scale Integration)*. SSI ICs contain 12 or fewer devices, such as logic gates. MSI ICs contain 13 to 99 devices, and VLSI ICs contain 100 or more devices. The typical CPU, such as a Pentium (Intel trademark) microprocessor, contains hundreds of thousands of devices. For a photo of a VLSI device or microchip, see *Microchip*.

Very Low Frequency (VLF) Radio frequencies in the range of 3 kHz to 300 kHz.

VGPL (Voice Grade Private Line) There are two grades of private line service, voice grade and data grade. Both can be analog (via modem) or digital. A private line is a communications path from one point to another that is not switched, or dial-up, it is a physical pair of wires or a virtual connection through a transport network (such as SONET or T1 carrier).

VHF (Very High Frequency) Radio frequencies in the range of 30 MHz to 300 MHz. VHF is utilized for broadcast television and FM radio.

Virtual Colocation The two types of colocations (also spelled *collocation*) are virtual and physical. A colocation is an interconnection agreement and a physical place where telephone companies hand-off calls and services to each other. This is usually done between a *CLEC (Competitive Local Exchange Carrier)* and an *RBOC (Regional Bell Operating Company)*. A virtual colocation is when telephone company A (the CLEC) requests that their phone company's network be connected to telephone company B's (the RBOC's) network. Telephone company B charges company A lots of money. Company B owns, installs and maintains the equipment. To company A the interconnection is virtual, because they never physically do anything to it when and after it is installed. Company B likes this because company A does not get free access to their premises.

Video Conferencing A video phone. A television connected to a telephone line that carries video and audio information. Some new companies specialize in this service. Some even have offices that contain video conferencing equipment that can be "rented" for meetings with people at other far away video-conferencing locations. PictureTel is a popular manufacturer of video-conferencing equipment.

Virtual DN (Virtual Directory Number) 1. Also called a *phantom directory number*. A directory number or extension on a PBX system that is used to attach a voice mailbox to. The virtual DN does not really have a telephone set, but the PBX system thinks it does, so it transfers calls to that DN, which are configured to be forwarded to a voice-mail system. A user of that DN can then dial into the voice-mail system, enter their extension, and receive their messages. 2. DNIS numbers are also referred to as virtual DN'S.

Virtual Tributary (VT) A virtual tributary is a communications channel or circuit that exists within another (larger) multiplexed communications channel. For example, a DS1 within a DS3 is a virtual tributary. Sometimes channels within an AT&T SLC96 carrier system (pair gain system for carrying 96 phone conversations on 8 pairs of wire) are referred to as *virtual pairs*.

VLF (Very Low Frequency) Radio frequencies in the range of 3 to 300 kHz.

Voice Band A reference to the frequency range of a human voice that is transferred on a telephone line. The range of the human voice is about 200 Hz to 12 kHz. The range of the voice transmitted over a POTS phone line is flat (an even reproduction) from 500 Hz to 3,500 Hz (3 kHz in bandwidth).

Voice Grade A reference to a POTS (Plain Old Telephone Service) telephone line, like the ones that are subscribed to by residential telephone company customers. Local telephone companies do not guarantee any transfer rate of data over these lines. So, if you can't transmit data at 28.8 Kb/s with your new modem, the phone company will probably ask you to subscribe to a switched 56K data line (which is more costly than a POTS line).

Voice Mail An answering machine system that integrates with a *PBX (Private Branch Exchange)* or key telephone system. Octel (now a

Voice Mail

part of Lucent Technologies) is a manufacturer of voice-mail systems that are used in business-office applications and in central-office applications for telephone companies to offer voice mail/voice messaging as a service to subscribers. Voice mail can also be purchased as a network interface card with software that runs on a PC. Shown is an Octel standalone voice-mail system.

Voice Recognition A reference to a computer or machines ability to recognize a specific individuals voice, like a fingerprint. This technology is frequently confused with speech recognition, which recognizes words, not distinct voices.

Voice Response Unit (VRU) Better known as *IVR (Interactive Voice Response)*. A telecommunications and data processing technology that interfaces a person to information held in a computer by using a phone line. If you have ever called your bank and entered your account number, a password, and a prompt so that a computerized voice can read back your bank-account balance, then you have used IVR.

Voice Grade Private Line (VGPL) There are two grades of private line service, voice grade and data grade. Both can be analog (via modem) or digital. A private line is a communications path from one point to another that is not switched, or dial-up, it is a physical pair of wires or a virtual connection through a transport network (such as SONET or T1 carrier).

Volt (V) The basic unit of electric force, pressure, or *EMF (Electromotive Force)*. In Ohm's Law formulas, *EMF* is used as the designator for voltage, *E*. The two main components of electricity are current (amperage) and voltage. Voltage is also referred to as the *potential energy* between two points in a circuit.

Voltage A reference to the amount of electric force, pressure, or *EMF (Electromotive Force)*. See *Volt*.

Voltage Drop A reference to the amount of voltage difference from one point in a circuit to another. A voltage drop is usually in reference to one component and is useful to know when troubleshooting electronic circuitry.

Volt Meter A meter used to verify the amount of voltage at a single point in a circuit (referenced to ground) or the amount of voltage potential between two different points in a circuit. Most meters for measuring voltage come in the form of a multimeter, which has a volt, ohm, and current (amp) meter incorporated into its design.

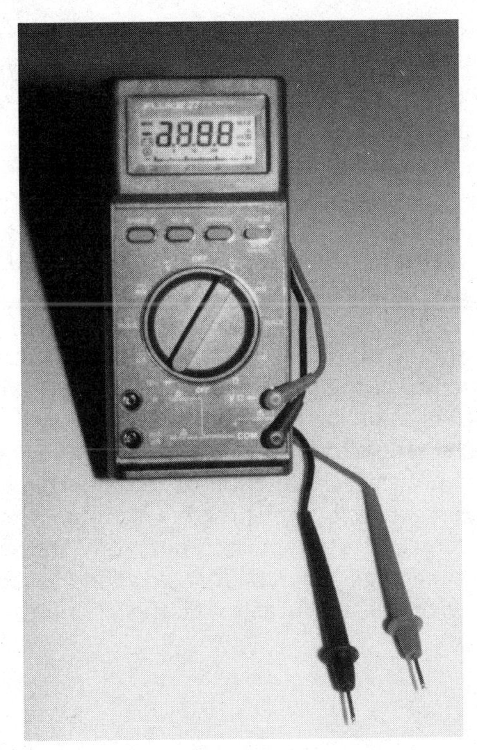

Volt Meter

Volume Unit Meter (VU Meter) See *VU Meter.*

VRC (Vertical Redundancy Check) A part of *Longitudinal Redundancy Checking (LRC)*. A method of checking for errors in communications/modem transmissions by combining vertical error checking and longitudinal error checking. A transmission device sends data in bytes, which are logically stacked on top of each other. The stack forms a block. The last bit of each line is used to form a check sequence. LRC is about 85% accurate in detecting and re-transmitting blocks containing errors. The newer method of error checking is *CRC (Cyclic Redun-dancy Checking)*. For a diagram, see *Vertical Redundancy Check*.

VT 1. Video terminal. 2. Virtual tributary. A virtual tributary is a communications channel or circuit that exists within another (larger) multiplexed communications channel. For example, a DS1 within a DS3 is a virtual tributary. Sometimes channels within an AT&T SLC96 carrier system (a pair-gain system for carrying 96 phone conversations on eight pairs of wire) are referred to as *virtual pairs*.

VT 100 A DEC device that consists of a monitor and a keyboard. It has no memory or programming, so it is called a *dumb terminal*. It is now possible to use a PC to act or communicate as a dumb I/O terminal. To use a PC in this application (such as to plug into a microwave link and boost its power), it must be equipped with terminal-emulation software. The microwave link device has its own microprocessor and only needs a device to communicate with; that device is usually a VT100 terminal. Terminal-emulation software allows your PC to "look" like a VT100 terminal to the microwave radio equipment.

VU Meter (Volume Unit Meter) The meter on tape recorders that moves while recordings are being made. The meter is used as a reference to make adjustments of the recording level (the strength of the signal that is sent to the tape). If the signal fed to the tape is strong enough to push the meter into the red, then the recording will be distorted. If the signal is not strong enough to push the meter, then the signal is not being recorded. Many cassette tape recorders today have LED VU meters. When recording messages for an *IVR (Integrated Voice Response)* system or other device, it is useful to watch the VU meter while recording.

Waffle Splice Closure A cable splice closure that is commonly used in air-pressure applications or shallow trenches, where a strong casing is needed to protect the splice. See also, *Air Pressure*.

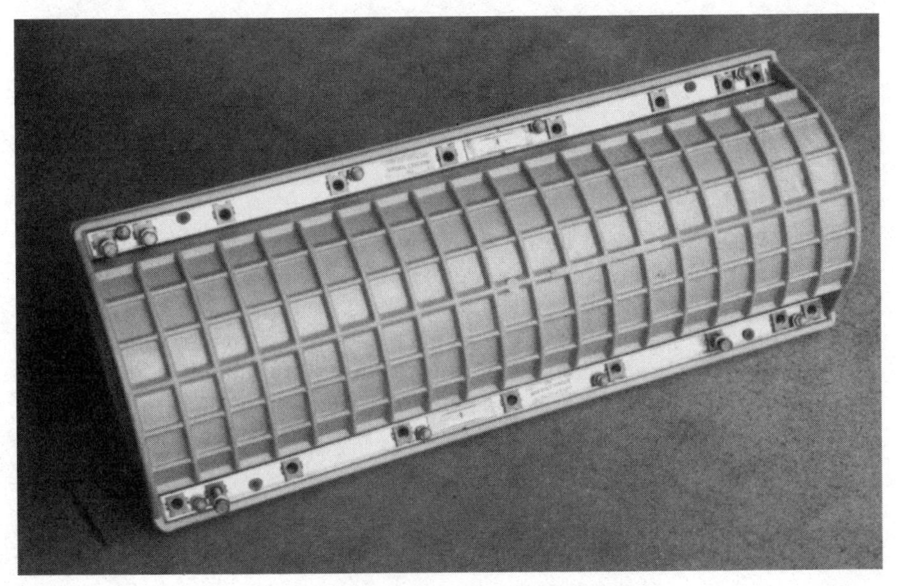

Air Pressure "Waffle" Splice Closure

WAN (Wide-Area Network) A network of computers or computing devices connected by telephone lines that extend beyond an area code's service area. An example of a WAN application is a computer that accesses another computer in another state to access information. Popular ways for computers to connect over long distances are by using a dial-up modem, a frame relay circuit, an ATM circuit, an ISDN circuit, or a 56K leased line. There are advantages and disadvantages to each of these services (which are offered by long-distance telephone companies). The faster and more reliable the service, the more expensive it is. Frame relay is rapidly becoming the most economical WAN protocol for applications that transfer data over long distances frequently.

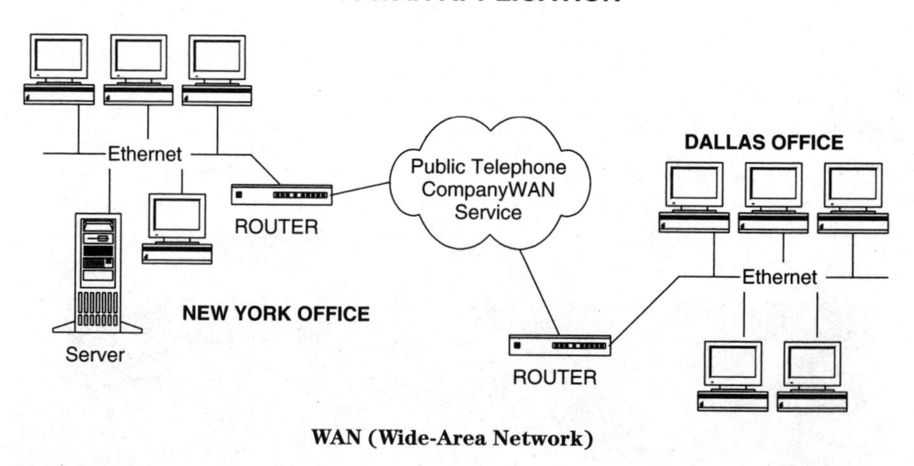

WAN (Wide-Area Network)

WATS (Wide-Area Telephone Service) A toll-free dialing service (800/888 lines) offered by telephone companies. In-WATS lines are priced and set up for incoming only calls, usually calls from a certain area. You can also subscribe to Out-WATS service as well. WATS can be for interstate and intrastate long distance. If you call an 800 number, you are most likely calling an in-WATS service line that a company has set up for customers. The time to start checking into WATS service is when your long distance to or from a specific area exceeds $200.00 per month.

Watt (W) The Unit of electrical power, represented as P in Ohm's Law formulas. Wattage is calculated by $P = I \times E$, where P is power in watts, I is current in amps, and E is voltage in volts. For example, if you have a light bulb that draws 1 amp at 100 volts, it would be a 100-watt light bulb.

The power consumed by the light bulb is radiated as heat and light. Just because a light bulb has a higher wattage rating does not mean that it is brighter, it could be hotter as well. Technically, one watt is equal to one joule per second. Another way to grasp the concept of a watt is to use the comparison that 746 watts is equal to one horsepower.

Wave Guide A device used to direct radio-frequency transmissions or light waves. Wave guides in radio transmitters look like high-tech plumbing. In lightwave applications, they are a small prism or optical fiber.

Wavelength The wavelength of a radio signal in meters is equal to 300,000,000 m/s (the speed of light) divided by the frequency in hertz. For example, the wavelength for an FM radio station's signal if they are at 96.3 MHz on the radio dial is equal to 300,000,000 m/s (96,300,000 Hz = 3.115 meters). It is useful to know wavelength in the design of radio antennas, which are made to be the same length or a fraction of the length of a radio signal's wavelength.

Wavelength Division Multiplexing (WDM) A way of increasing a fiber optic's capacity by using multiple colors of light. Each color of light has its own wavelength (and its own frequency). The electronic equipment on each end of the fiber can distinguish the different signals by their color (frequency/wavelength). In most applications today, each fiber optic in a communications network carries one light signal that is one pure color. In the future, fiber optic will be wavelength-division multiplexed to carry many transmission signals.

WDCS (Wide-Band Digital Cross-Connect System) Another name for *DCS (Digital Cross-Connect System)*.

WDM (Wavelength Division Multiplexing) See *Wavelength Division Multiplexing*.

Wet Circuit A T1 circuit is a wet circuit when its 135V DC battery voltage is present. When the T1 has the battery voltage removed via a *CSU/DSU (Channel Service Unit/Digital Service Unit)* it is a dry T1 circuit or more commonly known as a *DS1 (Digital Service Level 1)*.

Wet T1 See *Wet Circuit*.

White Board Also called a *mushroom board* or *peg board*. It is placed between termination blocks (such as 66M150 blocks) to provide a means of support for routing cross-connect wire.

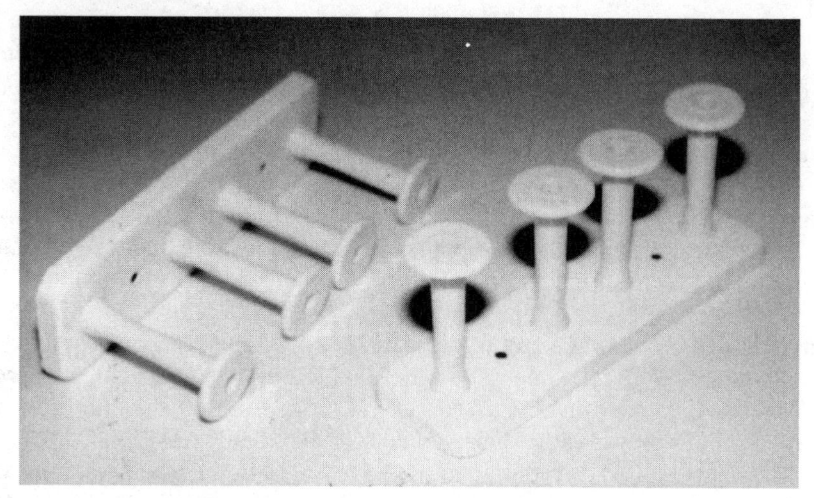

1/4 White Board "Mushroom Board" "Spool Board"

White Noise Random electrical noise, also called *ambient noise*. White noise is the sound you get from a TV or radio when it is not tuned to a station.

Wide-Area Network (WAN) See *WAN*.

Wide Band Another name for *Broadband*. Incorporating more than one channel into a communications transmission. T1 is a broadband communications protocol because it carries 24 conversations over four wires. Cable TV is also broadband because it carries many TV channels over one coax.

Wink Another name for a hookflash that is sent by PBX systems to telephone company central-office switches that signals a request for dial tone or other services.

Wink Start Signal Another name for a hookflash that is sent by PBX systems to telephone company central-office switches that signals a request for dial tone or other services. Wink start signaling is slowly being replaced by T1 out-of-band signaling, which is much faster, offers many more services, is less expensive, and is, of course, digital.

Wire Center A reference to a telephone company central office's geographical service area. The central office serves the area that its telephone wires (outside plant) reach to.

Wire Pair A reference to a two solid wires, twisted together, usually just called a *pair*. For more information, see *UTP*.

Wire Tap A device used to monitor telephone lines. With newer technology, the telephone company is capable of tapping or monitoring a telephone line with a stroke of a few keys. They can even set the telephone line to be monitored by a different telephone line, anywhere they want. Telephone companies (especially the Bell Companies) have extremely strict security guidelines regarding the monitoring of telephone conversations. They will not set up a tap or a monitor service without legal procedures being followed according to the laws of the area they are operating in. Watch-dogs are in place to be sure that telephone company employees do not monitor telephone lines when they are not supposed to. Some "spy" shops sell telephone line-tapping and recording devices, but I would not recommend the use of them because of strict laws regarding telephone privacy.

Wireless LAN A local-area network of computers and peripheral devices that communicates via radio signals or light waves (low-power laser beams). These systems are useful in situations where the cost of installing wiring between the devices is very expensive or for temporary/mobile applications.

Wire-Wrap Termination A type of twisted-pair wire termination used on DSX panels and other digital telecommunications services processing equipment. Wire-wrap termination is used because it requires very little space, compared to 110 blocks or 66M150 blocks.

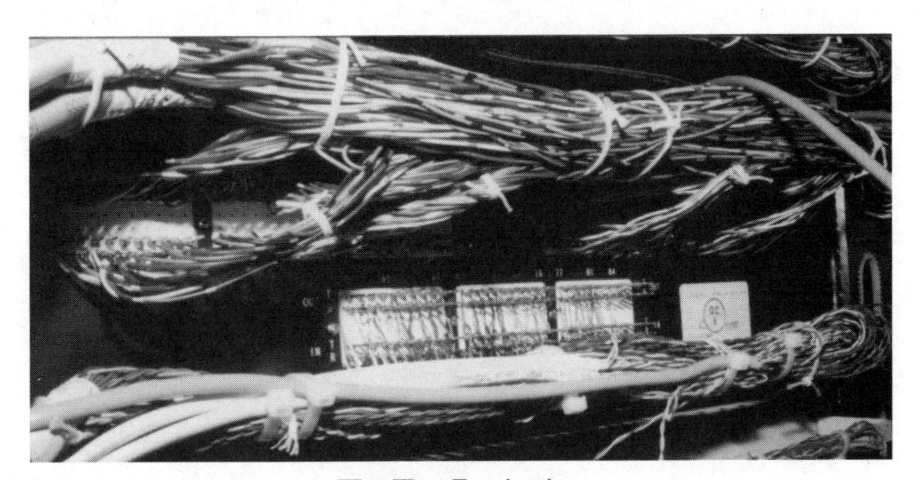

Wire-Wrap Termination

Word The number of bits that a microprocessor can extract from a bus and process at one time. Older computers and operating software recognize 8 bits as a word; newer computers recognize 64 bits as a word.

X.20 The recommended standard by the *ITU (International Telecommunications Union)* for asynchronous communications between modems on public-switched telephone networks (dial-up lines).

X.25 The recommended standard by the *ITU (International Telecommunications Union)* for packet network transmission and switching. X.25 is its own protocol; it requires its own equipment and should not be confused with its newer counterpart, frame relay.

X.61 The recommended standard by the *ITU (International Telecommunications Union)* for the data-user section of Signaling System 7 (SS7).

X.75 The recommended standard by the *ITU (International Telecommunications Union)* for the gateway (international) connection of X.25 packet networks.

X.121 The standard proposed (and widely used) addressing/numbering scheme for data networks by the *ITU (International Telecommunications Union)*. X.121 is utilized in the X.25 packet protocol.

X.130 The recommended standard by the *ITU (International Telecommunications Union)* for circuit set-up and tear-down times.

X.200 The recommended standard by the *ITU (International Telecommunications Union)* for the *OSI (Open Systems Interconnection)* model, and the functions and protocols of each layer.

X Standards Telecommunications standards that are developed and recommended by the *ITU-T (International Telecommunications Union Telecommunications).*

XC Cross Connect.

Xaga A type of "hot," buried splice enclosure for telephone cable that requires heating a special wrap that is placed around the splice. When the wrapping is heated, it shrinks into a sturdy casing. If it is installed properly, it is water proof. *Xaga* is a registered trademark of Raychem.

Xaga

Xmit Abbreviation for *transmit.*

Yagi Antenna A directional transmitting antenna. The Yagi antenna utilizes one or more director elements to "focus" a radiated signal in one direction or plane, and a reflector element that is placed at a wavelength distance 180 degrees from the driven element to null the transmitted signal in the backward direction.

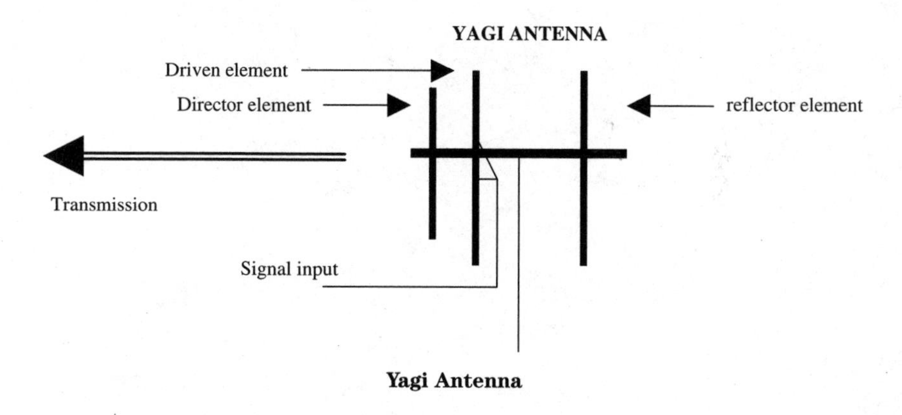

Yagi Antenna

Z

Zener Diode A diode that operates in the reversed biased/avalanche mode. When a zener diode is in this mode, it has a constant voltage from its anode to cathode. Zener diodes are used to regulate voltage, and they are available in many voltage values, such as 5 V, 12 V, 24 V, etc.

ZENER DIODE SCHEMATIC SYMBOL

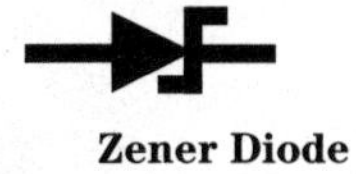

Zener Diode

Zero-Byte Time-Slot Interchange A method of line coding, where the quantity of zeros transmitted is reduced. The reduction is accomplished by recognizing bytes that are all zeros (eight consecutive bits) and replacing them with an alternate byte (or flag) that does not contain all zeros.

Zero Suppression To omit zeros from a number field.

Zulu Time Formerly known as *Greenwich Mean Time*. Time that is kept on an atomic 24-hour clock in France. The Zulu time standard is now more accurate by innovation of the cesium timing reference standard, which is the time-keeping element in "bits clocks" (timing devices used in central office nodes to synchronize SONET equipment). Zulu time is slang for *Universal Time Coordinated.*

Appendix

Vertical horizontal coordinates table for United States cities

City	V coordinate	H coordinate
Aberdeen SD	5992	5308
Akron, OH	5637	2472
Albany, NY	4639	1629
Albuquerque, NM	8549	5887
Allentown, PA	5166	1585
Altoona, PA	5460	1972
Amarillo,TX	8266	5076
Aneheim, CA	9250	7810
Appleton, WI	5589	3776
Asheville, NC	6749	2001
Atlanta, GA	7260	2083
Atlantic City, NJ	5284	1284
Augusta, GA	7089	1674
Augusta, ME	3961	1870
Austin, TX	9005	3996
Bakersfield, CA	8947	8060
Baltimore, MD	5510	1575
Baton Rouge, LA	8476	2874
Bellingham, WA	6087	8933
Billings, MT	6391	6790
Biloxi, MS	8296	2481
Binghamton, NY	4943	1837

City	V coordinate	H coordinate
Birmingham, AL	7518	2304
Bismarck, ND	5840	5736
Blacksburg, VA	6247	1687
Bloomington, IN	6417	2984
Boise, ID	7096	7869
Boston, MA	4422	1249
Bridgeport, CT	4841	1360
Buffalo, NY	5076	2326
Burlington, IA	6417	2984
Burlington, VT	4270	1808
Camden, NJ	5249	1453
Canton, OH	5676	2419
Carson City, NV	8139	8306
Casper, WY	6918	6297
Cedar Rapids, IA	6261	4021
Charleston, SC	7021	1281
Charlotte, NC	6657	1698
Chattanooga, TN	7098	2366
Cheyenne, WY	7203	5958
Chicago, Il	5986	3426
Cincinatti, OH	6263	2679
Clarksburg, WV	5865	2095
Clearwater, FL	8203	1206
Cleveland, OH	5574	2543
Columbia, SC	6901	1589
Columbus, OH	5872	2555
Concord, NH	4326	1426
Corpus Christi, TX	9475	3739
Dallas, TX	8436	4034
Danville, KY	6558	2561
Dayton, OH	6113	2705
Daytona Beach, FL	7791	1052
Denver, CO	7501	5899
Des Moines, IA	6471	4275
Detroit, MI	5536	2828
Dodge City, KS	7640	4958
Dubuque, IA	6088	3925
Duluth, MN	5352	4530
Eau Claire, WI	5698	4261
El Paso, TX	9231	5655
Fargo, ND	5615	5182
Farmingham, MA	4472	1249

City	V coordinate	H coordinate
Fayetteville, AR	7600	3872
Fayetteville, NC	6501	1385
Flagstaff, AZ	8746	2367
Flint, MI	5461	2993
Fort Collins, CO	7331	5965
Fort Lauderdale, FL	8282	557
Fort Wayne, IN	5942	2982
Fort Worth, TX	8479	4122
Frankfort, KY	6462	2634
Fresno, CA	8669	8239
Grand Forks, ND	5420	5300
Grand Island, NE	6901	4936
Grand Junction, CO	7804	6438
Grand Rapids, MI	5628	3261
Greeley, CO	7345	5895
Green Bay, WI	5512	3747
Hackensack, NJ	4976	1432
Harrisburg, PA	5363	1733
Hartford, CT	4687	1373
Helena, MT	6336	7348
Hot Springs, AR	7827	3554
Houston, TX	8938	3563
Huntsville, AL	7267	2535
Huron, SD	6201	5183
Indianapolis, IN	6272	2992
Iowa City, IA	6313	3972
Jackson, MS	8035	2880
Jacksonville, FL	7649	1276
Johnson City, TN	6595	2050
Joliet, Il	6088	3454
Joplin, MO	7421	4015
Kalamazoo, MI	5749	3177
Kansas City, MO	7027	4203
Kennewick, WA	6595	8391
Knoxville, TN	6801	2251
La Crosse, WI	5874	4133
Lansing, MI	5584	3081
Laredo, TX	9681	4099
Las Cruces, NM	9132	5742
Las Vegas, NV	8665	7411
Lawton, OK	8178	4451
Leesburg, VA	5634	1685

City	V coordinate	H coordinate
Lewiston, ME	4042	1391
Logan, UT	7367	7102
Long Beach, CA	9217	7856
Los Angeles, CA	9213	7878
Lubbock, TX	8596	4962
Lynchburg, VA	6093	1703
Macon, GA	7364	1865
Madison, WI	5887	3796
Madisonville, KY	6845	2942
Manchester, NH	4354	1388
Medford, OR	7503	8892
Memphis, TN	7471	3125
Meridian, MS	7899	2639
Miami, FL	8351	527
Milwaukee, WI	5788	3589
Minneapolis, MN	5777	4513
Missoula, MT	6336	7650
Mobile, AL	8167	2367
Montgomery, AL	7692	2247
Morgantown, WV	5764	2083
Morristown, NJ	5035	1478
Muncie, IN	6130	2925
Nashua, NH	4394	1356
Nashville, TN	7010	2710
Nassau, NY	4961	1355
New Brunswick, NJ	5085	1434
New Haven, CT	4792	1342
New London, CT	4700	1242
New Orleans, LA	8483	2638
New York City, NY	4977	1406
Newark, NJ	5015	1430
Norfolk, VA	5918	1223
North Bend, WA	6354	8815
Oakland, CA	8486	8695
Ogden, UT	7480	7100
Oklahoma City, OK	7947	4373
Omaha, NE	6687	4595
Orlando, FL	7954	1031
Paduca, KY	6982	3088
Pendleton, OR	6707	8326
Peoria, Il	6362	3592
Philadelphia, PA	5257	1501

City	V coordinate	H coordinate
Phoenix, AZ	9135	6748
Pine Bluff, AR	7803	3358
Pittsburgh, PA	5621	2185
Pocatello, ID	7146	7250
Portland, ME	4121	1384
Portland, OR	6799	8914
Poughkeepsie, NY	4821	1526
Providence, RI	4550	1219
Provo, UT	7680	7006
Pueblo, CO	7787	5742
Racine, WI	5837	3535
Raleigh, NC	6344	1436
Reading, PA	5258	1612
Redwood City, CA	8556	8682
Reno, NV	8064	8323
Richmond, VA	5906	1472
Roanoke, VA	6196	1801
Rochester, NY	4913	2195
Rock Island, Il	6276	3816
Sacramento, CA	8304	8580
Salt Lake City, UT	7576	7065
San Antonio, TX	9225	4062
San Bernardino, CA	9172	7710
San Diego, CA	9468	7629
San Francisco, CA	8492	8719
San Jose, CA	8583	8619
Santa Fe, NM	8389	5804
Santa Monica, CA	9227	7920
Santa Rosa, CA	8354	8787
Savannah, GA	7266	1379
Scranton, PA	5042	1715
Seattle, WA	6336	8896
Shreveport, LA	8272	3495
Sioux City, IA	6468	4768
Sioux Falls, SD	6279	4900
South Bend, IN	5918	3206
Spartanburg, SC	6811	1833
Spokane, WA	6247	8180
Springfield, Il	6539	3518
Springfield, MA	4620	1408
Springfield, MO	7310	3836
St. Joseph, MO	6913	4301

City	V coordinate	H coordinate
St. Paul, MN	5776	4498
Stamford, CT	4897	1388
Sunnyvale, CA	8576	8643
Syracuse, NY	4798	1990
Tallahassee, FL	7877	1716
Tampa, FL	8173	1147
Terre Haute, IN	6428	3145
Toledo, OH	5704	2820
Topeka, KS	7110	4369
Trenton, NJ	5164	1440
Troy, NY	4616	1633
Tulsa, OK	7707	4173
Tucson, AZ	9345	6485
Van Nuys, CA	9197	7919
Washington, DC	5622	1583
Westchester, NY	4912	1330
Wheeling, WV	5755	2241
Wichita, KS	7489	4520
Wilmington, DE	5326	1485
Winchester, KY	6441	2509
Winston-Salem, NC	6440	1710
Worchester, MA	4513	1330
Yakima, WA	6533	8607
Yuma, AZ	9385	7171

Appendix

B

Calling countries from the United States: country and city codes

Country	Code	City	Code
ALGERIA	213	city code not used	
AMERICAN SAMOA	684	city code not used	
ANDORRA	33	ALL CITIES	628
ARGENTINA	54	BUENOS ARIES	1
ARGENTINA	54	CORDOBA	1
ARGENTINA	54	LA PLATA	21
ARGENTINA	54	ROSARIO	41
ARUBA	297	ALL CITIES	8
ASCENSION ISL.	247	city code not used	
AUSTRALIA	61	ADELAIDE	8
AUSTRALIA	61	BRISBANE	7
AUSTRALIA	61	MELBOURNE	3
AUSTRALIA	61	SYDNEY	2
AUSTRIA	43	GRAZ	316
AUSTRIA	43	LINZ DONAU	732
AUSTRIA	43	VIENNA	1
BAHRAIN	973	city code not used	
BANGLEDESH	880	BARISAL	431
BANGLEDESH	880	CHITTAGONG	31
BANGLEDESH	880	DHAKA	2
BANGLEDESH	880	KHULNA	41
BELGIUM	32	ANTWERP	3

Country	Code	City	Code
BELGIUM	32	BRUSSELS	2
BELGIUM	32	GHENT	91
BELGIUM	32	LIEGE	41
BELIZE	501	BELIZE CITY	direct
BELIZE	501	COROZAL TOWN	04
BELIZE	501	PUNTA GORDA	07
BENIN	229	city code not used	
BOLIVIA	591	COCHABAMBA	42
BOLIVIA	591	LA PAZ	2
BOLIVIA	591	SANTA CRUZ	33
BRAZIL	55	BELE HORIZONTE	31
BRAZIL	55	RIO DE JANEIRO	21
BRAZIL	55	SAO PAULO	11
BRUNEI	673	BANDER SERI BEGAWAN	2
BRUNEI	673	KUALA BELAIT	3
BRUNEI	673	TUTONG	4
BULGARIA	359	PLOVDIV	32
BULGARIA	359	ROUSSE	82
BULGARIA	359	SOFIA	2
BULGARIA	359	VARNA	52
CAMEROON	237	city code not used	
CHILE	56	CONCEPCION	41
CHILE	56	SANTIAGO	2
CHILE	56	VALPARAISO	32
CHINA	86	BEJIENG (PEKING)	1
CHINA	86	FUZHOU	591
CHINA	86	GHUANGZHOU (CANTON)	20
CHINA	86	SHANGHAI	21
COLUMBIA	57	BARRANQUILLA	5
COLUMBIA	57	BOGATA	1
COLUMBIA	57	CALI	23
COLUMBIA	57	MEDELLIN	4
COSTA RICA	506	city code not used	
CYPRUS	357	LIMASSOL	51
CYPRUS	357	NICOSIA	2
CYPRUS	357	PAPHOS	61
CZECHOSLOVAKIA	42	BRATISLAVA	7
CZECHOSLOVAKIA	42	BRNO	5
CZECHOSLOVAKIA	42	HAVIROV	6994
CZECHOSLOVAKIA	42	OSTRAVA	69
CZECHOSLOVAKIA	42	PRAGUE	2
DENMARK	45	AALBORG	8

Country	Code	City	Code
DENMARK	45	AARHUS	6
DENMARK	45	COPENHAGEN	1
DENMARK	45	ODDENSE	7
ECUADOR	593	AMBATO	2
ECUADOR	593	CUENCA	7
ECUADOR	593	GUAYAQUIL	4
ECUADOR	593	QUITO	2
EGYPT	20	ALEXANDRIA	3
EGYPT	20	ASWAN	97
EGYPT	20	ASYUT	88
EGYPT	20	BENHA	13
EGYPT	20	CAIRO	2
EL SALVADOR	503	city code not used	
ETHIOPIA	251	ADDIS ABABA	1
ETHIOPIA	251	AKAKI	1
ETHIOPIA	251	ASMARA	4
ETHIOPIA	251	ASSAB	3
ETHIOPIA	251	AWASSA	6
FAEROE ISLANDS	298	city code not used	
FIJI ISLANDS	679	city code not used	
FINLAND	358	EPPO EBBO	15
FINLAND	358	HELSINKI	0
FINLAND	358	TAMMEFORS-TAMPERE	31
FINLAND	358	TURKU	21
FRANCE	33	LYON	7
FRANCE	358	MARSEILLE	91
FRANCE	358	NICE	93
FRANCE	358	PARIS	1
FRENCH ANTILLES	596	city code not used	
FRENCH GUIANA	594	city code not used	
FRENCH POLYNESIA	689	city code not used	
GABON	241	city code not used	
GAMBIA	220	city code not used	
GERMAN DEM. REP.	37	BERLIN	2
GERMAN DEM. REP.	37	DRESDEN	51
GERMAN DEM. REP.	37	LEIPZIG	41
GERMAN DEM. REP.	37	MAGDEBURG	91
GERMANY FED REP	49	BERLIN	30
GERMANY FED REP	49	BONN	228
GERMANY FED REP	49	FRANKFURT	69
GERMANY FED REP	49	MUNICH	89
GIBRALTAR	350	city code not used	

Country	Code	City	Code
GREECE	30	ATHENS	1
GREECE	30	IRAKLION	81
GREECE	30	LARISSA	41
GREECE	30	PIRAEUS PIREEFS	1
GREENLAND	299	GODTHAB	2
GREENLAND	299	SONDRE STROMFJORD	11
GREENLAND	299	THULE	50
GUADELOUPE	590	city code not used	
GUAM	671	city code not used	
GUANTANAMO BAY USN	5399	city code not used	
GUATEMALA	502	ANTIGUA	9
GUATEMALA	502	GUATEMALA CITY	2
GUATEMALA	502	QUEZALTENANGO	9
GUYANA	592	BARTICA	5
GUYANA	592	GEORGETOWN	2
GUYANA	592	NEW AMSTERDAM	3
HAITI	509	CAP-HAITIEN	3
HAITI	509	CAYES	5
HAITI	509	GONAIVE	2
HAITI	509	PORT AU PRINCE	1
HONDURAS	504	city code not used	
HONG KONG	852	HONG KONG	5
HONG KONG	852	KOWLOON	3
HONG KONG	852	NEW TERRITORIES	0
HUNGARY	36	BUDAPEST	1
HUNGARY	36	DERBRECEN	52
HUNGARY	36	GYOR	96
HUNGARY	36	MISKOLC	46
ICELAND	354	AKUREYRI	6
ICELAND	354	KEFLAVIC	2
ICELAND	354	REYKJAVIK	1
INDIA	91	BOMBAY	22
INDIA	91	CALCUTTA	33
INDIA	91	MADRAS	44
INDIA	91	NEW DELHI	11
INDONESIA	62	JAKARTA	21
INDONESIA	62	MEDAN	61
INDONESIA	62	SEMARANG	24
IRAN	98	ESFAHAN	31
IRAN	98	MASHAD	51
IRAN	98	TABRIZ	41
IRAN	98	TEHRAN	21

Country	Code	City	Code
IRELAND	353	CORK	21
IRELAND	353	DUBLIN	1
IRELAND	353	GALWAY	91
IRELAND	353	LIMERICK	61
ISRAEL	972	HAIFA	4
ISRAEL	972	JERUSALEM	2
ISRAEL	972	RAMAT GAN	3
ISRAEL	972	TEL AVIV	3
ITALY	39	FLORENCE	55
ITALY	39	GENOA	10
ITALY	39	MILAN	2
ITALY	39	NAPLES	81
ITALY	39	ROME	6
IVORY COAST	255	city code not used	
JAPAN	81	KYOTO	75
JAPAN	81	OSAKA	6
JAPAN	81	SAPORRO	11
JAPAN	81	TOKYO	3
JAPAN	81	YOKOHAMA	45
JORDAN	962	AMMAN	6
JORDAN	962	IRBID	2
JORDAN	962	JERASH	4
JORDAN	962	KARAK	3
JORDAN	962	MA'AN	3
KENYA	254	KISUMU	35
KENYA	254	MOMBASA	11
KENYA	254	NAIROBI	2
KENYA	254	NAKURU	37
KOREA	82	INCHEON	32
KOREA	82	PUSAN	51
KOREA	82	SEOUL	2
KOREA	82	TAEGU	53
KUWAIT	965	city code not used	
LESOTHO	266	city code not used	
LIBERIA	231	city code not used	
LIBYA	218	BENGHAZI	61
LIBYA	218	MISURATHA	51
LIBYA	218	TRIPOLI	21
LIBYA	218	ZAWAI	23
LEICHTENSTIEN	41	ALL CITIES	75
LUXEMBOURG	352	city code not used	
MACAO	853	city code not used	

Country	Code	City	Code
MALAWI	265	DOMASI	531
MALAWI	265	MAKWASA	474
MALAWI	265	ZOMBA	50
MALAYSIA	60	IPOH	5
MALAYSIA	60	JOHOR BAHRU	7
MALAYSIA	60	KAJANG	3
MALAYSIA	60	KUALA LUMPUR	3
MALTA	356	city code not used	
MARSHALL ISLANDS	692	EBEYE	871
MARSHALL ISLANDS	692	MAJURO	9
MICRONESIA	691	KOSREA	851
MICRONESIA	691	PONAPE	9
MICRONESIA	691	TRUK	8319
MICRONESIA	691	YAP	841
MIQUELON	508	city code not used	
MONACO	33	ALL CITIES	93
MOROCCO	212	AGADIR	8
MOROCCO	212	BENI-MELLAL	48
MOROCCO	212	CASSABLANCA	direct
MOROCCO	212	EL JADIDA	34
NAMBIA	264	GROOTFONTEIN	673
NAMBIA	264	KEETMANSHOOP	631
NAMBIA	264	MARIENTAL	661
NETHERLANDS	31	AMSTERDAM	20
NETHERLANDS	31	ROTTERDAM	10
NETHERLANDS	31	THE HAGUE	70
NETHERLANDS ANTILLES	599	BONAIRE	7
NETHERLANDS ANTILLES	599	CURACAO	9
NETHERLANDS ANTILLES	599	ST. EUSTATIUS	3
NETHERLANDS ANTILLES	599	ST. MAARTEN	5
NEW CALEDONIA	687	city code not used	
NEW ZEALAND	64	AUKLAND	9
NEW ZEALAND	64	CHRISTCHURCH	3
NEW ZEALAND	64	DUNEDIN	24
NEW ZEALAND	64	HAMILTON	71
NICARAGUA	505	CHINANDEGA	341
NICARAGUA	505	DIRIAMBA	42
NICARAGUA	505	LEON	311
NICARAGUA	505	MANAGUA	2
NIGERIA	234	LAGOS	1
NORWAY	47	BERGEN	5
NORWAY	47	OSLO	2

Country	Code	City	Code
NORWAY	47	STAVANGER	4
NORWAY	47	TRONDHEIM	7
OMAN	968	city code not used	
PAKISTAN	92	ISLAMABAD	51
PAKISTAN	92	KARICHI	21
PAKISTAN	92	LAHORE	42
PANAMA	507	city code not used	
PAPUA NEW GUINEA	675	city code not used	
PARAGUAY	595	ASUNCION	21
PARAGUAY	595	CONCEPCION	31
PERU	51	AREQUIPA	54
PERU	51	CALLAO	14
PERU	51	LIMA	14
PERU	51	TRUJILLO	44
PHILLIPINES	63	CEBU	32
PHILLIPINES	63	DAVAO	35
PHILLIPINES	63	MANILLA	2
POLAND	48	CRAKOW	12
POLAND	48	GDANSK	58
POLAND	48	WARSAW	22
PORTUGAL	351	COIMBRA	39
PORTUGAL	351	LISBON	1
PORTUGAL	351	PORTO	2
PORTUGAL	351	SETUBAL	65
QATAR	974	city code not used	
ROMANIA	40	BUCHAREST	0
ROMANIA	40	CLUJ-NAPOCA	51
ROMANIA	40	CONSTANTA	16
SAIPAN	670	ROTA ISL.	532
SAIPAN	670	SUSUPE CITY	234
SAIPAN	670	TINIAN ISL.	433
SAN MARINO	39	ALL CITIES	541
SAUDI ARABIA	966	HOFUF	3
SAUDI ARABIA	966	JEDDAH	2
SAUDI ARABIA	966	MAKKAH	2
SAUDI ARABIA	966	RIYADH	1
SENEGAL	221	city code not used	
SINGAPORE	65	city code not used	
SOUTH AFRICA	27	CAPE TOWN	21
SOUTH AFRICA	27	DURBAN	31
SOUTH AFRICA	27	JOHANNESBURG	11
SPAIN	34	BARCELONA	3

Country	Code	City	Code
SPAIN	34	MADRID	1
SPAIN	34	SEVILLE	54
SPAIN	34	VALENCIA	6
SRI LANKA	94	COLOMBO CENTRAL	1
SRI LANKA	94	KANDY	8
SRI LANKA	94	KOTTE	1
ST. PIERRE	508	city code not used	
SURINAME	597	city code not used	
SWAIZLAND	268	city code not used	
SWEDEN	46	GOTEBORG	31
SWEDEN	46	MALMO	40
SWEDEN	46	STOCKHOLM	8
SWEDEN	46	VASTERAS	21
SWITZERLAND	41	BASEL	61
SWITZERLAND	41	BERNE	31
SWITZERLAND	41	GENEVA	22
SWITZERLAND	41	ZURICH	1
TAIWAN	886	KAOHSIUNG	7
TAIWAN	886	TAINAN	6
TAIWAN	886	TAIPEI	2
TANZANIA	255	DAR ES SALAAM	51
TANZANIA	255	DODOMA	61
TANZANIA	255	MWANZA	68
TANZANIA	255	TANGA	53
THAILAND	66	BANGKOK	2
THAILAND	66	BURIRUM	44
THAILAND	66	CHANTHABURI	39
TOGO	228	city code not used	
TUNSIA	216	BIZERTE	2
TUNSIA	216	KAIROUAN	7
TUNSIA	216	MSEL BOURGUIBA	2
TUNSIA	216	TUNIS	1
TURKEY	90	ADANA	711
TURKEY	90	ANKARA	41
TURKEY	90	ISTANBUL	1
TURKEY	90	IZMIR	51
UGANDA	256	ENTEBBE	42
UGANDA	256	JINJA	43
UGANDA	256	KAMPALA	41
UGANDA	256	KYAMBOGO	41
UNITED ARAB EMIRATES	971	ABU DHABI	2
UNITED ARAB EMIRATES	971	AJMAN	6

Country	Code	City	Code
UNITED ARAB EMIRATES	971	AL AIN	3
UNITED ARAB EMIRATES	971	DUBAI	4
UNITED ARAB EMIRATES	971	SHARJAH	6
UNITED KINGDOM	44	BELFAST	232
UNITED KINGDOM	44	BIRMINGHAM	21
UNITED KINGDOM	44	GLASGOW	41
UNITED KINGDOM	44	LONDON	1
URAGUAY	598	CANELONES	332
URAGUAY	598	MERCEDES	532
URAGUAY	598	MONTEVIDEO	2
VATICAN CITY	39	ALL OF VAITICAN CITY	6
VENEZUELA	58	BARQUISIMETO	51
VENEZUELA	58	CARACAS	2
VENEZUELA	58	MARACAIBO	61
VENEZUELA	58	VALENCIA	41
YEMEN ARAB REPUBLIC	967	AMRAN	2
YEMEN ARAB REPUBLIC	967	SANAA	2
YEMEN ARAB REPUBLIC	967	TAIZ	4
YEMEN ARAB REPUBLIC	967	YARIM	4
YEMEN ARAB REPUBLIC	967	ZABID	3
YUGOSLAVIA	38	BELGRADE	11
YUGOSLAVIA	38	SARAJEVO	71
YUGOSLAVIA	38	ZAGREB	41
ZAIRE	243	KINSHASA	12
ZAIRE	243	LUBUMBASHI	222
ZAMBIA	260	CHINGOLA	2
ZAMBIA	260	KITWE	2
ZAMBIA	260	LUANSHYA	2
ZAMBIA	260	LUSAKA	1
ZAMBIA	260	NDOLA	26
ZIMBABWE	263	BULAWAYO	9
ZIMBABWE	263	HARARE	4
ZIMBABWE	263	MUTARE	20

Appendix

C

Area codes of the NANP listed by area

Location/State	City	Area Code
ALABAMA	BIRMINGHAM	205
ALABAMA	MONTGOMERY	334
ALASKA	ALL LOCATIONS	907
ARIZONA	PHOENIX	602
ARIZONA	TUSCON	520
CALIFORNIA	ANAHEIM	714
CALIFORNIA	BAKERSFIELD	805
CALIFORNIA	BARSTOW	760
CALIFORNIA	BEVERLY HILLS	310
CALIFORNIA	EUREKA	707
CALIFORNIA	FRESNO	209
CALIFORNIA	LONG BEACH	562
CALIFORNIA	LOS ANGELES	213
CALIFORNIA	MODESTO	209
CALIFORNIA	MONTEREY	408
CALIFORNIA	OAKLAND	510
CALIFORNIA	PALM SPRINGS	760
CALIFORNIA	PASADENA	818
CALIFORNIA	RIVERSIDE	909
CALIFORNIA	SACRAMENTO	916
CALIFORNIA	SAN BERNARDINO	909

Location/State	City	Area Code
CALIFORNIA	SAN DIEGO	619
CALIFORNIA	SAN FRANCISCO	415
CALIFORNIA	SAN JOSE	408
CALIFORNIA	SANTA BARBARA	805
CARRIBEAN ISLANDS	BAHAMAS	242
CARRIBEAN ISLANDS	BARBADOS	246
CARRIBEAN ISLANDS	BERMUDA	441
CARRIBEAN ISLANDS	MONTSERRAT	664
CARRIBEAN ISLANDS	ST. LUCIA	758
CARRIBEAN ISLANDS	TOBAGO	868
CARRIBEAN ISLANDS	TRINIDAD	868
COLORADO	ASPEN	970
COLORADO	AURORA	303
COLORADO	BOULDER	303
COLORADO	COLORADO SPRINGS	719
COLORADO	DENVER	303
COLORADO	DURANGO	970
COLORADO	GRAND JUNCTION	970
COLORADO	LEADVILLE	719
COLORADO	LONGMONT	303
COLORADO	PUEBLO	719
COLORADO	STEAMBOAT SPRINGS	970
CONNECTICUT	BRIDGEPORT	203
CONNECTICUT	HARTFORD	860
CONNECTICUT	NEW HAVEN	203
CONNECTICUT	STANFORD	203
DELAWARE	ALL LOCATIONS	302
DISTRICT OF COLUMBIA	WASHINGTON DC	202
FLORIDA	CLEARWATER	813
FLORIDA	FT. LAUDERDALE	954
FLORIDA	FT. MYERS	941
FLORIDA	GAINESVILLE	352
FLORIDA	JACKSONVILLE	904
FLORIDA	KEY WEST	305
FLORIDA	MAPLES	941
FLORIDA	MIAMI	305
FLORIDA	ORLANDO	407
FLORIDA	PENSACOLA	904
FLORIDA	SARASOTA	941
FLORIDA	ST. PETERSBURG	813
FLORIDA	TALLAHASSEE	904

Location/State	City	Area Code
FLORIDA	TAMPA	813
FLORIDA	WEST PALM BEACH	561
GEORGIA	ALBANY	912
GEORGIA	ATLANTA (METRO)	404
GEORGIA	ATLANTA (OUTSIDE 285 BELT)	770
GEORGIA	AUGUSTA	706
GEORGIA	COLUMBUS	706
GEORGIA	MARIETTA	770
GEORGIA	NORCROSS	770
GEORGIA	ROME	706
GEORGIA	SAVANNAH	912
HAWAII	ALL LOCATIONS	808
IDAHO	ALL LOCATIONS	208
ILLINOIS	ALTON	618
ILLINOIS	CAIRO	618
ILLINOIS	CHAMPAIGN - URBANA	217
ILLINOIS	CHICAGO	708
ILLINOIS	CHICAGO	773
ILLINOIS	CHICAGO	847
ILLINOIS	CHICAGO	312
ILLINOIS	CHICAGO (CENTRAL SUBURBS)	630
ILLINOIS	ELGIN	630
ILLINOIS	ELGIN	847
ILLINOIS	LA SALLE	815
ILLINOIS	MT. VERNON	618
ILLINOIS	PEORIA	309
ILLINOIS	ROCK ISLAND	309
ILLINOIS	ROCKFORD	815
ILLINOIS	SPRINGFIELD	217
ILLINOIS	WAUKEGAN	630
ILLINOIS	WAUKEGAN	847
INDIANA	EVANSVILLE	812
INDIANA	GARY	219
INDIANA	HAMMOND	219
INDIANA	INDIANAPOLIS	317
INDIANA	KOKOMO	317
INDIANA	MICHIGAN CITY	219
INDIANA	SOUTH BEND	219
IOWA	COUNCIL BLUFFS	712
IOWA	DAVENPORT	319
IOWA	DES MOINES	515

Location/State	City	Area Code
IOWA	DUBUQUE	319
IOWA	SIOUX CITY	712
KANSAS	DODGE CITY	316
KANSAS	LAWRENCE	913
KANSAS	SALINA	913
KANSAS	TOPEKA	913
KANSAS	WICHITA	316
KENTUCKY	ASHLAND	606
KENTUCKY	FRANKFORT	502
KENTUCKY	LOUISVILLE	502
KENTUCKY	PADUCAH	502
KENTUCKY	SHELBYVILLE	502
KENTUCKY	WINCHESTER	606
LOUISIANA	BATON ROUGE	504
LOUISIANA	LAKE CHARLES	318
LOUISIANA	NEW ORLEANS	504
LOUISIANA	SHREVEPORT	318
MAINE	ALL LOCATIONS	207
MARYLAND	ANNAPOLIS	410
MARYLAND	BALTIMORE	410
MARYLAND	HAGERSTOWN	301
MARYLAND	ROCKVILLE	301
MASSACHUSETTS	BOSTON	617
MASSACHUSETTS	FALL RIVER	508
MASSACHUSETTS	NEW BEDFORD	508
MASSACHUSETTS	PITTSFIELD	413
MASSACHUSETTS	SPRINGFIELD	413
MASSACHUSETTS	WORCESTER	508
MICHIGAN	ANN ARBOR	313
MICHIGAN	BATTLE CREEK	616
MICHIGAN	BAY CITY	517
MICHIGAN	DEARBORN	313
MICHIGAN	DETROIT	313
MICHIGAN	FLINT	810
MICHIGAN	GRAND RAPIDS	616
MICHIGAN	JACKSON	517
MICHIGAN	KALAMAZOO	616
MICHIGAN	LANSING	517
MICHIGAN	MARQUETTE	906
MICHIGAN	PONTIAC	810
MICHIGAN	SAULT STE. MARIE	906

Location/State	City	Area Code
MICHIGAN	SOUTHFIELD	810
MICHIGAN	TROY	810
MINNESOTA	DULUTH	218
MINNESOTA	MINNEAPOLIS	612
MINNESOTA	ROCHESTER	507
MINNESOTA	ST. PAUL	612
MINNESOTA	ST. CLOUD	320
MISSISSIPPI	ALL LOCATIONS	601
MISSOURI	COLUMBIA	314
MISSOURI	JEFFERSON CITY	573
MISSOURI	JOPLIN	417
MISSOURI	KANSAS CITY	816
MISSOURI	SPRINGFIELD	417
MISSOURI	ST. JOSEPH	816
MISSOURI	ST. LOUIS	314
MONTANA	ALL LOCATIONS	406
NEBRASKA	LINCOLN	402
NEBRASKA	NORTH PLATTE	308
NEBRASKA	OMAHA	402
NEBRASKA	SCOTTSBLUFF	308
NEVADA	ALL LOCATIONS	702
NEW HAMPSHIRE	ALL LOCATIONS	603
NEW JERSEY	ATLANTIC CITY	609
NEW JERSEY	CAMDEN	609
NEW JERSEY	HACKENSACK	201
NEW JERSEY	JERSEY CITY	201
NEW JERSEY	NEW BRUNSWICK	908
NEW JERSEY	NEWARK	201
NEW JERSEY	PATERSON	201
NEW JERSEY	TRENTON	609
NEW JERSEY	VINELAND	609
NEW MEXICO	ALL LOCATIONS	505
NEW YORK	ALBANY	518
NEW YORK	BINGHAMTON	607
NEW YORK	BRONX	718
NEW YORK	BRONX CEL AND PAGER	917
NEW YORK	BROOKLYN	718
NEW YORK	BROOKLYN CEL AND PAGE	917
NEW YORK	BUFFALO	716
NEW YORK	ELMIRA	607
NEW YORK	HEMPSTEAD	516

Location/State	City	Area Code
NEW YORK	LONG ISLAND	516
NEW YORK	MANHATTAN	212
NEW YORK	MANHATTAN CEL AND PAGE	917
NEW YORK	NIAGARA FALLS	716
NEW YORK	PEEKSKILL	914
NEW YORK	POUGHKEEPSIE	914
NEW YORK	QUEENS	718
NEW YORK	QUEENS CEL AND PAGE	917
NEW YORK	ROCHESTER	716
NEW YORK	SCHENECTADY	518
NEW YORK	STATEN ISLAND	718
NEW YORK	STATEN ISLAND CEL PAGE	917
NEW YORK	SYRACUSE	315
NEW YORK	TROY	518
NEW YORK	UTICA	315
NEW YORK	WHITE PLAINS	914
NEW YORK	YONKERS	914
NORTH CAROLINA	ASHEVILLE	704
NORTH CAROLINA	CHARLOTTE	704
NORTH CAROLINA	DURHAM	914
NORTH CAROLINA	FAYETTEVILLE	910
NORTH CAROLINA	GREENSBORO	910
NORTH CAROLINA	RALEIGH	919
NORTH CAROLINA	WINSTON-SALEM	910
NORTH DAKOTA	ALL LOCATIONS	701
OHIO	AKRON	330
OHIO	CANTON	330
OHIO	CINCINNATI	513
OHIO	CLEVELAND	216
OHIO	COLUMBUS	614
OHIO	DAYTON	513
OHIO	LORAIN	216
OHIO	STEUBENVILLE	614
OHIO	TOLEDO	419
OHIO	YOUNGSTOWN	330
OKLAHOMA	ENID	405
OKLAHOMA	OKLAHOMA CITY	405
OKLAHOMA	TULSA	918
OREGON	ASTORIA	503
OREGON	BEND	541
OREGON	CORVALLIS	541

Location/State	City	Area Code
OREGON	EUGENE	541
OREGON	MEDFORD	541
OREGON	PENDLETON	541
OREGON	PORTLAND	503
OREGON	SALEM	503
PENNSYLVANIA	ALLENTOWN	610
PENNSYLVANIA	ALTOONA	814
PENNSYLVANIA	ERIE	814
PENNSYLVANIA	HARRISBURG	717
PENNSYLVANIA	PHILADELPHIA	215
PENNSYLVANIA	PITTSBURGH	412
PENNSYLVANIA	READING	610
PENNSYLVANIA	SCRANTON	717
PENNSYLVANIA	WILKES-BARRE	717
RHODE ISLAND	ALL LOCATIONS	401
SOUTH CAROLINA	CHARLESTON	803
SOUTH CAROLINA	COLUMBIA	803
SOUTH CAROLINA	FLORENCE	803
SOUTH CAROLINA	GREENVILLE	864
SOUTH CAROLINA	SPARTANBURG	864
SOUTH DAKOTA	ALL LOCATIONS	605
TENNESSEE	CHATTANOOGA	423
TENNESSEE	JOHNSON CITY	423
TENNESSEE	KNOXVILLE	423
TENNESSEE	MEMPHIS	901
TENNESSEE	NASHVILLE	615
TEXAS	ABILENE	915
TEXAS	AMARILLO	806
TEXAS	AUSTIN	512
TEXAS	BEAUMONT	409
TEXAS	BROWNSVILLE	210
TEXAS	CORPUS CHRISTI	512
TEXAS	DALLAS	214
TEXAS	EL PASO	915
TEXAS	FORT WORTH	817
TEXAS	GALVESTON	409
TEXAS	HOUSTON	713
TEXAS	HOUSTON CEL AND PAGE	281
TEXAS	LAREDO	210
TEXAS	LUBBOCK	806
TEXAS	SAN ANTONIO	210

Location/State	City	Area Code
TEXAS	TYLER	903
TEXAS	WACO	817
UTAH	CEDAR CITY	435
UTAH	LOGAN	435
UTAH	OGDEN	801
UTAH	PROVO	435
UTAH	SALT LAKE CITY	801
UTAH	ST. GEORGE	435
VERMONT	ALL LOCATIONS	802
VIRGINIA	ALEXANDRIA	703
VIRGINIA	ARLINGTON	703
VIRGINIA	CHARLOTTESVILLE	804
VIRGINIA	NEWPORT NEWS	804
VIRGINIA	NORFOLK	757
VIRGINIA	RICHMOND	804
VIRGINIA	ROANOKE	540
WASHINGTON	OLYMPIA	360
WASHINGTON	SEATTLE	206
WASHINGTON	SPOKANE	509
WASHINGTON	TACOMA	206
WASHINGTON	VANCOUVER	360
WASHINGTON	WALLA WALLA	509
WASHINGTON	YAKIMA	509
WASHINGTON DC	ALL LOCATIONS	202
WEST VIRGINIA	ALL LOCATIONS	304
WISCONSIN	EAU CLAIRE	715
WISCONSIN	FOND DE LAC	414
WISCONSIN	GREEN BAY	414
WISCONSIN	MADISON	608
WISCONSIN	MILWAUKEE	414
WISCONSIN	RACINE	414
WISCONSIN	WAUSAU	715
WYOMING	ALL LOCATIONS	307

Appendix

D

Area codes of the NANP listed by number

Location/State	City	Area Code
NEW JERSEY	HACKENSACK	201
NEW JERSEY	JERSEY CITY	201
NEW JERSEY	NEWARK	201
NEW JERSEY	PATERSON	201
WASHINGTON DC	ALL LOCATIONS	202
DISTRICT OF COLUMBIA	WASHINGTON DC	202
CONNECTICUT	BRIDGEPORT	203
CONNECTICUT	NEW HAVEN	203
CONNECTICUT	STANFORD	203
ALABAMA	BIRMINGHAM	205
WASHINGTON	SEATTLE	206
WASHINGTON	TACOMA	206
MAINE	ALL LOCATIONS	207
IDAHO	ALL LOCATIONS	208
CALIFORNIA	FRESNO	209
CALIFORNIA	MODESTO	209
TEXAS	BROWNSVILLE	210
TEXAS	LAREDO	210
TEXAS	SAN ANTONIO	210
NEW YORK	MANHATTAN	212
CALIFORNIA	LOS ANGELES	213

Location/State	City	Area Code
TEXAS	DALLAS	214
PENNSYLVANIA	PHILADELPHIA	215
OHIO	CLEVELAND	216
OHIO	LORAIN	216
ILLINOIS	CHAMPAIGN - URBANA	217
ILLINOIS	SPRINGFIELD	217
MINNESOTA	DULUTH	218
INDIANA	GARY	219
INDIANA	HAMMOND	219
INDIANA	MICHIGAN CITY	219
INDIANA	SOUTH BEND	219
CARRIBEAN ISLANDS	BAHAMAS	242
CARRIBEAN ISLANDS	BARBADOS	246
TEXAS	HOUSTON CEL AND PAGE	281
MARYLAND	HAGERSTOWN	301
MARYLAND	ROCKVILLE	301
DELAWARE	ALL LOCATIONS	302
COLORADO	AURORA	303
COLORADO	BOULDER	303
COLORADO	DENVER	303
COLORADO	LONGMONT	303
WEST VIRGINIA	ALL LOCATIONS	304
FLORIDA	KEY WEST	305
FLORIDA	MIAMI	305
WYOMING	ALL LOCATIONS	307
NEBRASKA	NORTH PLATTE	308
NEBRASKA	SCOTTSBLUFF	308
ILLINOIS	PIORIA	309
ILLINOIS	ROCK ISLAND	309
CALIFORNIA	BEVERLY HILLS	310
ILLINOIS	CHICAGO	312
MICHIGAN	ANN ARBOR	313
MICHIGAN	DEARBORN	313
MICHIGAN	DETROIT	313
MISSOURI	COLUMBIA	314
MISSOURI	ST. LOUIS	314
NEW YORK	SYRACUSE	315
NEW YORK	UTICA	315
KANSAS	DODGE CITY	316
KANSAS	WICHITA	316
INDIANA	INDIANAPOLIS	317

Location/State	City	Area Code
INDIANA	KOKOMO	317
LOUISIANA	LAKE CHARLES	318
LOUISIANA	SHREVEPORT	318
IOWA	DAVENPORT	319
IOWA	DUBUQUE	319
MINNESOTA	ST. CLOUD	320
OHIO	AKRON	330
OHIO	CANTON	330
OHIO	YOUNGSTOWN	330
ALABAMA	MONTGOMERY	334
FLORIDA	GAINESVILLE	352
WASHINGTON	OLYMPIA	360
WASHINGTON	VANCOUVER	360
RHODE ISLAND	ALL LOCATIONS	401
NEBRASKA	LINCOLN	402
NEBRASKA	OMAHA	402
GEORGIA	ATLANTA (METRO)	404
OKLAHOMA	ENID	405
OKLAHOMA	OKLAHOMA CITY	405
MONTANA	ALL LOCATIONS	406
FLORIDA	ORLANDO	407
CALIFORNIA	MONTEREY	408
CALIFORNIA	SAN JOSE	408
TEXAS	BEAUMONT	409
TEXAS	GALVESTON	409
MARYLAND	ANNAPOLIS	410
MARYLAND	BALTIMORE	410
PENNSYLVANIA	PITTSBURGH	412
MASSACHUSETTS	PITTSFIELD	413
MASSACHUSETTS	SPRINGFIELD	413
WISCONSIN	FOND DE LAC	414
WISCONSIN	GREEN BAY	414
WISCONSIN	MILWAUKEE	414
WISCONSIN	RACINE	414
CALIFORNIA	SAN FRANCISCO	415
MISSOURI	JOPLIN	417
MISSOURI	SPRINGFIELD	417
OHIO	TOLEDO	419
TENNESSEE	CHATTANOOGA	423
TENNESSEE	JOHNSON CITY	423
TENNESSEE	KNOXVILLE	423

Location/State	City	Area Code
UTAH	CEDAR CITY	435
UTAH	LOGAN	435
UTAH	ST. GEORGE	435
CARRIBEAN ISLANDS	BERMUDA	441
KENTUCKY	FRANKFORT	502
KENTUCKY	LOUISVILLE	502
KENTUCKY	PADUCAH	502
KENTUCKY	SHELBYVILLE	502
OREGON	ASTORIA	503
OREGON	PORTLAND	503
OREGON	SALEM	503
LOUISIANA	BATON ROUGE	504
LOUISIANA	NEW ORLEANS	504
NEW MEXICO	ALL LOCATIONS	505
MINNESOTA	ROCHESTER	507
MASSACHUSETTS	FALL RIVER	508
MASSACHUSETTS	NEW BEDFORD	508
MASSACHUSETTS	WORCESTER	508
WASHINGTON	SPOKANE	509
WASHINGTON	WALLA WALLA	509
WASHINGTON	YAKIMA	509
CALIFORNIA	OAKLAND	510
TEXAS	AUSTIN	512
TEXAS	CORPUS CHRISTI	512
OHIO	CINCINNATI	513
OHIO	DAYTON	513
IOWA	DES MOINES	515
NEW YORK	HEMPSTEAD	516
NEW YORK	LONG ISLAND	516
MICHIGAN	BAY CITY	517
MICHIGAN	JACKSON	517
MICHIGAN	LANSING	517
NEW YORK	ALBANY	518
NEW YORK	SCHENECTADY	518
NEW YORK	TROY	518
ARIZONA	TUSCON	520
VIRGINIA	ROANOKE	540
OREGON	BEND	541
OREGON	CORVALLIS	541
OREGON	EUGENE	541
OREGON	MEDFORD	541

Location/State	City	Area Code
OREGON	PENDLETON	541
FLORIDA	WEST PALM BEACH	561
CALIFORNIA	LONG BEACH	562
MISSOURI	JEFFERSON CITY	573
MISSISSIPPI	ALL LOCATIONS	601
ARIZONA	PHOENIX	602
NEW HAMPSHIRE	ALL LOCATIONS	603
SOUTH DAKOTA	ALL LOCATIONS	605
KENTUCKY	ASHLAND	606
KENTUCKY	WINCHESTER	606
NEW YORK	BINGHAMTON	607
NEW YORK	ELMIRA	607
WISCONSIN	MADISON	608
NEW JERSEY	ATLANTIC CITY	609
NEW JERSEY	CAMDEN	609
NEW JERSEY	TRENTON	609
NEW JERSEY	VINELAND	609
PENNSYLVANIA	ALLENTOWN	610
PENNSYLVANIA	READING	610
MINNESOTA	MINNEAPOLIS	612
MINNESOTA	ST. PAUL	612
OHIO	COLUMBUS	614
OHIO	STEUBENVILLE	614
TENNESSEE	NASHVILLE	615
MICHIGAN	BATTLE CREEK	616
MICHIGAN	GRAND RAPIDS	616
MICHIGAN	KALAMAZOO	616
MASSACHUSETTS	BOSTON	617
ILLINOIS	ALTON	618
ILLINOIS	CAIRO	618
ILLINOIS	MT. VERNON	618
CALIFORNIA	SAN DIEGO	619
ILLINOIS	CHICAGO (CENTRAL SUBURBS)	630
ILLINOIS	ELGIN	630
ILLINOIS	WAUKEGAN	630
CARRIBEAN ISLANDS	MONTSERRAT	664
NORTH DAKOTA	ALL LOCATIONS	701
NEVADA	ALL LOCATIONS	702
VIRGINIA	ALEXANDRIA	703
VIRGINIA	ARLINGTON	703
NORTH CAROLINA	ASHEVILLE	704

Location/State	City	Area Code
NORTH CAROLINA	CHARLOTTE	704
GEORGIA	AUGUSTA	706
GEORGIA	COLUMBUS	706
GEORGIA	ROME	706
CALIFORNIA	EUREKA	707
ILLINOIS	CHICAGO	708
IOWA	COUNCIL BLUFFS	712
IOWA	SIOUX CITY	712
TEXAS	HOUSTON	713
CALIFORNIA	ANHEIM	714
WISCONSIN	EAU CLAIRE	715
WISCONSIN	WAUSAU	715
NEW YORK	BUFFALO	716
NEW YORK	NIAGARA FALLS	716
NEW YORK	ROCHESTER	716
PENNSYLVANIA	HARRISBURG	717
PENNSYLVANIA	SCRANTON	717
PENNSYLVANIA	WILKES-BARRE	717
NEW YORK	BRONX	718
NEW YORK	BROOKLYN	718
NEW YORK	QUEENS	718
NEW YORK	STATEN ISLAND	718
COLORADO	COLORADO SPRINGS	719
COLORADO	LEADVILLE	719
COLORADO	PUEBLO	719
VIRGINIA	NORFOLK	757
CARRIBEAN ISLANDS	ST. LUCIA	758
CALIFORNIA	BARSTOW	760
CALIFORNIA	PALM SPRINGS	760
GEORGIA	ATLANTA (OUTSIDE 285 BELT)	770
GEORGIA	MARIETTA	770
GEORGIA	NORCROSS	770
ILLINOIS	CHICAGO	773
UTAH	OGDEN	801
UTAH	PROVO	801
UTAH	SALT LAKE CITY	801
VERMONT	ALL LOCATIONS	802
SOUTH CAROLINA	CHARLESTON	803
SOUTH CAROLINA	COLUMBIA	803
SOUTH CAROLINA	FLORENCE	803
VIRGINIA	CHARLOTTESVILLE	804

Location/State	City	Area Code
VIRGINIA	NEWPORT NEWS	804
VIRGINIA	RICHMOND	804
CALIFORNIA	BAKERSFIELD	805
CALIFORNIA	SANTA BARBARA	805
TEXAS	AMARILLO	806
TEXAS	LUBBOCK	806
HAWAII	ALL LOCATIONS	808
MICHIGAN	FLINT	810
MICHIGAN	PONTIAC	810
MICHIGAN	SOUTHFIELD	810
MICHIGAN	TROY	810
INDIANA	EVANSVILLE	812
FLORIDA	CLEARWATER	813
FLORIDA	ST PETERSBURG	813
FLORIDA	TAMPA	813
PENNSYLVANIA	ALTOONA	814
PENNSYLVANIA	ERIE	814
ILLINOIS	LA SALLE	815
ILLINOIS	ROCKFORD	815
MISSOURI	KANSAS CITY	816
MISSOURI	ST. JOSEPH	816
TEXAS	FORT WORTH	817
TEXAS	WACO	817
CALIFORNIA	PASADENA	818
ILLINOIS	CHICAGO	847
ILLINOIS	ELGIN	847
ILLINOIS	WAUKEGAN	847
CONNECTICUT	HARTFORD	860
SOUTH CAROLINA	GREENVILLE	864
SOUTH CAROLINA	SPARTANBURG	864
CARRIBEAN ISLANDS	TOBAGO	868
CARRIBEAN ISLANDS	TRINIDAD	868
TENNESSEE	MEMPHIS	901
TEXAS	TYLER	903
FLORIDA	JACKSONVILLE	904
FLORIDA	PENSACOLA	904
FLORIDA	TALLAHASSEE	904
MICHIGAN	MARQUETTE	906
MICHIGAN	SAULT STE. MARIE	906
ALASKA	ALL LOCATIONS	907
NEW JERSEY	NEW BRUNSWICK	908

Location/State	City	Area Code
CALIFORNIA	RIVERSIDE	909
CALIFORNIA	SAN BERNARDINO	909
NORTH CAROLINA	FAYETTEVILLE	910
NORTH CAROLINA	GREENSBORO	910
NORTH CAROLINA	WINSTON-SALEM	910
GEORGIA	ALBANY	912
GEORGIA	SAVANAH	912
KANSAS	LAWRENCE	913
KANSAS	SALINA	913
KANSAS	TOPEKA	913
NORTH CAROLINA	DURHAM	914
NEW YORK	PEEKSKILL	914
NEW YORK	POUGHKEEPSIE	914
NEW YORK	WHITE PLAINS	914
NEW YORK	YONKERS	914
TEXAS	ABILENE	915
TEXAS	EL PASO	915
CALIFORNIA	SACRAMENTO	916
NEW YORK	BRONX CEL AND PAGER	917
NEW YORK	BROOKLYN CEL AND PAGE	917
NEW YORK	MANHATTAN CEL AND PAGE	917
NEW YORK	QUEENS CEL AND PAGE	917
NEW YORK	STATEN ISLAND CEL PAGE	917
OKLAHOMA	TULSA	918
NORTH CAROLINA	RALEIGH	919
FLORIDA	FT. MYERS	941
FLORIDA	MAPLES	941
FLORIDA	SARASOTA	941
FLORIDA	FT. LAUDERDALE	954
COLORADO	ASPEN	970
COLORADO	DURANGO	970
COLORADO	GRAND JUNCTION	970
COLORADO	STEAMBOAT SPRINGS	970

Appendix

E

Binary, decimal, and hexadecimal conversions

b2=binary, b10=decimal, b16=hexadecimal

b10	b2	b16
0	0	0
1	1	1
2	10	2
3	11	3
4	100	4
5	101	5
6	110	6
7	111	7
8	1000	8
9	1001	9
10	1010	A
11	1011	B
12	1100	C
13	1101	D
14	1110	E
15	1111	F
16	10000	10
17	10001	11
18	10010	12
19	10011	13

b10	b2	b16
20	10100	14
21	10101	15
22	10110	16
23	10111	17
24	11000	18
25	11001	19
26	11010	1A
27	11011	1B
28	11100	1C
29	11101	1D
30	11110	1E
31	11111	1F
32	100000	20
33	100001	21
34	100010	22
35	100011	23
36	100100	24
37	100101	25
38	100110	26
39	100111	27
40	101000	28
41	101001	29
42	101010	2A
43	101011	2B
44	101100	2C
45	101101	2D
46	101110	2E
47	101111	2F
48	110000	30
49	110001	31
50	110010	32
51	110011	33
52	110100	34
53	110101	35
54	110110	36
55	110111	37
56	111000	38
57	111001	39
58	111010	3A
59	111011	3B
60	111100	3C
61	111101	3D

b10	b2	b16
62	111110	3E
63	111111	3F
64	1000000	40
65	1000001	41
66	1000010	42
67	1000011	43
68	1000100	44
69	1000101	45
70	1000110	46
71	1000111	47
72	1001000	48
73	1001001	49
74	1001010	4A
75	1001011	4B
76	1001100	4C
77	1001101	4D
78	1001110	4E
79	1001111	4F
80	1010000	50
81	1010001	51
82	1010010	52
83	1010011	53
84	1010100	54
85	1010101	55
86	1010110	56
87	1010111	57
88	1011000	58
89	1011001	59
90	1011010	5A
91	1011011	5B
92	1011100	5C
93	1011101	5D
94	1011110	5E
95	1011111	5F
96	1100000	60
97	1100001	61
98	1100010	62
99	1100011	63
100	1100100	64
101	1100101	65
102	1100110	66
103	1100111	67

b10	b2	b16
104	1101000	68
105	1101001	69
106	1101010	6A
107	1101011	6B
108	1101100	6C
109	1101101	6D
110	1101110	6E
111	1101111	6F
112	1110000	70
113	1110001	71
114	1110010	72
115	1110011	73
116	1110100	74
117	1110101	75
118	1110110	76
119	1110111	77
120	1111000	78
121	1111001	79
122	1111010	7A
123	1111011	7B
124	1111100	7C
125	1111101	7D
126	1111110	7E
127	1111111	7F
128	10000000	80
129	10000001	81
130	10000010	82
131	10000011	83
132	10000100	84
133	10000101	85
134	10000110	86
135	10000111	87
136	10001000	88
137	10001001	89
138	10001010	8A
139	10001011	8B
140	10001100	8C
141	10001101	8D
142	10001110	8E
143	10001111	8F
144	10010000	90
145	10010001	91

b10	b2	b16
146	10010010	92
147	10010011	93
148	10010100	94
149	10010101	95
150	10010110	96
151	10010111	97
152	10011000	98
153	10011001	99
154	10011010	9A
155	10011011	9B
156	10011100	9C
157	10011101	9D
158	10011110	9E
159	10011111	9F
160	10100000	A0
161	10100001	A1
162	10100010	A2
163	10100011	A3
164	10100100	A4
165	10100101	A5
166	10100110	A6
167	10100111	A7
168	10101000	A8
169	10101001	A9
170	10101010	AA
171	10101011	AB
172	10101100	AC
173	10101101	AD
174	10101110	AE
175	10101111	AF
176	10110000	B0
177	10110001	B1
178	10110010	B2
179	10110011	B3
180	10110100	B4
181	10110101	B5
182	10110110	B6
183	10110111	B7
184	10111000	B8
185	10111001	B9
186	10111010	BA
187	10111011	BB

b10	b2	b16
188	10111100	BC
189	10111101	BD
190	10111110	BE
191	10111111	BF
192	11000000	C0
193	11000001	C1
194	11000010	C2
195	11000011	C3
196	11000100	C4
197	11000101	C5
198	11000110	C6
199	11000111	C7
200	11001000	C8
201	11001001	C9
202	11001010	CA
203	11001011	CB
204	11001100	CC
205	11001101	CD
206	11001110	CE
207	11001111	CF
208	11010000	D0
209	11010001	D1
210	11010010	D2
211	11010011	D3
212	11010100	D4
213	11010101	D5
214	11010110	D6
215	11010111	D7
216	11011000	D8
217	11011001	D9
218	11011010	DA
219	11011011	DB
220	11011100	DC
221	11011101	DD
222	11011110	DE
223	11011111	DF
224	11100000	E0
225	11100001	E1
226	11100010	E2
227	11100011	E3
228	11100100	E4
229	11100101	E5

b10	b2	b16
230	11100110	E6
231	11100111	E7
232	11101000	E8
233	11101001	E9
234	11101010	EA
235	11101011	EB
236	11101100	EC
237	11101101	ED
238	11101110	EE
239	11101111	EF
240	11110000	F0
241	11110001	F1
242	11110010	F2
243	11110011	F3
244	11110100	F4
245	11110101	F5
246	11110110	F6
247	11110111	F7
248	11111000	F8
249	11111001	F9
250	11111010	FA
251	11111011	FB
252	11111100	FC
253	11111101	FD
254	11111110	FE
255	11111111	FF

Appendix

F

Color codes

The twisted-pair cable color code

This system has tip colors, ring colors, group colors, and binder colors. The twisted-pair color code is used to identify the number of a pair within a cable. For example, a pair in the Orange-Yellow binder group that is Yellow-Green is pair 418 (O-Y+Y-O). The twisted-pair color code in our telephone network is:

Pair codes for pairs 1 to 25 within binders.

Pair #	Tip color	Ring color	Abbreviation
1	white	blue	W-BL
2	white	orange	W-O
3	white	green	W-G
4	white	brown	W-BR
5	white	slate	W-S
6	red	blue	R-BL
7	red	orange	R-O
8	red	green	R-G
9	red	brown	R-BR
10	red	slate	R-S
11	black	blue	BK-BL
12	black	orange	BK-O
13	black	green	BK-G
14	black	brown	BK-BR
15	black	slate	BK-S
16	yellow	blue	Y-BL
17	yellow	orange	Y-O
18	yellow	green	Y-G
19	yellow	brown	Y-BR

20	yellow	slate	Y-S
21	violet	blue	V-BL
22	violet	orange	V-O
23	violet	green	V-G
24	violet	brown	V-BR
25	violet	slate	V-SL

A *binder* is a plastic ribbon wrapped around 25 pairs of wire or 600 pairs of wire. Binder codes are for identifying groups of 25 or 600. The binder code is the same as the pair code, only the group colors and the pair colors are reversed (e.g., white-blue becomes blue-white), so when the abbreviated colors are put together, they are easier to understand. Binder codes are good for 600 pairs of wire. After that, each group of 600 pairs gets an additional plastic binder.

Binder color code

Binder#	Binder color	Abbreviation	Pairs in binder
1	blue white	BL-W	1–25
2	orange white	O-W	26–50
3	green white	G-W	51–75
4	brown white	BR-W	76–100
5	slate white	S-W	101–125
6	blue red	BL-R	126–150
7	orange red	O-R	151–175
8	green red	G-R	176–200
9	brown red	BR-R	201–225
10	slate red	S-R	226–250
11	blue black	BL-BK	251–275
12	orange black	O-BK	276–300
13	green black	G-BK	301–325
14	brown black	BR-BK	326–350
15	slate black	S-BK	351–375
16	blue yellow	BL-Y	376–400
17	orange yellow	O-Y	401–425
18	green yellow	G-Y	426–450
19	brown yellow	BR-Y	451–475
20	slate yellow	S-Y	476–500
21	blue violet	BL-V	501–525
22	orange violet	O-V	526–550
23	green violet	G-V	551–575
24	brown violet	BR-V	576–600

Large cable binders:

Pair 1 to 600, surrounded by a white binder
Pair 601 to 1200, surrounded by a red binder
Pair 1201 to 1800, surrounded by a white binder

Fiber-optic color code

Groups of 12 fibers are placed in loose tube buffers:

Fiber #	Color
1	blue
2	orange
3	green
4	brown
5	slate
6	white
7	red
8	black
9	yellow
10	violet
11	aqua (light blue)
12	pink

The resistor color code

Resistors have four color bands on them. They are regarded to as the first, second, third, and fourth bands. The first band is the closest to one side of the resistor and the following bands count to the inside. The first band indicates the first integer of the value of resistance. The second band indicates the second integer of the value of resistance. The third band indicates a multiplier or number of zeros to be placed after the first two band numbers. A diagram is the easiest way to demonstrate this code.

Color	band 1	band 2	×	band 3
Black	0	0	×	1
Brown	1	1	×	10
Red	2	2	×	100
Orange	3	3	×	1000
Yellow	4	4	×	10,000
Green	5	5	×	100,000
Blue	6	6	×	1,000,000
Violet	7	7	×	10,000,000
Gray	8	8	×	100,000,000
White	9	9		none

Index

B

C

D

I

J

K

P

S

U

V

W

X

Y

Z

About the Author

Jade Clayton is a communications engineer for American Express. He designs and coordinates communication services that involve multiple telecommunication companies around the world. Clayton, who wrote the definitions in the McGraw-Hill Illustrated Telecom Dictionary, has over 12 years of experience in almost all facets of the telecommunications industry, including public network traffic switching, private line services, SONET network engineering, IVR maintenance, inside and outside plant construction, cable TV, PBX configuration and administration, LAN/WAN installation, electronic engineering, cellular/PCS tower and site construction, broadcast radio, terrestrial microwave engineering and installation, call center ACD management, and voice mail systems and public telephone network services installation and maintenance.